抽線琢鍼圖

This third edition published in 2007
by André Deutsch
an imprint of the Carlton Publishing Group
20 Mortimer Street
London W1T 3JW

10 9 8 7 6 5 4 3 2 1

First published in Great Britain in 1986 by Patrick Stephens Ltd

A catalogue record for this book is available from the British Library.

ISBN-13: 978-0-233-00202-6

Designer: Michelle Pickering
Art Director: Clare Baggaley
Picture Research: Steve Behan, Paul Langan
Project Editor: Gareth Jones
Cartographer: Bill Smuts
Production: Lisa Moore

Printed and bound in Dubai

THE GENIUS OF
CHINA
BY ROBERT TEMPLE
INTRODUCTION BY DR JOSEPH NEEDHAM FRS, FBA
3,000 YEARS OF SCIENCE, DISCOVERY & INVENTION
ANDRE DEUTSCH

CONTENTS

DEDICATION

TO DR SONG JIAN,
WITH DEEP GRATITUDE FOR
HIS FRIENDSHIP. HIS QUALITIES
OF CHARACTER, INTELLECTUAL
BRILLIANCE, AND PRACTICAL GRASP
OF PUBLIC AFFAIRS, HAVE MADE
HIM ONE OF THE MOST LOVED
AND ADMIRED OF ALL THE
LEADERS OF CHINA.

PREFACE

Dr Joseph Needham, Fellow of the Royal Society in Britain, a famous biochemist and historian of science and an old friend of the Chinese people, devoted his later life to the study of the history of Chinese science and technology and to encouraging friendship between the British and Chinese people. From the middle of the last century, Dr Needham gave his full strength to doing a conscientious job of classifying and researching the rich ancient Chinese traditions of science and technology, and he achieved great success. *Science and Civilization in China*, which is about 20 million words in length, is an unprecedented contribution to the world history of science and technology. With the correct views and thoughtful insights of a Western scholar, he came to a systematic and comprehensive conclusion about the glorious achievements of ancient Chinese science and technology through this series of works. His works fully affirm the great contribution of ancient Chinese civilization to world civilization, and erect a bridge of mutual communication and understanding between the two major Eastern and Western systems. His work has exerted tremendous and profound influence in respect of enhancing the comprehensive understanding of China by other countries, facilitating the scientific and cultural exchange between East and West, and promoting research into Chinese science and technology.

Robert Temple is a British scholar who loves the Chinese people and has a deep interest in and admiration for traditional Chinese culture and scientific achievements. Through comparative study, he came to a surprising conclusion: Possibly more than half of the basic inventions and discoveries upon which the 'modern world' rests come from China. Therefore, he came to the profound realization that civilization is an inseparable entirety and that the modern technological world is a joint creation of Eastern and Western civilisations, and that both East and West should acknowledge and respect China's contribution. In order to acquaint western readers with a general knowledge of ancient Chinese science and technology, he wrote his book *The Genius of China* (the Chinese edition is called *The Spirit of Chinese Invention: 100 World Firsts*). It selected a hundred World First cases either chronologically or according to scientific criteria from the rich and varied Chinese scientific traditions and made a concrete and concise introduction with pictures. Most of the materials are from Dr Needham's *Science and Civilization in China*, other works of Joseph Needham, and other unpublished typescripts and materials. Temple's work was praised by Dr Needham, who therefore wrote a warm Foreword for it.

The Genius of China is a book with rich content, clear expression and excellence both in pictures and text. It is scientific, knowledgeable and interesting, which is highly suitable for young people. Our new century is an era of the knowledge-based economy. The intense competition between nations is, to be exact, a competition for national quality. Our core target of deepening educational reform and advancing quality education in an all-round way is speedily cultivating the highly-qualified personnel who possess an innovative spirit and can demonstrate capability. General Secretary Jiang Zemin has stressed repeatedly that 'innovation is the spirit of a nation and the endless momentum for a nation's prosperity'. He has also said 'The Chinese nation should occupy its proper place in the field of world high-technology'. Our young people are expected to bear firmly in mind General Secretary Jiang's instructions, undertake the mission of the times, always regard the building of the state's prosperity and the people's happiness as their duty. They should be determined to have the lofty aspirations of loving their homeland and going all out from childhood to make their country strong, to be the 'New People' (having ideals, morality, knowledge and discipline) and dedicate their wisdom and capability to their country, and to help to realize the great rejuvenation of the Chinese nation. Reading this book, young people could not only acquire a relatively clear understanding of the splendid history of science and technology of their motherland, but also draw stimulation in the course of studying and carrying on the long-standing and well-established innovative spirit and courage of the Chinese nation. Therefore *The Genius of China*, written by the friend of China, Robert Temple, is a good teaching material with a unique perspective to assist in quality education for young Chinese people.

MADAME CHEN ZHILI
State Councillor of China for
Education, Science, Technology and Culture

ABOVE (1) The Chinese invention of the stirrup led to the development of the game of polo. This watercolour painting on silk by Li-Lin, made c. 1635, was probably based on a Yuan Dynasty (1280–1368) original. (Victoria and Albert Museum.)

FOREWORD

I should like to give a warm welcome to this book by Robert Temple. It is, in its own way, a brilliant distillation of my *Science and Civilisation in China*, published by the Cambridge University Press, a work which will be complete in some twenty-five volumes and of which fifteen have now appeared or are passing through the press.

The extraordinary inventiveness, and insight into nature, of ancient and medieval China raises two fundamental questions. First, why should they have been so far in advance of other civilizations; and second, why aren't they now centuries ahead of the rest of the world? We think it was a matter of the very different social and economic systems between China and the West, as I will explain more fully in a moment. Modern science arose only in Europe in the seventeenth century when the best method of discovery was itself discovered; but the discoveries and inventions made then and thereafter depended in so many cases on centuries of previous Chinese progress in science, technology and medicine.

Perhaps I should describe how I became involved in all this. Coming to Cambridge in 1918, intending to read medicine, I was trained primarily as a biochemist, and specialized in the connections of biochemistry with embryology, producing *Chemical Embryology* in 1931, and *Biochemistry and Morphogenesis* in 1942. But I was always interested in the history of science, and in a way Charles Singer did as much for me as Frederick Gowland Hopkins. My *History of Embryology* came out in 1934.

Three years later the laboratories in which I worked received several Chinese scientists, some intending to do research leading to the doctorate. We became great friends, and this had two effects upon me: first, I found that the better I got to know them the more exactly like my own their minds were, which raised in acute form the question of why modern science had begun only in Europe. Second, I learnt the Chinese language; and I did so as a labour of love, which is quite a different thing from going through the mill of the Oriental Studies Tripos. By the time that one of them, now long my chief collaborator, Lu Gwei-Djen, left in 1940, we had decided that 'something absolutely must be done' about the history of science, technology and medicine in traditional Chinese culture.

Then in 1942 I was asked to go to China, as an envoy from the Royal Society, and this led to my staying there 'for the duration' of the Second World War as Scientific Counsellor at the British Embassy in Chungking. Since the post involved a great deal of travelling, I had ample opportunities for learning about what had happened in Chinese history; wherever I went there was usually someone, a medical doctor, a mathematician or an engineer, who was deeply interested in how the subject had developed in his own culture, and was able to tell me what to read, what books to buy if possible, and so on. Thus I gained a remarkable orientation, which perhaps could hardly have been achieved in any other way. After the war was over I put in two years helping to build up Unesco by organizing the Division of Natural Sciences.

When I returned to Cambridge in 1948 the project of *Science and Civilisation in China* began in earnest, with the help of my first collaborator Wang Ching-Ning. Although I was still Sir William Dunn Reader in Biochemisty, and therefore had a good many lectures to give to the advanced students, we made good progress, and the first volume of the series came out in 1954. Looking back at it now, I feel that in a task of this sort it is very important not to know too much, but yet to be in possession of a boundless enthusiasm for the Chinese people and their achievements over the ages. Now some fifteen volumes have been published, and the whole set is expected to consist of at least twenty-five, so we have come a long way since we started.

And gradually what a cave of glittering treasures was opened up! My friends among the older generation of sinologists had thought that we should find nothing – but how wrong they were. One after another, extraordinary inventions and discoveries clearly appeared in Chinese literature, archeological evidence or pictorial witness, often, indeed generally, long preceding the parallel, or adopted, inventions and discoveries of Europe. Whether it was the array of binomial coefficients, or the standard method in interconversion of rotary and longitudinal motion, or the first of all clockwork escapements, or the ploughshare of malleable cast iron, or the beginnings of geo-botany and soil science, or cutaneous-visceral reflexes, or the finding of smallpox inoculation – wherever one looked there was 'first' after 'first'.

Francis Bacon had selected three inventions, paper and printing, gunpowder, and the magnetic compass, which had done more, he thought, than any religious conviction, or any astrological influence, or any conqueror's achievements, to transform completely the modern world and mark it off from antiquity and the Middle Ages. He regarded the origins of these inventions as 'obscure and inglorious' and he died without ever knowing that all of them were Chinese. We have done our best to put this record straight.

Chauvinistic Westerners, of course, always try to minimize the indebtedness of Europe to China in antiquity and the Middle Ages, but often the circumstantial evidence is compelling. For example, the first blast furnaces for cast iron, now known to be Scandinavian of the late eighth century AD, are of closely similar form to those of the previous century in China; while as late as the seventeenth century AD the magnetic compasses of surveyors and astronomers pointed south, not north, just as the compasses of China had always done. In many cases, however, we cannot as yet detect the capillary channels through which knowledge was conveyed from East to West. Nevertheless we have always adopted the very reasonable assumption that the longer the time elapsing between the appearance of a discovery or invention in one part of the world, and its appearance later on in some other part of the world far away, the less likely is it that the new thing was independently invented or discovered.

But all these things being agreed, a formidable question then presents itself. If the Chinese were so advanced in antiquity and the Middle Ages, how was it that the Scientific Revolution, the coming of *modern* science into the world, happened only in Europe? This is what we call the 'sixty-four thousand dollar question', and it may be remembered that it was precisely this problem which presented itself to me so forcefully when I first met the Chinese scientists who came to Cambridge in 1937. The fact is that in the seventeenth century we have to face a package deal; the Scientific Revolution was accompanied both by the Protestant Reformation and by the rise of capitalism, the ascendancy of the entrepreneurial bourgeoisie. Distinctively *modern* science, which then developed, was a mathematization of hypotheses about nature, combined with relentless experimentation. The sciences of all the ancient and medieval worlds had had an indelibly ethnic stamp, but now nature was addressed for the first time in a universal and international language, the precise and quantitative idiom of mathematics, a tongue which every man and woman, irrespective of colour, creed or race, can use and master if given the proper training. And to the technique of experiment the same applies. It was like the merchant's universal standard of value. How one looks at the primary causative factor in all this depends on one's own background; if one is a theologian one probably thinks that the liberation of the Reformation was responsible, if one is an old-fashioned scientist, one naturally thinks that the scientific movement occurred first and powered all the others, and if one is a Marxist, one certainly thinks that the economic and social changes bear the main responsibility.

One factor which must have great relevance here is the undeniable circumstance that the feudalism of Europe and China were fundamentally different. European feudalism was military-aristocratic: the peasantry were governed by the knights in their manors, and they in turn were subject to the barons in their castles, while the king in his palace ruled over all. In time of war he needed the help of the lower ranks in the feudal hierarchy who were bound to rally to him with stated numbers of men-at-arms. How different was the feudalism of China, long very justifiably described as bureaucratic. From the time of the first emperor, Ch'in Shih Huang Ti, onwards (third century BC), the old hereditary feudal houses were gradually attacked and destroyed, while the king or emperor (as he soon became) governed by the aid of an enormous bureaucracy, a civil service unimaginable in extent and degree of organization to the petty kingdoms of Europe. Modern research is showing that the bureaucratic organization of China in its earlier stages strongly helped science to grow; only in its later ones did it forcibly inhibit further growth, and in particular prevented a break-through which has occurred in Europe. For example, no other country in the world at the beginning of the eighth century AD could have set up a meridian arc survey stretching from south to north some 2500 miles. Nor could it have mounted an expedition at that time to go and observe the stars of the southern hemisphere to within 20° of the south celestial pole. Nor indeed would it have wanted to.

It may well be that a similar pattern will appear in the future when the history of science, technology and medicine, for all the great classical literary cultures, such as India or Sri Lanka, comes to be written and gathered in. Europe has entered into their inheritance, producing an ecumenical universal science and technology valid for every man and woman on the face of the earth. One can only hope that the shortcomings of the distinctively European traditions in other matters will not debauch the non-European civilizations. For example, the sciences of China and of Islam never dreamed of divorcing science from ethics, but when at the Scientific Revolution the final cause of Aristotle was done away with, and ethics chased out of science, things became very different, and more menacing. This was good in so far as it clarified and discriminated between the great forms of human experience, but very bad and dangerous when it opened the way for evil men to use the great discoveries of modern science and activities disastrous for humanity. Science needs to be lived alongside religion, philosophy, history and esthetic experience; alone it can lead to great harm. All we can do today is to hope and pray that the unbelievably dangerous powers of atomic weapons, which have been put into the hands of human beings by the development of modern science, will remain under control by responsible men, and that maniacs will not release upon mankind powers that could extinguish not only mankind, but all life on earth.

JOSEPH NEEDHAM

THE WEST'S DEBT TO CHINA

One of the greatest untold secrets of history is that the 'modern world' in which we live is a unique synthesis of Chinese and Western ingredients. Possibly more than half of the basic inventions and discoveries upon which the 'modern world' rests come from China. And yet few people know this. Why?

The Chinese themselves are as ignorant of this fact as Westerners. From the seventeenth century, the Chinese became increasingly dazzled by European technological expertise, having experienced a period of amnesia regarding their own achievements. When the Chinese were shown a mechanical clock by Jesuit missionaries, they were awestruck, forgetting that it was they who had invented mechanical clocks in the first place!

It is just as much a surprise for the Chinese as for Westerners to realize that *modern* agriculture, *modern* shipping, the *modern* oil industry, *modern* astronomical observatories, *modern* music, decimal mathematics, paper money, umbrellas, fishing reels, wheelbarrows, multi-stage rockets, guns, underwater mines, poison gas, parachutes, hot-air balloons, manned flight, brandy, whisky, the game of chess, printing, and even the essential design of the steam engine, all came from China.

Without the importation from China of nautical and navigational improvements such as ships' rudders, the compass and multiple masts, the great European Voyages of Discovery could never have been undertaken. Columbus would not have sailed to America, and Europeans would never have established colonial empires.

Without the importation from China of the stirrup, to enable them to stay on horseback, knights of old would never have ridden in their shining armour to aid damsels in distress; there would have been no Age of Chivalry. And without the importation from China of guns and gunpowder, the knights would not have been knocked from their horses by bullets which pierced the armour, bringing the Age of Chivalry to an end.

Without the importation from China of paper and printing, Europe would have continued for much longer to copy books by hand. Literacy would not have become so widespread. Johannes Gutenberg did *not* invent movable type. It was invented in China. William Harvey did *not* discover the circulation of the blood in the body. It was discovered – or rather, always assumed – in China. Isaac Newton was *not* the first to discover his First Law of Motion. It was discovered in China.

These myths and many others are shattered by our discovery of the true Chinese origins of many of the things, all around us, which we take for granted. Some of our greatest achievements turn out to have been not achievements at all, but simple borrowings. Yet there is no reason for us to feel inferior or downcast at the realization that much of the genius of mankind's advance was Chinese rather than European. For it is exciting to realize that the East and the West are not as far apart in spirit or in fact as most of us have been led, by appearances, to believe, and that the East and the West *are already combined* in a synthesis so powerful and so profound that it is all-pervading. Within this synthesis we live our daily lives, and from it there is no escape. The modern world *is* a combination of Eastern and Western ingredients which are inextricably fused. The fact that we are largely unaware of it is perhaps one of the greatest cases of historical blindness in the existence of the human race.

Why are we ignorant of this gigantic, obvious truth? The main reason is surely that the Chinese themselves lost sight of it. If the very originators of the inventions and discoveries no longer claim them, and their memory of them has faded, why should their inheritors trouble to resurrect their lost claims? Until our own time, it is questionable whether many Westerners even wanted to know the truth. It is always more satisfying to the ego to think that we have reached our present position alone and unaided, that we are the masters of all abilities and crafts.

The discovery of the truth is a result of incidents in the life of the distinguished scholar Dr Joseph Needham, author of the great work *Science and Civilisation in China*. In 1937, aged 37, Needham was one of the youngest Fellows of the Royal Society and a biochemist of considerable distinction at Cambridge. He had already published many books, including the definitive history of embryology. One day he met and befriended some Chinese students, including a young lady from Nanking named Lu Gwei-Djen, whose father had passed on to her his unusually profound knowledge of the history of Chinese science. Needham began to hear tales of how the Chinese had been the true discoverers of this and that important thing, and at first he could not believe it. But as he looked further into it, evidence began to come to light from Chinese texts, hastily translated by his new friends.

Needham became obsessed with the subject. Not knowing a word of Chinese, he set about learning the language. In 1942 he was sent to China as Scientific Counsellor to the British Embassy in Chungking. He was able to travel all over China, learn the language thoroughly, meet men of science, and accumulate vast quantities of priceless ancient Chinese science books. These were flown back to Britain by the Royal Air Force and today form the basis of the finest library, outside China, on the history of Chinese science, technology and medicine, at the Needham Research Institute in Cambridge. After the War, Needham was among those who 'put the "s" into Unesco', having persuaded that organization to concern itself with science as well as education and culture. He became Unesco's first Assistant Director General for the natural sciences.

In July 1946 Needham stated in a lecture to the China Society in London that: 'What is really very badly needed is a proper book on the history of science and technology in China, especially with reference to the social and economic background of Chinese life. Such a book would be by no means academic, but would have a wide bearing on the general history of thought and ideas.'

When he returned to Cambridge, where he eventually became Master of Caius College for many years, Joseph went ahead and wrote the work which he had envisaged, except that it was very academic and impenetrable to the ordinary educated reader. The result, *Science and Civilisation in China*, became a huge multi-volume project, envisaged eventually in 36 volumes (at least 24 are now available). Since Joseph's death, further volumes in the series have been issued by a number of specialist collaborators. This was a process which had begun even while Joseph was alive, with the appearance of the excellent volume on agriculture, written by a then young, intrepid sinologist, Francesca Bray, under Joseph's occasional supervision.

Gwei-Djen died tragically before Joseph, leaving him emotionally bereft, but he continued working right up to his death. One day when Gwei-Djen was still alive, I pointed to an ornate sealed gate at Caius and said to Joseph: 'What is that, and why is it so tightly shut?' He said: 'That gate is only opened at the inauguration of a new Master, or when one dies, for his funeral. One day they'll carry out me through there.' When, years later, I passed through the gate behind his coffin, I sadly recalled his comment.

Joseph never lost his early vision of a work which was 'by no means academic', as he had originally promised. He had always wanted to make his work accessible in every possible way. Therefore, when I approached him in 1984 with the suggestion that I write a popular book for the general reader based upon his half-century's labours, he agreed more readily than at that time I could understand. He and Gwei-Djen told me that they strongly approved of some things I had published about the Shang Dynasty, the I Ching and such matters, and liked the way I wrote about such abstruse subjects for the ordinary reader without sacrificing scholarly accuracy.

Although Joseph did not personally like Professor Derk Bodde, under whom I had studied Chinese philosophy, my academic background was considered acceptable because Joseph knew of Bodde's high standards. As far as Joseph and Gwei-Djen were concerned, those writings of mine proved to them that I was qualified for the task, and the only thing that remained was for Joseph to make the hard decision to relinquish the task himself, which was first announced in 1946.

I have taken certain minor liberties which must be pointed out to those readers who may consult Needham's own volumes. I have used the convention, which he avoids, of BC and AD for dates, substituting them in my quotations for his plus and minus signs. I have ironed out various passages, particularly translations from the Chinese, by eliminating Chinese words, occasional parentheses, and specialized matter which does not concern the general reader. I have also, at Dr Needham's own suggestion, eliminated the extra letter 'h' in Chinese words which he had introduced as a substitution for the aspirate apostrophe. Hence, his *chhien* becomes *ch'ien*, etc. The system of transliteration used in this book is thus the pure Wade-Giles system. The Pinyin system which has been adopted by the Chinese government and newspapers around the world in recent years is not suitable, for it would have made reference to Needham's own volumes impossible to the non-specialist.

This book has purposely been prepared without footnotes or other scholarly accessories. Many volumes have continued to appear in the *Science and Civilisation in China* series, and the list of those in print should always be consulted by anyone wishing to go more deeply into certain specific subjects. The main aim of this book has been to make Needham's work accessible to the general non-specialist reader, whilst providing an overview for specialists. In preparing the book, I used many typescripts of unpublished material, discussions with Joseph and Gwei-Djen, proofs and oral and written accounts of material that had not yet been published. My account of porcelain was done entirely without the assistance of any material by Joseph, as he never wrote about that subject at all. Those collaborators, such as H. T. Huang, who were generous in helping me in my efforts have been specially acknowledged for it, for which see the *Author's Acknowledgements* (see pages 286–7).

In the 1946 lecture which was so prophetic of his future activities, Dr Needham went on to say:

> **I personally believe that all Westerners, all people belonging to the Euro-American civilization, are subconsciously inclined to congratulate themselves, feeling with some self-satisfaction that, after all, it was Europe and its extension into the Americas which developed modern science and technology. In the same way I think that all my Asian friends are subconsciously inclined to a certain anxiety about this matter, because their civilization did not, in fact, develop modern science and technology.**

We need to set this matter right, from both ends. And I can think of no better single illustration of the folly of Western complacency and self-satisfaction than the lesson to be drawn from the history of agriculture. Today, a handful of Western nations have grain surpluses and feed the world. When Asia starves, the West sends grain. We assume that Western agriculture is the very pinnacle of what is possible in the productive use of soil for the growth of food. But we should take to heart the astonishing and disturbing fact that the European agricultural revolution, which laid the basis for the Industrial Revolution, came about only because of the importation of Chinese ideas and inventions. The growing of crops in rows, intensive hoeing of weeds, the 'modern' seed drill, the iron plough, the mouldboard to turn the ploughed soil, and efficient harnesses were all imported from China. Before the arrival from China of the trace harness and collar harness, Westerners choked their horses with straps round their throats. Although ancient Italy could produce plenty of grain, it could not be trans-ported overland to Rome for lack of satisfactory harnesses. Rome depended on shipments of grain by sea from places like Egypt. As for sowing methods – probably over half of Europe's seed was wasted every year before the Chinese idea of the seed drill came to the attention of Europeans. Countless millions of farmers throughout European history broke their backs and their spirits by ploughing with ridiculously poor ploughs, while for two thousand years the Chinese were enjoying their relatively effortless method. Indeed, until two centuries ago, the West was so backward in agriculture compared to China, that the West was the Underdeveloped World in comparison to the Chinese Developed World. The tables have now turned. But for how long? And what an uncomfortable realization it is that the West owes its very ability to eat today to the adoption of Chinese inventions two centuries ago.

It would be better if the nations and the peoples of the world had a clearer understanding of each other, allowing the mental chasm between East and West to be bridged. After all they are, and have been for several centuries, intimate partners in the business of building a world civilization. The technological world of today is a product of both East and West to an extent which until recently no one had ever imagined. It is now time for the Chinese contribution to be recognized and acknowledged, by East and West alike. And, above all, let this be recognized by today's schoolchildren, who will be the generation to absorb it into their most fundamental conceptions about the world. When that happens, Chinese and Westerners will be able to look each other in the eye, knowing themselves to be true and full partners.

R.K.G. TEMPLE

Part 1
AGRICULTURE

ROW CULTIVATION AND INTENSIVE HOEING

SIXTH CENTURY BC

Growing crops in rows, and taking care to weed them thoroughly, may seem to us to be obvious and necessary processes. But they were not practised in Europe until the eighteenth century. As late as 1731, the agricultural propagandist Jethro Tull was trying to persuade European farmers to adopt what he called 'horse-hoeing husbandry', which involved growing crops in rows and hoeing them thoroughly.

The Chinese were doing this at least by the sixth century BC, and were thus a good 2200 years in advance of the West in one of the most sensible aspects of agriculture. A treatise of the third century BC, *Master Lu's Spring and Autumn Annals*, tells us: 'If the crops are grown in rows they will mature rapidly because they will not interfere with each other's growth. The horizontal rows must be well drawn, the vertical rows made with skill, for if the lines are straight the wind will pass gently through.'

At first, the seed was sown by hand along ridges, in a ridge-and-furrow pattern. By the first century BC at the latest, the multi-tube seed drill greatly increased the rate of sowing in rows, and with this went the intensive hoeing techniques which were pioneered by the Chinese. About the sixth or fifth century BC, cast-iron hoes were commonly available in China, after the unique advances the Chinese had made in their working of metals. A good iron hoe could be expected to last ten years for a hard-working farmer. Hoes and all agricultural tools took on a much longer life in the third century BC when the Chinese developed a malleable (non-brittle) form of cast iron. Then in about the first century BC, an improved design of hoe became widely available. Known as the swan-neck hoe, it was capable of weeding round plants without damaging them, and it had a variety of interchangeable blades. It was a splendid technological advance.

There was an ancient Chinese proverb: 'There are three inches of moisture on the end of a hoe.' And there is no doubt that careful hoeing does wonderfully conserve soil moisture. This was enormously important in north China, which is dry, often windy, and where the main crops are wheat and millet. Rice, which most Westerners think is found all over China, is mostly grown in the south, which is a quite different agricultural region.

From the *Treatise on Agriculture* of the great agriculturalist Wang Chen, published in 1313 AD, we have this colourful picture of how hand-hoeing was traditionally practised by the poorer farmers:

> **In the villages of the North they frequently form hoeing societies, generally of ten families. First they hoe the fields of one family which provides all the rest with food and drink, then the other families follow in turn over the ten-day period.... This is a quick and pleasant way of performing the task of hoeing, and if one family should fall ill or meet with an accident the others will help them out. The fields are free from weeds and so the harvests are always bountiful. After the autumn harvest the members of the society contribute bowls of wine and pigs' trotters for a celebratory feast.**

OPPOSITE (2) The fundamental agricultural innovation of sowing (and hoeing) crops in rows has been practised in China for 2500 years, but in the West it has only been practised for one tenth of that time – before that no-one seems to have thought of it.

ABOVE (3) Intensive hoeing and row cultivation of crops originated in China in or before the sixth century BC. Here we see the technique in operation during the Han Dynasty (207 BC–220 AD).

The more prosperous farmers did not have to rely on such methods, for they had animal-drawn hoes, which were obviously much quicker. The first type seems to have been a kind of plough without a mouldboard, having two sharp, pointed shares. This was dragged along, with the shares going each side of the ridge where the crop was growing in a row. It cut the weeds away from each side of the ridge, deepened the furrows or irrigation trenches, and further banked up the soil around the roots of the plants. This implement is mentioned in Liu Hsi's book, *Expositor of Names*, in the second century AD. Such horse hoes came with single or double blades, which could be adapted for ridging or for more shallow hoeing. By medieval times, an improvement called a 'goose-wing' was attached to the horse hoe. This consisted of two wide flaps which further increased the deepening of the trench and the piling of the soil around the roots. Each hoeing by the horse hoe saved several hoeings by hand, though it was common for a farmer to follow behind afterwards and hand-hoe a small number of intractable weeds, by way of finishing off the job.

THE IRON PLOUGH

SIXTH CENTURY BC

Of all the advantages which China had for centuries over the rest of the world, the greatest was perhaps the superiority of its ploughs. Nothing underlines the backwardness of the West more than the fact that for thousands of years, millions of human beings ploughed the earth in a manner which was so inefficient, so wasteful of effort, and so utterly exhausting, that this deficiency of sensible ploughing may rank as mankind's single greatest waste of time and energy.

Only the Chinese freed themselves from the tyranny of bad ploughs. And, when the Chinese plough was finally brought to Europe, it was copied and led directly to the European agricultural revolution (in combination with the growing of crops in rows and the use of the seed drill, also adopted from Chinese practice). Since the agricultural revolution of Europe is generally thought to have led to the Industrial Revolution, and to the West's superior power over the rest of the world, it is ironic that the basis of it all came from China, and was not by any means indigenous to Europe.

The most basic and universal form of plough is called an 'ard'. It has a shallow ploughshare and makes only a slight furrow, so is sometimes preferred in areas of continual winds and thin, dry soil. Ards can still be seen at work in Spain, for example. We have pictorial representations of these implements dating back to the third millennium BC at Uruk (in present-day Iraq). They were often made entirely of wood, and thus most have not survived.

Archeological evidence for early ploughs is very scanty in China. However, archaic Chinese writings from the fourteenth century BC give evidence for ploughing.

Triangular stone ploughshares for ards have been excavated in China which go back as far as the fourth or even early fifth millennium BC. Ox-drawn ards were therefore in use in China from neolithic times. Some sixteenth-century BC bronze ploughshares for true ploughs (more exactly, turn-ploughs) have been excavated in Tonkin in Vietnam, a region with which China had trade contacts at that time. Most Chinese ploughshares, however, seem to have been of wood at this time, and consequently have not survived.

By the sixth century BC, iron ploughshares became available in China, in the form either of iron laid over wood, or of solid iron. These were the first iron ploughs in the world. They were attached in a far better way than were ard shares in the West. Greek and Roman ard shares were usually simply tied onto the bottom of the sole with bits of rope; the two types, called 'stangle shares' and 'sleeve shares', were both flimsy and insecure compared to the Chinese ones, even when made of iron.

Improved iron supplies and casting techniques in China by the third century BC led to the design of ploughshares called *kuan*. At this period, the Chinese developed a malleable (non-brittle) cast iron which was far sturdier for use in agriculture. From the beginning, these *kuan* were advanced in their design, with a central ridge ending in a sharp point to cut the soil and wings which sloped gently up towards the centre to throw the soil off the plough and reduce friction. From about the time iron ploughs came into circulation, the bow-framed ard began to be ousted, in all but the lightest soils and windiest localities, by a heavier and more efficient square-framed turn-plough. Such proper ploughs could be used in heavier soils. These ploughs also made possible the working of much virgin land which could not have been ploughed by ards, which were too light and feeble. Heavy and waterlogged soils now became capable of proper cultivation.

By the first century BC, ploughshares attained widths of over 6 inches, and were capable of making really worthwhile furrows, scoured deep in the earth on each side of ample, wide ridges.

BELOW (4) A stone relief of the second century AD, from Yeng-tzu-shan in the southern province of Szechuan, showing a typical ancient Chinese plough drawn by an ox.

ABOVE (5) A traditional Chinese iron plough in use today. When these were introduced from China into Holland and England in the seventeenth century, they sparked the European agricultural revolution.

By the fourth century BC at the latest, the frame-plough was being officially promoted by government officials and literati. Nowhere else in the world at this time were there ploughs to compare with the Chinese ones. The sturdy, square frames, strong, heavy, well-designed shares and the new mouldboards were all factors well in advance of anywhere else. But perhaps of greater importance still was the use of an adjustable strut which precisely regulated the ploughing depth by altering the distance between the blade and the beam.

The new control this gave to the farmer meant that the plough could now be altered to suit whatever type of soil he encountered, from season to season, for different weather conditions and for different crops. The plough became a versatile tool indeed. For farmers, this was like going from the bow and arrow to the gun. The Romans could only adjust the depth of furrow by leaning more or less heavily on the beam – both a clumsy and an exhausting means of control. This was mostly the case all through medieval times in Europe.

By the second century BC in China, large numbers of private foundries for casting iron farming tools existed all over China. By 100 BC, the imperial government had established huge state foundries in most provinces. Iron was in fairly wide general use among the populace – so iron cooking-pots were quite common for the ordinary person. There was no shortage of these advanced iron ploughs, and they were not the rare possessions of rich people, as were the early European seed drills for the first two centuries of their use.

By the first or second centuries BC, four different kinds of mouldboard were widely available for ploughs. The mouldboard is of crucial importance. It is the twisted piece of the plough, above soil level, which guides the ploughed-up earth gently to one side, where it falls in a neat ridge and does not clog up the works. There was a smooth connection between the mouldboard and the share. There were different shapes and angles, so that the soil could be turned in different ways, landing in different patterns. Some of the earliest mouldboards already had the principles enunciated in 1784 by James Small, the

Scottish pioneer of scientific plough design, who wrote (unaware that he had been anticipated by 2200 years):

> **The back of the sock [share] and mouldboard shall make one continued fair surface without any interruption or sudden change. The twist, therefore, must begin from nothing at the point of the sock, and the sock and the mouldboard must be formed by the very same rule.**

The Chinese knew, too, that the extra weight of an iron ploughshare and mouldboard were more than compensated by the dramatic reduction in friction of ploughing. Arthur Young wrote in 1797, two millennia later:

> **It appears that the weight of the plough is of little consequence, very contrary to common ideas.... The weight of the plough is the least part of the horse's labour; the great object is the resistance met with in the cohesion of the earth; lightness does nothing to overcome this; it is effected by just proportions only.**

A good mouldboard design turns the clods right over, smoothly and with minimum friction, so that with a good plough, ploughing becomes like running a knife through butter, and just as a heavier knife will go through butter better, so a heavier plough will make a finer and deeper furrow with less trouble, if it has the proper design.

In Europe, mouldboards were completely unknown until late medieval times, and even then, they were extremely crude in their design. They were just flat bits of wood stuck on to the plough to provide an angled surface against which the upturned soil could collide and be deflected. The curved mouldboard, adopted from the outset in China, did not appear in Europe until the eighteenth century, and the lack of it probably caused more hardship to farmers than any other single factor. They had to stop repeatedly and scrape caught-up mud and weeds off their ploughs because there was no smooth connecting surface between the sharp, pointed share which cut the earth and the crude angled board which sent the clods to one side. The increased friction meant that huge multiple teams of oxen were required, whereas Chinese ploughmen could make do with a single ox, and rarely more than two.

Europeans had to pool their resources and waste valuable time and money in getting hold of six to eight oxen to plough the simplest field. This also meant leaving aside much more land for grazing, so that there was less for crops to produce food for human beings. It is no exaggeration to say that China was in the position of America and Western Europe today, and Europe was in the position of, say, Morocco. There was simply no comparison between the primitive and hopeless agriculture of

BELOW (FIG. 1) A drawing of the iron plough as described in Lu Kuei-Meng's book *The Classic of the Plough* of 880. The relevant parts are: (9) the ploughshare which cuts the soil; (8) the mouldboard above, which turned the soil over neatly as the plough moved forward; (6) the mouldboard brace; (7) the wooden nose of the sole, or slade, onto which the iron ploughshare was snugly fitted; (12) the beam; and (13) the beginning of the 'whippletree', which was attached to each side of the beast or beasts pulling the plough.

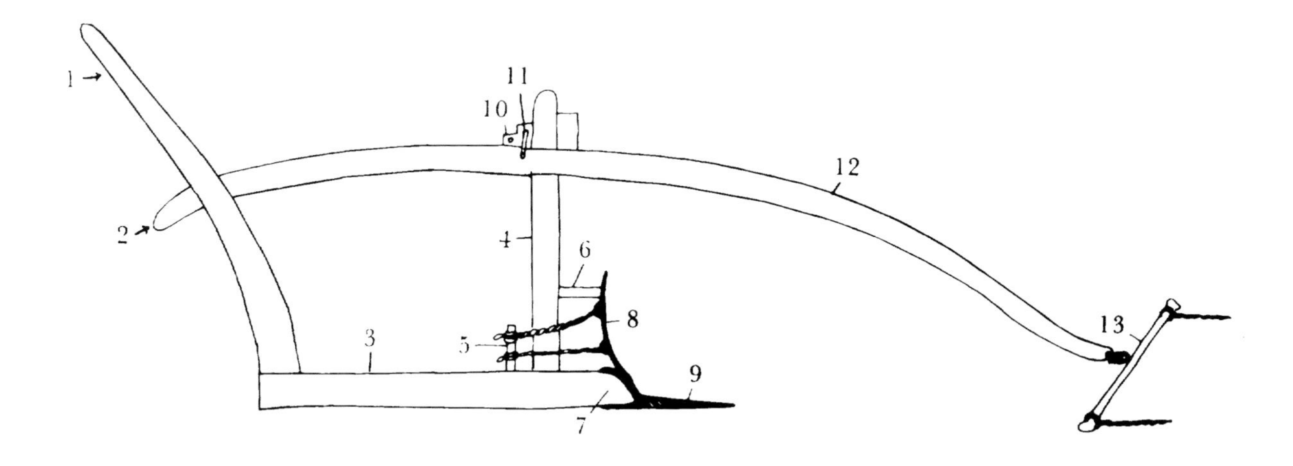

Europe before the eighteenth century and the excellent and advanced agriculture of China after the fourth century BC.

Chinese ploughs, with mouldboards, were brought to Holland in the seventeenth century by Dutch sailors. And because the Dutch were hired by the English to drain the East Anglian fens and Somerset moors at that time, they brought with them their Chinese ploughs, which came to be called 'Rotherham ploughs'. Thus, the Dutch and the English were the first to enjoy efficient ploughs in Europe. Another name for the Chinese design was the 'bastard Dutch plough'. It was extremely successful on wet, boggy land, and it was soon realized that it would be just as successful on ordinary land.

From England it spread to Scotland, and from Holland it spread to America and France. By the 1770s it was the cheapest and best plough available. Western designers adapted and improved it considerably over the succeeding decades. James Small's plough of 1784 was a step forward, and the various nineteenth-century ploughs of J. Allen Ransome were further improvements. Steel frames were adopted, and the modern plough was born in the nineteenth century as a result of these improved Chinese ploughs. There was no single more important element in the European agricultural revolution. When we reflect that only two hundred years have elapsed since Europe suddenly began to catch up with and then surpassed Chinese agriculture, we can see what a thin temporal veneer overlies our assumed Western superiority in the production of food.

EFFICIENT HORSE HARNESSES

FOURTH AND THIRD CENTURIES BC

THE TRACE HARNESS

China was the only ancient civilization to develop efficient horse harnesses. There were none in ancient Europe. Thus through most of man's history he has been severely handicapped by the lack of an efficient means of harnessing horsepower for transport. This had an enormous effect on the course of history.

From earliest times until the eighth century AD in the West (and, as we shall see, much earlier in China), the only means of harnessing horses was by the 'throat-and-girth harness'. It was an absurd method since the strap across the throat meant that the horse was choked as soon as he exerted himself. Yet for thousands of years, nobody could think of anything better. As long as man was restricted to the use of this pathetic harness, horsepower was all but useless for transport by cart. Even individual riders could half-strangle their mounts at a gallop. The confusion and slaughter of cavalry battles must have been much increased by inadequate harnessing. Long-distance rides would have been seriously impeded, no matter how good the horse and rider, by the fact that the poor horse was being not merely tired but also choked half to death. If ever the feebleness of human ingenuity has been displayed, it is by the fact that mankind was prepared to put up with the throat-and-girth harness for millennia.

Those who read about ancient Rome are often struck by the importance attached to the shipping of grain from Egypt. Without Egyptian grain, Rome

BELOW (FIG. 2) The three main forms of horse harness. (a) The throat-and-girth harness of Western antiquity, which severely choked the horse. (b) The trace harness, with its breast strap, which was in use in China by the fourth century BC. The horse can exert itself with this harness, since the pressure is on the chest bone (sternum) rather than the throat. The load is therefore borne by the horse's skeletal system rather than its windpipe. (c) The collar harness. From the third century BC the Chinese used this greater refinement. Again, the load is borne by the skeletal system by means of the pull on the chest bone (sternum). The collar is padded to avoid chafing the horse's skin.

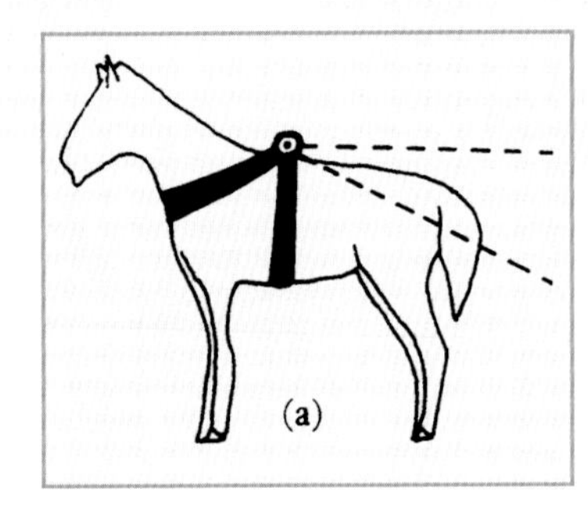

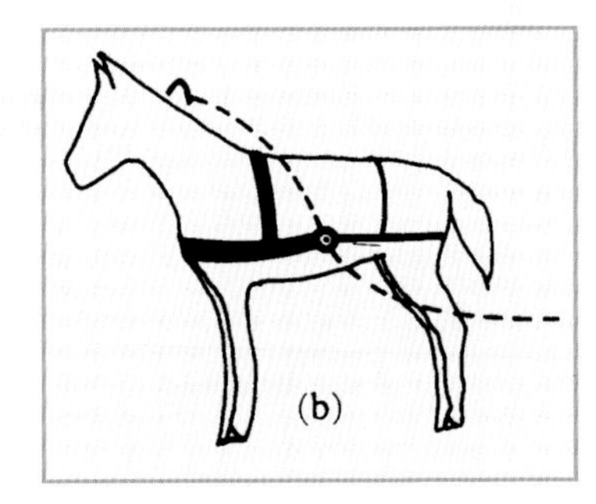

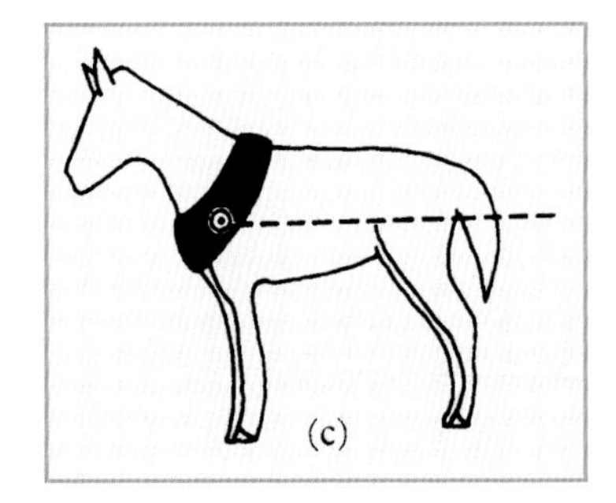

ABOVE (6) A Han Dynasty relief of about the first century BC, showing the use of the trace harness, with its breast strap, to pull a typical carriage of that date. A precursor of the umbrella protects the passengers from sun and rain.

must starve. But why? What was wrong with grain grown in Italy, one asks? Why was Rome dependent on ships from Egypt in order to be able to eat? The answer is simply that there was no horse harness capable of making it possible for Italian grain to be transported to Rome. We often overlook such technological factors when we seek to interpret events in the ancient world.

In about the fourth century BC the Chinese made a great breakthrough. A lacquered box of the period bears a painting which shows a yoke across the horse's chest, from which traces connect it to the chariot shafts. Although this cannot be considered a truly satisfactory harness, it shows that the throat-and-girth harness was abandoned in favour of a band across the breast of the horse. Soon, the hard yoke across the breast was also abandoned and replaced by the obviously more satisfactory breast strap, commonly called the 'trace harness'. There is no longer a strap across the horse's throat; the weight of the load is borne by the horse's chest and collar bones.

Experiments have been carried out to establish the relative efficiency of the different types of harness. Two horses harnessed in the throat-and-girth fashion can pull a load of half a ton. But a single horse in a collar harness (described below) can easily pull a ton-and-a-half. With a trace harness, the efficiency is only slightly less. As Needham says: 'The throat-and-girth harness would not have been able, therefore, to draw modern vehicles, even when empty.' The vehicles of the Greeks and Romans had to be so light, generally carrying no more than two people if they were for passengers, that effective transport by horse was impossible.

Needham suggests two factors which may have led the Chinese to invent the trace harness. There was the motivation of the Chinese, Mongols and Huns, living on the edge of the Gobi Desert where they were always getting stuck in the sand, from which horses using the throat-and-girth harness could not extricate them. Secondly, there was the use of human hauliers. Man's own experience of hauling, for example, canal boats upstream, meant that he was quickly aware of the inadequacy of a rope round the neck. Common sense dictated that the chest and collar bones should bear the weight. Therefore, the breast strap for the horse may well have been inspired by the breast strap used by humans.

The trace harness seems to have arrived in Europe by way of Central Asia. The Avars invaded Hungary from the East in 568 AD, and it is thought that they brought the trace harness with them. The same tribe brought the stirrup to Europe (see page 101). The harness spread to the Magyars, Bohemians, Poles and Russians. Evidence for trace harnesses has been found by archeologists in graves of the seventh to the tenth centuries. By the eighth century, the trace harness had made its way across Europe and it appears for the first time in a Western depiction on an Irish monument.

ABOVE (7) The collar harness in use in modern China.

The Vikings also acquired it. Various illustrations of it occur, notably in the Bayeux Tapestry of 1130, where it is shown in a ploughing scene.

THE COLLAR HARNESS

The most efficient harness is the collar harness. It may be seen in the diagram in Figure 2. In the beginning, this harness effectively overcame a deficiency in the horse's anatomy and supplied a horse with a feature of the ox. An ox has a perfectly horizontal spine, with a hump more or less above the shoulders. A yoke can be fitted there with great ease, enabling considerable weights to be pulled. But a horse's neck is on an upward slope and has no hump. The earliest Chinese collar harness provided the horse with an artificial 'hump', to which a yoke was then attached. In other words, the horse was transformed into an ox-substitute by the collar and the hump it created at the top of the horse's neck. The collar was padded to avoid rubbing and causing sores on the horse's back.

The earliest evidence for the collar harness in China may be seen in a rubbing from an ancient brick, showing the collar harness on three horses pulling a chariot. It dates from some time between the fourth and first centuries BC. Therefore, we must consider the collar harness as having been invented in China by the first century BC at the latest. This is a full thousand years before its appearance in Europe a century after the trace harness.

After some time, it was found by the Chinese that the collar could be used in another and simpler way: traces could be attached from the sides of the collar directly to the vehicle. It is this form of the collar harness which is used today all round the world.

In connection with the collar harness of the modern form, the Chinese invented the 'whippletree', which is an attachment to a vehicle. If two horses pull a modern cart, the collars and traces lead to the whippletree. The earliest evidence of the whippletree in China goes back to the third century AD, in use with oxen.

One other factor in the invention of the collar harness might have been that similar collars had for some time been used to place baggage onto Bactrian camels. China had a Camel Corps and considerable familiarity with large numbers of the animals by the second century BC at the latest. The camel pack-saddle at that time was a felt-padded, horseshoe-shaped wooden ring, and, with some modifications, could have been used on a horse.

THE ROTARY WINNOWING FAN

SECOND CENTURY BC

The Chinese were about two thousand years ahead of the West in their approach towards the winnowing of grain, the means used to separate out husks and stalks from the grain after harvest and threshing. The easiest method goes back even before the cultivation of crops: the grain is thrown up into the air, preferably in a strong wind, so that the chaff is blown away while the grain falls down to the ground. Later, winnowing baskets were used, which required dextrous handling. With the right kind of rhythmic wrist movement, one can separate the heavy grain from the chaff, which is gradually tipped over the edge of the basket, leaving the grain behind. Later still, the winnowing sieve was introduced.

But the Chinese were not satisfied with waiting for a strong wind for the tossing method, or with the slow and laborious basket and sieve methods. By the second century BC they had made a brilliant invention: the rotary winnowing fan. Models of them have been found in ancient tombs, made of pottery and with miniature working parts.

Grain was put into a hopper and was subjected to continuous streams of air which came from a crank-operated fan. The fan had a large air inlet behind it, and was set at the end of a broad sloping tunnel leading towards the grain. The air from the fan blew the chaff away and out through a vent. The grain then fell down into a chaff-free pile beneath. One type of rotary winnowing fan was portable. This was an important development because the original machine was expensive, but could now be rented out, so enabling owners to recoup their costs. Another type was operated, not by a small crank handle suitable for one person, but by a treadle connected to a crank which left the operator's hands free to carry out other tasks at the same time.

The primary use of rotary winnowing fans was generally in the south of China, in connection with rice. Although apparently invented and first used in the north with wheat and millet, after several centuries the use shifted southwards and the device, for various economic reasons, was largely forgotten in the north. Many farmers reverted to the traditional tossing and hand-sifting of grain because they could not afford the fans.

A poem by the poet Mei Yao-Ch'en, or Mei Sheng-Yu, survives from the eleventh century, celebrating the rotary winnowing fan:

There on the threshing floor stands
the wind-maker,
Not like the feeble round fans of
the dog-days,
But wood-walled and fan-cranked,
a cunning contrivance,
He blows in his tempest all the
coarse chaff away,
Easy the work for those manning
the handles –
No call to wait for the weather, the breezes
To free the fine grain from its husks,
that our fathers
Needed for tossing their baskets on high.

The rotary winnowing fan was exported to Europe, brought there by Dutch sailors between 1700 and 1720. Apparently they had obtained them from the Dutch settlement of Batavia in Java, Dutch East Indies. The Swedes imported some from south China at about the same time and the Jesuits had taken several to France from China by 1720.

LEFT (8) A traditional rotary winnowing fan in use in modern China, in the southern province of Yunnan.

ABOVE (9) A glazed pottery model from the Han Dynasty (207 BC–220 AD), showing a grain pounder and a rotary grain mill. The figure on the right is turning the crank handle of a rotary winnowing fan, which was invented in the second century BC in China, but did not reach Europe until two thousand years later. Grain was poured into the open receptacle at the top for winnowing inside the machine. (The Seattle Art Museum, Eugene Fuller Memorial Collection.)

Until the beginning of the eighteenth century, no rotary winnowing fans existed in the West. Until then, tossing into the air with a shovel and sifting in winnowing-baskets were the primary techniques, though roughly fanning with canvases, blankets, and so on was occasionally practised from at least the early sixteenth century. But this rudimentary fanning was very rare and only adopted by the most professional farmers. It is estimated that the most advanced winnowing technique in common use in Europe before the eighteenth century was the winnowing basket. It could yield about 99 pounds of winnowed grain per hour, if done by an expert. But in the eighteenth century the Swedes studied a Chinese rotary winnowing fan which they had transported to Gothenburg, and discovered it could process an astonishing seventeen barrels of grain per day. European engineers were not slow to improve on the Chinese design, adapt it to European grain sizes, and even (something which the Chinese never did) combine it with threshing by machine as well.

Once more, we see the Chinese giving the West one of the most essential tools for the Western agricultural revolution. We might note, by way of a footnote, that the actual traditional Chinese rotary winnowing fan, though developed and improved immensely in the West, still survives in its basic form in the Third World countries of today, where it is found to be cheaper and more practical than modern Western versions.

THE MULTI-TUBE ('MODERN') SEED DRILL

SECOND CENTURY BC

It may come as a surprise to those who are unfamiliar with the history of Western agriculture to learn that the West had no seed drills until the sixteenth century AD. Until the seed drill was adopted, broadcasting of seed by hand was practised. This was appallingly wasteful, and it was common for as much as half the crop to have to be saved for sowing the next year. Of those seeds which then germinated, many fell into hollows in the ground, with the resulting plants all being clumped together, competing for moisture, light and nutrients. Also, proper weeding was out of the question because it was impossible to get at the weeds.

Although it never made its way to Europe, the Sumerians of the Middle East had a primitive single-tube seed drill 3500 years ago. But it was the multi-tube seed drill invented by the Chinese in the second century BC (and adopted also in India) which made possible the efficient sowing of crop seed for the first time in history. The drill is pulled along behind the horse, ox, or mule and dribbles the seed at a controlled rate into straight rows.

Small iron seed drill shares have been excavated in China, dating from about the second century BC. A government official named Chao Kuo introduced the seed drill to the metropolitan area of the capital in 85 BC. We read of this in a surviving fragment of a book called *On Government*:

> **Three ploughshares were all drawn by one ox, with one man leading it, dropping the seed and holding the drill simultaneously. Thus, 100 *mu* could be sown in a single day....**

The later agriculturalist, Wang Chen, gives more details:

> **The *huo* ... or sowing shares are the shares ... that trace the drill, like a triangular ploughshare but smaller, with a high ridge down the centre, 4 inches long and 3 inches wide. They are inserted into the two holes at the back of the seed-drill's feet and bound tightly to the crosspiece. The share bites 3 inches or so deep into the soil, and the seed dribbles down through the foot of the drill, so it is sown very deep in the soil and the yield is improved. Soil tilled with a seed-drill looks as if it had been gone over with a very small plough.**

The Chinese system was at least ten times as efficient as the European one, and could be up to thirty times as efficient, in terms of harvest yield. And this was the case for seventeen or eighteen hundred years. Through all those centuries, China was so far in advance of the West in terms of agricultural productivity that the contrast, if the two halves of the world had only been able to see it, was rather like the contrast today between what is called the 'developed world' and what is called the 'developing world'.

The inspiration for the first Western seed drills came from China. But because seed drills were mostly used in north China, far from the ports of south China frequented by Europeans, actual specimens were not transported to Europe for examination. What came over was a rather imprecise account and description of the device. Word of mouth combined with inadequate descriptions and pictures in Chinese books meant that the frustrated Europeans receiving this news could not quite fully understand it. They were therefore forced to reinvent the seed drill. As a result, European seed drills were based on principles quite different from the Chinese principles. This was a case of 'stimulus diffusion' – the transmission of an idea without the accompanying details of construction. So the Europeans finally got their seed drills, but at the cost of working it all out from scratch.

The earliest European seed drill was patented by the Venetian Senate in 1566; its inventor was Camillo Torello. The earliest for which we have detailed descriptions was that of Tadeo Cavalina of Bologna, in 1602; it was very primitive. The first really sound seed drill in Europe was that developed by Jethro Tull. This drill was first produced soon after 1700, and descriptions of it published in 1731. However, seed drills of this and successive types were both expensive and unreliable, as well as fragile. Sturdy, sound seed drills became available in quantity in Europe only in the mid-nineteenth century. Until that time, Jethro Tull's vision of an agricultural revolution was somewhat delayed in its full form.

There had been a very fine seed drill in the eighteenth century in Europe, invented by James Sharp, but it was only for single-row sowing and was too small, so that its perfect functioning did not attract sufficient interest. Basically, it was a lack of engineering skill which made European seed drills essentially ineffective and uneconomical until the middle of the nineteenth century. Thus, two centuries of knowledge of seed drills were wasted, due to the failure to exploit the principles properly.

OPPOSITE (10) The multi-tube seed drill was in use in China by the second century BC. Europeans had no seed drills at all until the sixteenth century. Here we see spring wheat being sown in north China, in an engraving published in 1742 in *Compendium of Works and Days*, compiled by O-Erh-T'ai.

北耕兼種圖
麥粟粱皆用此具
種子
鐵尖
鐵尖

Part 2
ASTRONOMY AND CARTOGRAPHY

RECOGNITION OF SUNSPOTS AS SOLAR PHENOMENA

FOURTH CENTURY BC

In the West, the heavens were supposed to be so perfect that no such thing as a sunspot could be thought possible. Most of the sunspots seen in the West before the seventeenth century were explained away as transits of the Sun by the planets Mercury and Venus. The theory of 'perfection of the Heavens' forbade the admission of any imperfections on the surface of the Sun. Consequently, it was assumed that these 'blemishes' were planets or small invisible satellites.

The Chinese suffered from no such preconceived insistence on 'perfection'. Since sunspots are sometimes large enough to be seen by the naked eye, the Chinese naturally saw them. The earliest surviving record we have of their observations would seem to be some remarks by one of the three known early astronomers in China. He was Kan Te, who lived in the fourth century BC. He and two contemporaries, Shih Shen and Wu Hsien, drew up the first great star catalogues. Their work was fully comparable to that of the Greek Hipparchos, though two centuries earlier.

Kan Te appeared to be referring to sunspots when he spoke of solar eclipses which began at the centre of the Sun and spread outwards. Although he was incorrect, this has the merit of being eminently reasonable, and showed that Kan Te recognized that the sunspots were solar phenomena, and characteristics of the Sun which partly darkened its surface.

The next indication of a sunspot observation dates from 165 BC. We are told in a much later encyclopedia, *The Ocean of Jade*, that in that year, the Chinese character Wang appeared in the Sun. This was therefore a sunspot which appeared not round, but shaped like a cross with a bar drawn across the top and the bottom. The astronomer D.J. Schove accepts this as the world's earliest precisely dated sunspot. The recording of sunspot observations in the voluminous official imperial histories of China commenced on 10 May 28 BC. But systematic Chinese observations of sunspots probably began at the latest by the fourth century BC, and only the loss of much literature of that time denies us more specific information.

There are reasons for believing that the early natural philosopher Tsou Yen, the loss of whose writings is so much lamented, included sunspot observation in his school's curriculum. If the works of Tsou Yen could ever be recovered from one of the Han tombs, as have works on medicine and other subjects, a huge gap would be filled in our knowledge of early science.

Most people today believe that sunspots were first observed in the West by Galileo, who is also supposed to have been the first person to 'invent' or at least use the telescope. Neither belief is true. Galileo most certainly did not invent the telescope, though he gave it prominence, and courageously advocated its use to study the heavens. As for the observation of sunspots, the earliest clear reference to them so far found in Western literature is in Einhard's *Life of Charlemagne*, of about

OPPOSITE (11) The Chinese had recognized that sunspots were features on the surface of the Sun by the fourth century BC. Until the sixteenth century Western observers thought they were objects in intervening space. This illustration of sunspots is from a Chinese manuscript by the Ming Emperor Chu Kao-Chih entitled *Essay on Astronomical and Meteorological Presages*. Written in the year 1425, this work was never printed, but the manuscript survives. This painting was done by the Emperor himself. (Cambridge University Library.)

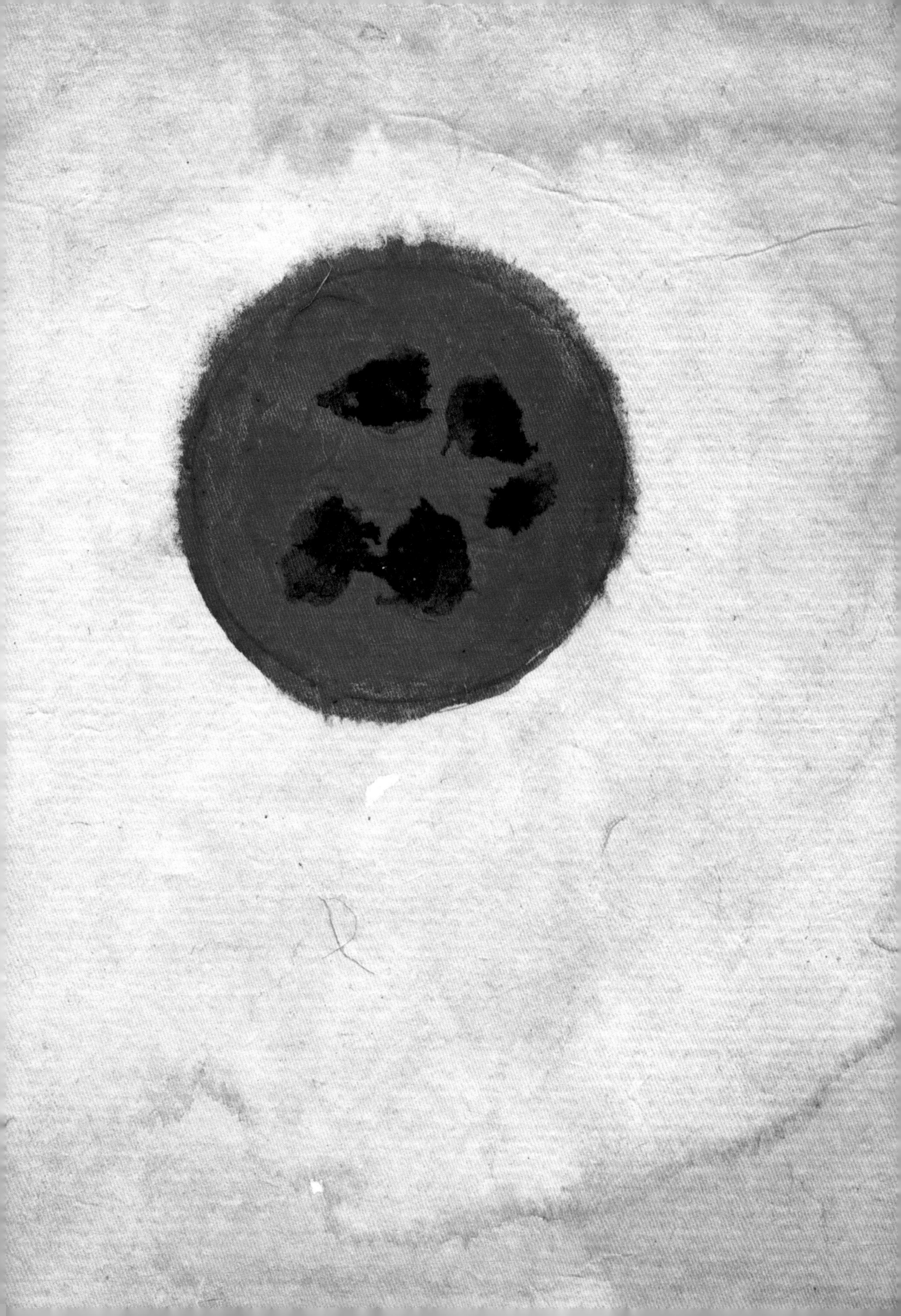

807 AD, or eight centuries before Galileo's first observation of sunspots in 1610.

Later sunspot observations in the West were made by the Arab Abu al-Fadl Ja'far ibn al-Muqtafī in 840 AD, by Ibn Rushd about 1196, and by Italian observers around 1457. Galileo's priority in the seventeenth century was disputed by the Jesuit Christopher Scheiner in Holland, by Fabricius in Germany, and by Thomas Hariot in England, all of whom seem to have seen the sunspots before Galileo did.

Needham has counted the numbers of sunspot observations in the official histories between 28 BC and 1638 AD, and has found 112 instances. There are also hundreds of notices of sunspots in other Chinese books during the centuries, but no one has ever had the time or stamina to collect them all together in a body. These Chinese records are the oldest and longest continuous series of such observations in the world. Many of the specific descriptions are full of interest. Extensive, though incomplete, lists of them exist in various old Chinese encyclopedias, and some of these have been published in English. The extreme importance of these lists may be seen from the fact that modern astronomers have tried to analyze them to determine cycles in the occurrence of sunspots. It is well known that a period of about eleven years appears during which sunspots seem to increase, decrease, and then begin to increase again.

Since these cycles have an effect upon the Earth's ionosphere and weather ('magnetic storms' in the atmosphere are related to sunspot periods), the more we know about these the better. By studying the most complete available list of the Chinese material, the Japanese astronomer Shigeru Kanda believed he had detected a 975-year cycle in sunspots. If the Sun has a period of 975 years, that may have important implications for weather cycles on our planet, and directly affect the futures of us all.

QUANTITATIVE CARTOGRAPHY

SECOND CENTURY AD

The science of map-making took a great step forward when Chang Heng invented quantitative cartography in the second century AD. Chang was the inventor of the first seismograph (page 177) and one of China's leading scientific figures. He first applied the grid system to maps so that positions, distances and itineraries could be calculated and studied in a more scientific way. The Chinese tradition of applying grids to maps eventually developed so far that by the Middle Ages schematic grid-maps were actually appearing with only the grid and names, omitting the map itself. (See Plate 13 (page 34).) Geographical positions on such extreme mathematical abstractions were determined by counting the X and Y co-ordinates of the grid. This was quantitative cartography of a kind more suitable to a computer, since it entirely dispensed with images.

Chang Heng's own works on cartography are lost. He wrote a book entitled *Discourse on New Calculations* which apparently laid the groundwork for the mathematical use of the grid with maps; another of his books seems to have borne the title *Bird's-Eye Map*; and we know that he presented a map to the emperor in 116 AD. The most important piece of evidence we have regarding his role, however, is the statement in the official history of the Han Dynasty that he 'cast a network of co-ordinates about heaven and earth, and reckoned on the basis of it'.

In the third century, Chang had a successor in the scientific study of cartography. He was P'ei Hsiu, who in 267 was appointed Minister of Works by the first emperor of the Chin Dynasty. The official history quotes the following from his preface to a great map in eighteen sheets which he presented to the emperor:

> **The origin of maps and geographical treatises goes far back into former ages. Under the three dynasties [Hsia, Shang and Chou] there were special officials for this. Then, when the Han people sacked Hsien-yang, Hsiao Ho collected all the maps and documents of the Ch'in. Now it is no longer possible to find the old maps in the secret archives, and even those which Hsiao Ho found are missing; we only have maps, both general and local, from the later Han time. None of these employs a graduated scale and none of them is arranged on a rectangular grid. Moreover, none of them gives anything like a complete representation of the celebrated mountains and the great rivers; their arrangement is very rough and imperfect, and one cannot rely on them. Indeed some of them contain absurdities, irrelevancies, and exaggerations, which are not in accord with reality, and which should be banished by good sense....**

In making a map there are six principles observable:
1) The graduated divisions, which are the means of determining the scale to which the map is to be drawn.
2) The rectangular grid (of parallel lines in two dimensions), which is the way of depicting the correct relations between the various parts of the map.
3) Pacing out the sides of right-angled triangles, which is the way of fixing the lengths of derived distances (i.e., the third side of the triangle which cannot be walked over).
4) Measuring the high and the low.
5) Measuring right angles and acute angles.
6) Measuring curves and straight lines. These last three principles are used according to the nature of the terrain, and are the means by which one reduces what are really plains and hills to distances on a plane surface.

If one draws a map without having graduated divisions, there is no means of distinguishing between what is near and what is far. If one has graduated divisions, but no rectangular grid or network of lines, then while one may attain accuracy in one corner of the map, one will certainly lose it elsewhere (i.e., in the middle, far from guiding marks). If one has a rectangular grid, but has not worked upon the principle of pacing out the sides of right-angled triangles, then when it is a case of places in difficult country, among mountains, lakes or seas (which cannot be traversed directly by the surveyor), one cannot ascertain how they are related to one another.... But if we examine a map which has been prepared by the combination of all these principles, we find that a true scale representation of the distances is fixed by the graduated divisions.... When the principle of the rectangular grid is properly applied, then the straight and the curved, the near and the far, can conceal nothing of their form from us.

P'ei Hsiu's actual map has not survived. Note that it was placed in the secret archives. This was not exceptional, for throughout history, and especially in China, the possession of superior maps was the key to political and military success, analogous to having advanced strategic weapons today. Shen Kua, in his *Dream Pool Essays* of 1086, gives the following illuminating story:

In the Hsi-Ning reign-period [1068 to 1077 AD] ambassadors came from Korea bringing tribute. In every *hsien* city or provincial capital which they passed through they asked for local maps, and these were made and given to them. Mountains and rivers, roads, escarpments and defiles, nothing was omitted. When they arrived at T'iehchow they asked for maps, as usual, but Ch'en Shu-Kung, who later became Prime Minister but was at that time in charge of the military guard at Yangchow, played a trick on them. He said that he would like to see all the maps of the two Chekiang provinces with which they had been furnished, so that he could copy them for what was now wanted, but when he got hold of them, he burnt them all, and made a complete report on the affair to the emperor.

It is easy to understand how so many early maps did not survive; they were not copied, and were frequently destroyed. Their information was simply too dangerous to risk its falling into the wrong hands. However, two splendid maps of the eleventh century carved in stone survive. Both are preserved in the Pei Lin Museum at Sian, in China. One of them, 'Map of the Tracks of Yü the Great', has a rectangular grid laid over it. It is generally superior to the other map, and especially so with regard to the coastal details; also it includes the Shantung peninsula, which the other has omitted. But the other incorporates much more accurate information about the south-western rivers. The two maps therefore each display a regional bias.

Another great map is that of Chu Ssu-Pen, prepared between 1311 and 1320. This map existed only in manuscript for two centuries but was finally printed in about 1555 in the *Enlarged Terrestrial Atlas* in an edition by Lo Hung-Hsien (1504-64). Lo says of the map:

Chu Ssu-Pen's map was prepared by the method of indicating the distance by a network of squares, and thus the actual geographic picture was faithful. Hence, even if one divided the map and put it together again, the individual parts in the east and west fitted faultlessly together.... His map was seven feet long and therefore inconvenient to unroll; I have therefore now arranged it in book form on the basis of its network of squares.

和寧
火州
土魯番
哈密
沙州
赤斤
阿端
合黎河
曲先
罕東
西寧
永昌
鎮番
莊浪
應理
寧夏
固原
平涼
慶陽
鳳翔
會
鞏昌
臨洮
疊
貴德
五嶺白大庾始安
臨賀桂陽揭陽
松潘
江源
龍安
漢中
漢源
保寧
順慶
重慶
四川
成都
朶甘思
烏思藏
番
黎
馬湖
敘州
越巂邛部
建昌
永寧
順寧
麗江
瀾滄江
姚安
大理
蒙化
景東
鎮沅
楚雄
武定
雲南
曲靖
烏撒
烏蒙
東川
畢節
普定
貴州
貴陽
都勻
思南
石阡
鎮遠
思州
黎平
銅仁
南寧
臨安
昆明
廣西
元江
車里
老撾
孟艮
孟養
孟定
廣南
太平
鎮安
思明
安南
雲遠
蒙光
通西
可難
騰衝
南甸
鎮西
雲龍
干崖
隴川
緬甸
江頭城
古城
八百媳婦
木邦
大古剌

沙漠
泰寧
合蘭府
開平
應昌
兀良哈
大寧
河套
東勝
大同
宣府
北京順天
薊鎮
山海關
遼陽
鴨綠江
朝鮮
榆林
延安
山西太原
平陽
真定
保定
河間
山東濟南
青州
萊州
登州
陝西西安
河南開封
歸德
淮安
揚州
南陽
鄖陽
襄陽
鳳陽
廬州
安慶
黃州
承天
荊州
漢陽
湖廣武昌
南京應天
鎮江
蘇州
松江
湖州
浙江杭州
寧波
紹興
台州
溫州
處州
金華
嚴州
徽州
池州
九江
南康
饒州
廣信
江西南昌
岳州
常德
長沙
辰州
衡州
寶慶
袁州
撫州
建昌
邵武
建寧
延平
福建福州
興化
泉州
漳州
汀州
南安
南雄
韶州
惠州
潮州
永州
平樂
廣西桂林
梧州
肇慶
高州
廣東廣州
雷州
瓊州
五指
昌國
日本
琉球

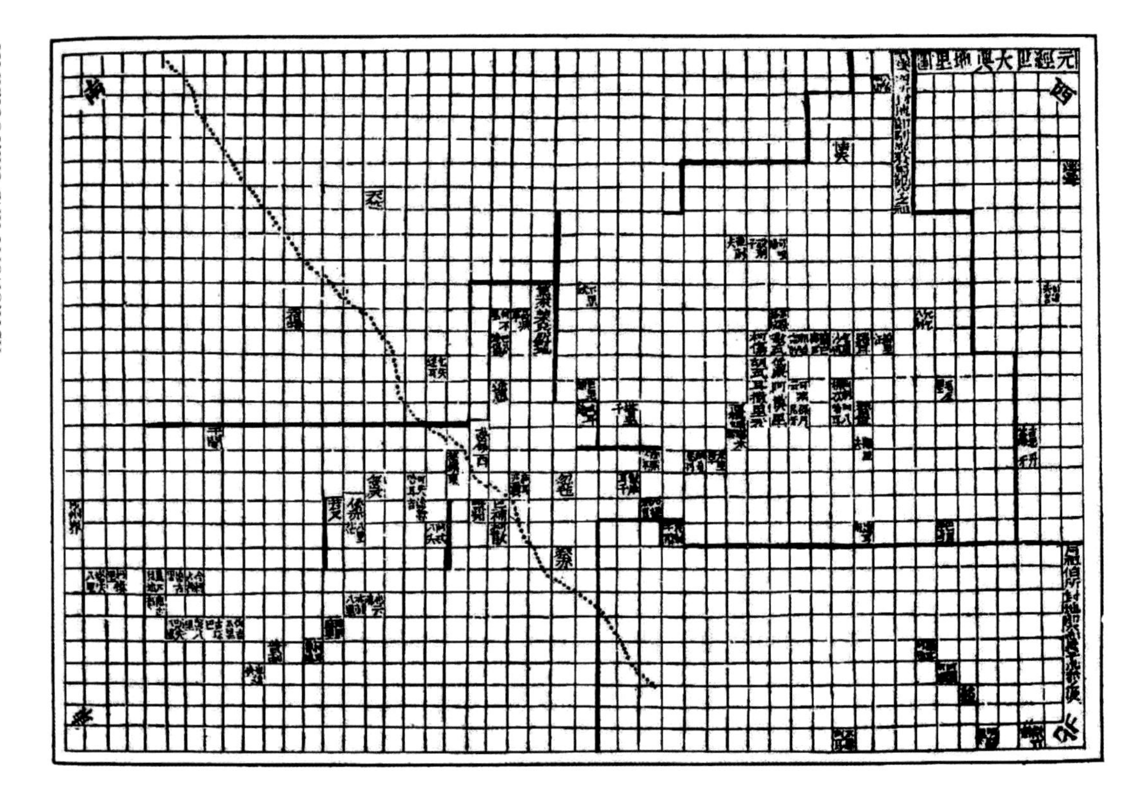

ABOVE (13) The schematic grid has entirely usurped the pictorial aspect of this map of the north-western part of China, which has North in the bottom right-hand corner. Places are simply marked off where they lie on the grid, and other surface features are ignored. This is known as the 'Mongolian Style' of cartography, from the fact that it was practised in 1329 under the Mongol (Yuan) Dynasty of China. The map is from the book of that date, *History of Institutions of the Yuan Dynasty*.

Thus we see the grid used successfully to reduce the size of the map in a manner analogous to photographic reduction. (See Plate 12 (page 32–3).)

What was happening in Europe while all this was going on? After the time of Ptolemy (c. 120-170 AD), Western map-making degenerated under the influence of religion to a point scarcely credible. For instance, a well-known Western world map from a manuscript of 1150 is so pathetically inaccurate that it hardly even rates the description of 'map'. It was only during the Renaissance that respectable maps once more began to be found in Europe. Fourteenth-century sailing charts (called portolans) were of good quality, but reasonable maps of larger areas did not appear in Europe until the fifteenth century – thirteen hundred years after Chang Heng introduced scientific, quantified cartography in China.

DISCOVERY OF THE SOLAR WIND

SIXTH CENTURY AD

Comet tails always point away from the Sun, blown that way by the 'solar wind'. The Chinese were history's most noted observers of comets. The computation of approximate orbits for about forty comets appearing before 1500 have been based almost entirely upon Chinese records of their sightings. Comet movements were described with such precision by the Chinese that many precise trajectories across the sky can be drawn on a star map, simply from reading an ancient Chinese text. This report of 1472 describes one comet's movements:

> Suddenly it went to the north, touched the star 'Right Conductor' (in Boötes), and swept through the 'Enclosure' of stars in Virgo, Coma Berenices and Leo, touching *nu* Coma Berenices, 2629 Coma Berenices, E Leonis, and 2567 Leonis. Its tail now pointed directly toward the west. It swept transversely across a to k Comae Berenices.... On a *chia-mao* day its tail had greatly lengthened. It extended from east to west across the heavens. The comet then proceeded northwards, covering about 28°, touched *iota*, *theta* and *chi* Boötes, swept through the Great Bear, and passed near the three small stars at the north of Canes Venatici and *chi* Ursae Majoris....

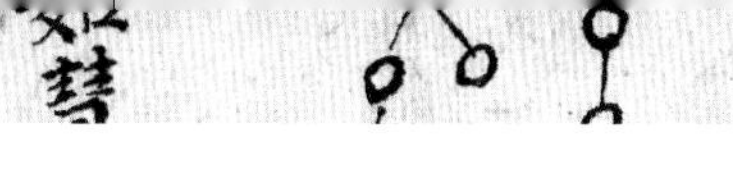

It will be noticed with what attention the precise position and direction of the tail of each comet was studied. (The stars named here are obviously the modern equivalents, the Chinese names being omitted.)

From at least the seventh century AD, and probably the sixth, the Chinese observations of comet tails had been refined enough to establish the principle that comet tails always point away from the Sun. We find a clear statement of this principle in the official history of the T'ang Dynasty for 635 AD: 'In general, when a comet appears in the morning, its tail points towards the west, and when it appears in the evening, its tail points towards the east. This is a constant rule. If the comet is north or south of the Sun, its tail always points following the same direction as the light radiating from the Sun.'

The astronomers of that time already realized that comets shine by reflected light, like the Moon. It is now known that comet tails always point away from the Sun because they are so tenuous that the force of the 'solar wind' pushes them away into that position. In other words, radiation from the Sun has sufficient force to act upon them as a wind would do.

Did the Chinese in the sixth century AD merely discover this empirically, or go so far as to formulate the notion of the 'solar wind' itself? There is no unambiguous evidence, but on balance it is highly likely that the Chinese did not even need to formulate the theory of the 'solar wind'. It would have been so congenial to their underlying assumptions about the Universe that it would probably have been taken for granted as the explanation as soon as they had formulated (as in the text of 635 AD above) the principle that comet tails followed the same direction as the Sun's radiation. For Chinese literature is replete with countless references to the *ch'i* of the Sun's radiation. This concept of *ch'i*, which is essentially untranslatable, can in this context be thought of as the 'emanative or radiative force' coming from the Sun. To the ancient Chinese, it would have been obvious, believing as they did in this *ch'i*, that the Sun's *ch'i* was strong enough to blow the tails of comets away from the Sun, as in a high wind. And that the Chinese conceived of space as being full of strong forces, we can see in our account of manned flight with kites (page 191).

BELOW (14) Comets (or 'broom stars') are depicted here in relation to different star groupings in the sky. These are illustrations from a manuscript of an unpublished work entitled *True Canon of Understanding the Mysteries* (*T'ung-hsüan Ching*). By careful observation of the tails of comets over the millennia, the Chinese were able – no later than the sixth century AD – to postulate the existence of a solar wind which blew the tails away from the sun. The text on the right says: 'If a comet invades the Ti, Fang, or Hsin stars, there will be floods and war in the area of the Sung State. The people there will suffer great vicissitudes, and this will happen within three years.' In the middle: 'If a comet invades the Wei star or Chih star, there will be serious drought for one year, and a shortage of food production in the region of the Yan State, and this will happen very soon.' On the left: 'If a comet invades the Hsü and Wei stars, it indicates that there will be a poor harvest and trouble for the people, within one year. This sign never fails.' (Collection of Robert Temple.)

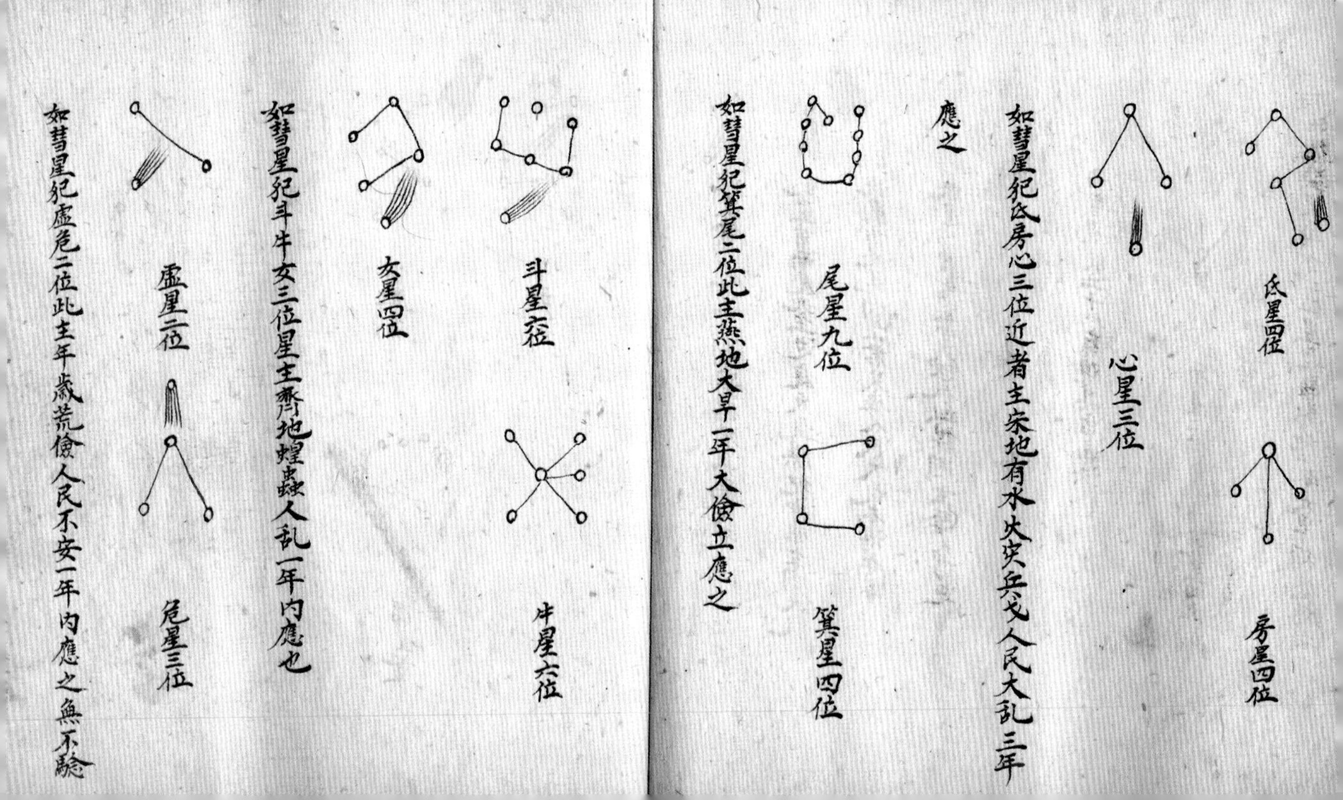

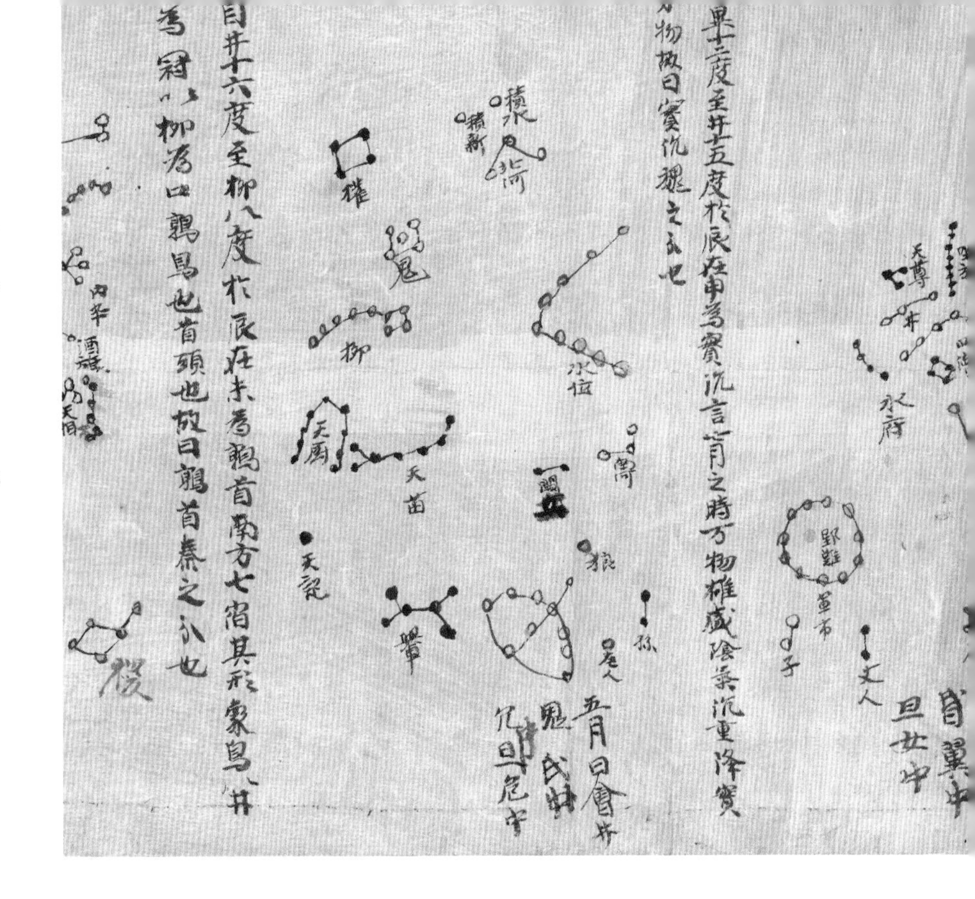

RIGHT (15) A Chinese star map drawn on a 'Mercator' projection, c.940, discovered by Sir Aurel Stein at Tunhuang. Here we see three hour-angle segments of the sky. Among the stars and constellations shown are Orion, Canis Major, Canis Minor, Lepus, Cancer and Hydra. They have different configurations in Chinese astronomy; Sirius, the Bow Star, is the target at which the bow and arrow at bottom left are pointing. Some stars are differently coloured from others, to indicate that they were discovered, or at least commented upon, by different astronomers. (British Library, London.)

THE MERCATOR MAP-PROJECTION

TENTH CENTURY AD

The Mercator map-projection is the map of the world most commonly seen on the walls of classrooms, in which Greenland is enormous and the north and south polar regions appear larger than Europe and America.

The projection is a cylindrical projection; that is, if one inserted a transparent globe of the Earth in the centre of a hollow cylinder and then turned on a light-bulb inside the globe, the features of the Earth's surface thrown onto the cylinder would be those of the Mercator projection. The equator is a straight line running across the middle of the Mercator projection and only the features near the equator are anything like their proper shapes. The higher up and lower down on the globe, the more the features are distorted by being cast onto portions of the cylinder which are further away. The projection is virtually useless for land travel, but is very popular at sea because it has the peculiar feature that a navigational course drawn on it comes out as a straight line, whereas with other maps such courses are arcs.

The Mercator projection originated, in Europe, as the presumed invention of Gerardus Mercator, which is the Latin name of Gerhard Kremer (1512-94), a Flemish mathematician and geographer. He published the first map on 'Mercator's projection', a navigation map, in 1568.

The cylindrical projection, however, was used by the Chinese centuries before Mercator. A manuscript star map dating from about 940 AD is preserved in the British Library (see Plate 15 (above)). This map presents the celestial globe (that is, the sky portrayed as a globe) as projected onto a surface by the cylindrical projection technique. The Chinese divided the sky into twenty-eight sections called *hsiu*, which were rather like the sections of an orange. They were 'lunar mansions' (stages of the moon's progress through the sky) with the pole at their centre. In this star map, the *hsiu* are represented as long rectangles centred on the equator and very distorted towards the poles.

A century-and-a-half later, Su Sung (see page 121-2) published further Mercator-style map-projections in his book *New Design for a Mechanized Armillary Sphere and Celestial Globe*, published in 1094. One of these had a straight line running across the middle as the equator and an arc above it, the ecliptic. The rectangular boxes of the lunar mansions are clearly seen, with the stars near the equator being more tightly packed together and those near the poles spread further apart. Su Sung published two star maps on the 'Mercator projection' and two on polar projections. These are the world's oldest published star maps of any kind.

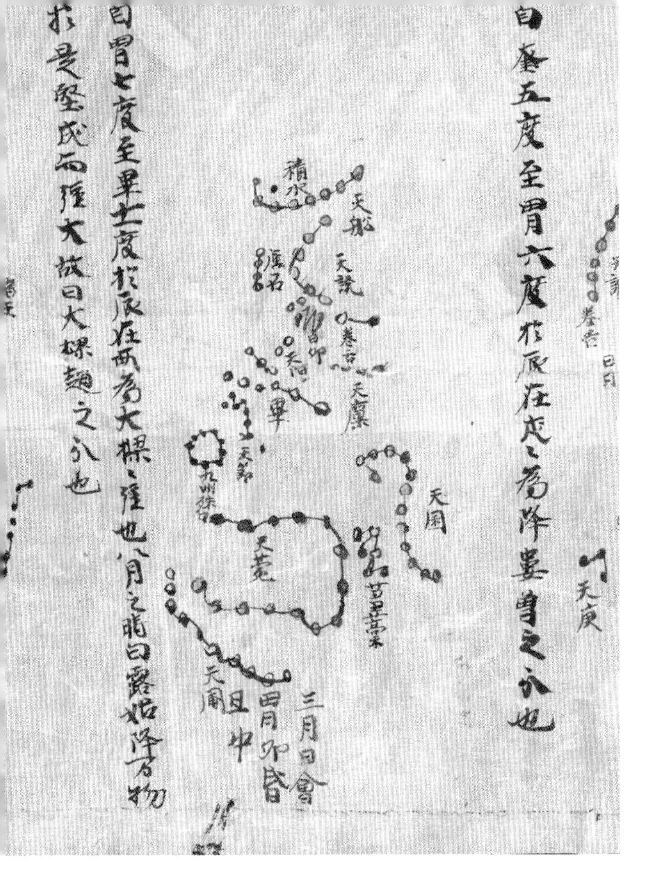

EQUATORIAL ASTRONOMICAL INSTRUMENTS

THIRTEENTH CENTURY AD

Modern astronomical observatories derive from a Chinese, not a European tradition, which makes an understanding of sky positions easier. They are oriented and mounted according to what is known as the equatorial system of astronomy. This is traditionally Chinese, and it goes back to at least 2400 BC. It takes the equator as the horizontal circle around the side of the instrument, and the pole as the top point. This may seem simple and obvious, but it was not the system used by our own European ancestors. In our tradition, which is called 'ecliptic', the two horizontal circles which were of importance were not the equator but the horizon and the ecliptic (the circle described by the Sun's motion in the sky, which is the same plane as the Earth's orbit around the Sun). Whereas Europeans more or less ignored the equator, the Chinese largely ignored the horizon and the ecliptic. But when making astronomical observations, it came to be realized in seventeenth-century Europe that the Chinese system of equatorial astronomy was more convenient and showed greater promise. So it was adopted by Tycho Brahe (1576–1601) and his successors and is still the basis of modern astronomy today.

The Chinese system was really very simple. Everything was conceived of as radiating from the celestial pole, as if it were the point where the stem of an orange were attached. The sky was then divided up into twenty-eight sections rather like orange segments, known as *hsiu*, or lunar mansions. Each one of these *hsiu* contained certain star constellations which were known and given names. Since the pole star and the stars near it never set beneath the horizon at any time during the year (whereas most stars do), the Chinese gave greatest attention to them, and by noticing

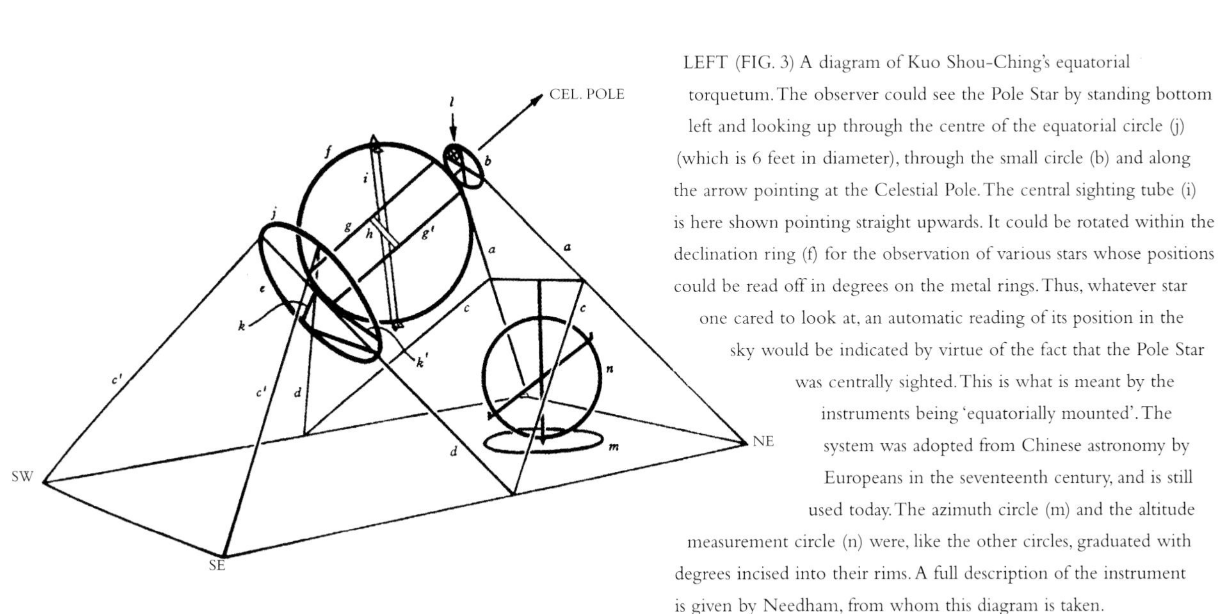

LEFT (FIG. 3) A diagram of Kuo Shou-Ching's equatorial torquetum. The observer could see the Pole Star by standing bottom left and looking up through the centre of the equatorial circle (j) (which is 6 feet in diameter), through the small circle (b) and along the arrow pointing at the Celestial Pole. The central sighting tube (i) is here shown pointing straight upwards. It could be rotated within the declination ring (f) for the observation of various stars whose positions could be read off in degrees on the metal rings. Thus, whatever star one cared to look at, an automatic reading of its position in the sky would be indicated by virtue of the fact that the Pole Star was centrally sighted. This is what is meant by the instruments being 'equatorially mounted'. The system was adopted from Chinese astronomy by Europeans in the seventeenth century, and is still used today. The azimuth circle (m) and the altitude measurement circle (n) were, like the other circles, graduated with degrees incised into their rims. A full description of the instrument is given by Needham, from whom this diagram is taken.

where the stars at the top of a sky segment were, they could then precisely specify where the stars at the bottom of the same sky segment were, even though they might be invisible beneath the horizon.

To do this sort of thing with precision required instruments. The Chinese had such superior expertise in metal casting, having after all invented cast iron (see page 44), that they made large and impressive instruments of bronze and iron. These would take the form of huge metal rings precisely graduated with the degrees of the circle. Different rings representing different sky-circles would then be joined together at the two points where they crossed one another, forming what looked like the skeletons of spheres. These we call armillary spheres, from the Latin word *armilla*, meaning 'bracelet'. One ring would obviously represent the equator. Another would represent what is called the meridian, which is a great sky-circle that passes directly over one's head and also through the pole.

These instruments also had sighting-tubes, through which one could peer at particular stars. The sighting-tube was moved along the equator ring until a star was found. Then one counted the number of degrees marked on that ring back to the meridian ring, which stood up from it vertically. As soon as the degrees had been counted, the exact position of the star along the equator would become clear and one could tell what sky segment it was in. By such means as these, star maps were drawn with great precision, and positions of stars were recorded. The sky became not a maze of points of light, but a sensibly ordered arrangement of constellations. And to make sense of the night sky is, after all, what astronomy is.

Armillary rings still exist today in the form of the mountings which allow the great modern telescopes to be oriented to particular points in the sky. But instead of mere sighting-tubes today, we have the additional aids of lenses or mirrors to peer at our stars, so we can not only find the star positions, but stare closely at the stars, magnified greatly, and observe things about them that are not apparent to the unaided eye. In the past, finding the positions of stars was an end in itself; today it forms the basic framework enabling us to carry out further studies.

The Chinese astronomer Keng Shou-Ch'ang introduced the first permanently mounted equatorial armillary ring in 52 BC, and the astronomers Fu An and Chia Kuei in 84 AD added a second ring to show the ecliptic. Chang Heng (who also invented the first seismograph – see page 177) added a ring for the

LEFT (16) The Suchow Planisphere, or map of the night sky, dating from 1193 AD, drawn for the heir to the throne, who would rule as Emperor Ning Tsung (1195–1224), by his imperial tutor Huang Shang. In 1247, this planisphere was carved in stone to preserve it forever; this image is a rubbing of that stone (only ten such rubbings were ever authorized). The inner circle which matches the exterior circle is the equator, called 'the red road which encircles the heart of heaven.... With regard to the two poles it occupies exactly the mid-distance between the south and the north and constitutes the heart of Heaven, wherein dwells the central spirit'. The eccentric inner circle is the ecliptic ('the yellow road'). The parallel pair of wiggly lines drawn from the bottom right upwards, sweeping across to the top left, represent the Milky Way ('the river of heaven'). The lengthy text of the Suchow Planisphere states that there are 1565 named stars in the night sky (although only 1440 of them are represented on this planisphere). Between 4 and 5 o'clock is the constellation of Canis Major, known to the Chinese as the Bow Star and depicted by a bow and arrow. (Chinese depictions of constellations are wholly different from those in the West). The many lines radiating outwards from the centre of this planisphere, which divide the sky into unequal sections, are indications of the boundaries of what the Chinese called the 'lunar mansions'. These form the basis of a highly complex theoretical division of the sky which is fundamental to traditional Chinese astronomy. I also have in my collection a very old and large original painted planisphere which omits the ecliptic and is wholly equatorial, but it is somewhat damaged and there are portions which have been obliterated, so it was not suitable for reproduction here. (Collection of Robert Temple.)

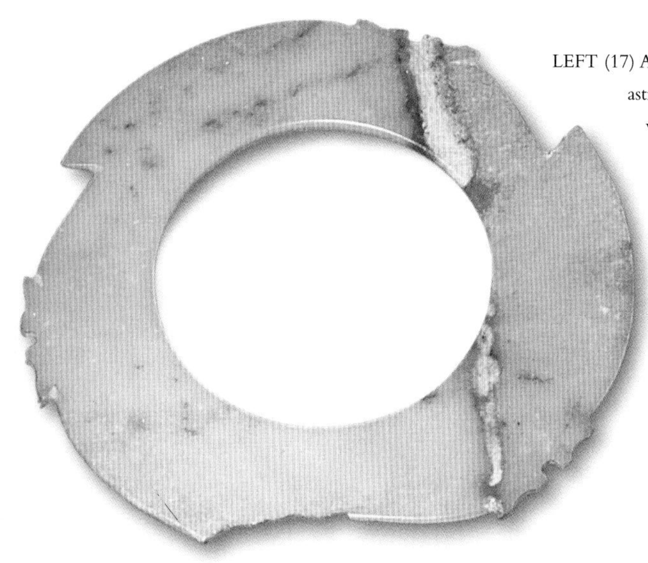

LEFT (17) A jade circumpolar constellation template. An ancient Chinese astronomer would have held this up at approximately arm's length, with the Pole Star in the centre. The jagged outer edges of the circle would then have perfectly matched the pattern of stars surrounding the Pole Star: that is, a star would twinkle in each tiny notch in the jade. These templates varied with time because of the shifting positions of the Stars. Although the date of this one has not been determined, similar ones have been dated to 1000 BC and 600 BC. Unique to China, they were used to find orientations in the sky in the study of the constellations, and were among the earliest and most primitive equatorial astronomical instruments. (Victoria and Albert Museum, London.)

meridian in 125 AD, as well as one for the horizon. By that date, the true armillary sphere could be said to exist in its full form. But Chang Heng was not yet satisfied. He made non-observational armillary spheres rotate by water power, about 132 AD. He used a water wheel powered by a constant pressure-head of water in a clepsydra (water-clock mechanism) to rotate his sphere slowly. Here is a description in an official history:

> **Though many have discoursed upon the theory of the heavens, few have been as well acquainted with the principles of the *yin* and the *yang* as Chang Heng ... Chang Heng made his bronze armillary sphere and set it up in a closed chamber, where it rotated by the force of flowing water. Then, the order having been given for the doors to be shut, the observer in charge of it would call out to the watcher on the observatory platform, saying the sphere showed that such and such a star was just rising, or another star just culminating, or yet another star just setting. Everything was found to correspond with the phenomena like the two halves of a tally.**

Another passage in the official history tells us:

> **In the time of the emperor Shun Ti (126–144 AD) Chang Heng constructed a computational armillary, which included the inner and outer circles, the south and north celestial poles, the ecliptic and the equator, the twenty-four fortnightly periods, the stars within [i.e., north of] and beyond [i.e., south of] the twenty-eight *hsiu*, and the paths of the sun, moon and five planets. The instrument was rotated by the water of a clepsydra [lit. dripping water] and was placed inside a closed chamber above a hall. The transits, risings and settings of the heavenly bodies shown on the instrument in the chamber corresponded with those in the actual heavens, following the motion of the trip-lug and the turning of the auspicious wheel.**

This water-powered rotating instrument was a tremendous tool for demonstrating (and computing) the movements of the heavenly bodies. It must have given astronomers of the time a sense of intellectual power. Chang Heng is also said to have constructed one which was actually for observation of the stars – obviously not inside a chamber – which was also water-powered. However, full descriptions of it do not survive.

An astronomical instrument which is one step up on the armillary sphere is the torquetum, invented by the Arabs. With this instrument, all the various rings and so on are not nested together in a single sphere, but are mounted at various different parts of a set of struts, in a way more convenient and advanced than is allowed by the constraints of the single sphere. A non-equatorial form of the torquetum was transmitted by the Arabs to the Chinese. In Plate 18 (opposite) and Figure 3 (page 37) we see the great metal equatorial torquetum of the astronomer Kuo Shou-Ching of 1270 which was called the 'Simplified Instrument'. This was because it was made purely equatorial, and had all the Arab ecliptic components left out. We are fortunate that this great instrument survives, for the Jesuits melted down many such metal instruments in China as scrap. The base plate of this instrument measures 18 by 12 feet, the revolving meridian ring in the centre is 6 feet across, and its sighting-tube is thus the height of a man.

Kuo had taken the torquetum of the Arabs and adapted it to the equatorial system of the Chinese. Needham says of this huge machine that 'it constitutes the precursor of all equatorial mountings of telescopes', and of Kuo's accomplishment:

> ...though Arabic influence may have been responsible for suggesting its construction, Kuo adapted it to the specific character of Chinese astronomy, namely equatorial coordinates. And in so doing, he fully anticipated the equatorial mounting so widely used for modern telescopes.

Needham believes that some knowledge of this reached Tycho Brahe in Denmark three centuries later, and led to Brahe's taking up equatorial astronomy for his instruments. As for the transmission of the idea from China, Needham believes it came from the Arabs to Gemma Frisius in 1534, and from him to Tycho Brahe. And through him and his successor, Johannes Kepler, modern European astronomy came to be equatorial in the Chinese manner. Since Kuo Shou-Ching in thirteenth-century China, M.C. Johnson emphasizes that: 'Actually our present-day equatorial mounting has made no further essential advance.' Here indeed is a Chinese contribution to an aspect of modern Western science which is barely appreciated.

BELOW (18) Kuo Shou-Ching's famous 'Simplified Instrument' of 1270, at the Purple Mountain Observatory in Nanking, seen from the east. It was originally constructed for use at Linfen in Shansi, but was moved to Nanking during the Ming Dynasty, when it was no longer appreciated that the difference of 3¾° in latitude rendered it useless. At its original latitude, the Pole Star would have appeared at the precise point required for the use of this equatorially mounted instrument. Cast in bronze and weighing several tons, the declination ring (centre) is 6 feet across. See Figure 3, page 37.

Part 3 ENGINEERING

SPOUTING BOWLS AND STANDING WAVES

FIFTH CENTURY BC

The strange objects known as spouting bowls, which date from the fifth century BC or earlier, are some of the most impressive artifacts of ancient Chinese science. They are bronze bowls which are so precisely proportioned that when just the right amount of water is placed in them, and the two handles are rubbed at the correct rate, water spouts appear. Needham has witnessed spouts 3 inches high being generated in this way. Apparently the spouts can be teased up to 3 feet high by an expert.

The vibrations set up in the bronze by rhythmic rubbing of the bowl with the hands is supposed to cause the precise frequencies necessary to generate what are called 'standing waves', whose only movement is up and down. The spouts of water rise from these waves.

A bronze spouting bowl is, in effect, the equivalent of a vibrating string on a musical instrument; its vibrations, like those of the string, have certain points, called nodes, where the vibration drops to zero. If a coin is placed on the correct node of a vibrating piano wire, the note will continue to sound because that precise point on the wire is in any case motionless. At the nodes, therefore, within the substance of the bronze bowl, standing waves are formed when the bowl is set vibrating.

The four nodes of the bowl were somehow determined in advance, so the bowl could be cast with decorations pointing at them – in this case, lines emanating from the mouths of fish, an amusing pictorial guide to the nodes, which determine where the real spouts will occur, generated as the spouts are in the areas between the nodal points.

These spouting bowls, which are triumphs of precision casting, are evidence of total control of the dimensions and proportions which determine nodal points in vibrating bronze objects. We may regard the spouting bowls as a specialized form of bell adapted to demonstrate the actions of sound upon liquids. The effects with liquids other than water are not known. The bowls may have had a utilitarian purpose, for example to give an index of viscosity, or they may have been used to perform a sophisticated test for adulteration of some precious liquid.

The subject of standing waves is one of the most intriguing questions in physics. Looked at from one point of view, a standing wave is a product of the collision of two equal and opposite waves. There is a modern mathematical theory showing clearly that only certain wavelengths may exist in standing waves inside vessels. Perhaps the ancient Chinese had some empirical grasp of this theory.

A standing wave is extraordinary by any standards, for it balances the normal dispersive forces which would destroy it in most circumstances with other, peculiar, counteracting forces which give the wave continued cohesion and allow it to prolong its existence. A standing wave is therefore a profoundly apt model for ancient Chinese concepts of 'the mean', or 'following the Tao'. It is also an entity

OPPOSITE (19) A photograph taken by Cecil Beaton of a boy demonstrating a bronze spouting bowl at the Chinese Temple at North Hot Springs. The handles at either side are rubbed rapidly. As long as there is no grease on the hands, and the water is at exactly the right level in the bowl, a strange humming sound is given off and water spouts appear from standing waves set up in the liquid. Such spouts can apparently spurt as high as 3 feet in the air.

created entirely out of its surrounding medium without the addition of any other distinguishing substance, and is therefore a perfect example of what the Chinese believed all things to be: everything emanated from the Tao and dissolved back into the Tao, like a standing wave in a spouting bowl.

The Chinese believed the Tao to be 'the way of the Universe', which was always manifested from within rather than imposed from without by a god standing outside the Universe (a Western concept utterly alien to all Chinese thought). The spout spurting up from one of these bowls is a form teased out of uniform and featureless fluid by the mysterious forces of resonance and sound – the cosmic forces which led to the creation and dissolution of all forms in that great sea, the Universe.

CAST IRON
FOURTH CENTURY BC

Blast furnaces for cast iron are now known to have existed in Scandinavia by the late eighth century AD, but many readers will be amazed to learn that cast iron was not widely available in Europe before 1380. The Chinese, however, practised the technique from at least the fourth century BC. What were the reasons for the Chinese superiority? There were a number of factors. China had good refractory clays for the construction of the walls of blast furnaces. The

RIGHT (20) This cast iron pagoda in Jining, Shantung Province, was built in 1105. It was cast layer by layer in octagonal sections, and stands 78 feet high. There are several such cast iron pagodas in China, some even older than this one.

Chinese also knew how to reduce the temperature at which the iron would melt. They threw in something which they called 'black earth', which contained much iron phosphate. If up to 6 per cent of phosphorus is added in this way to an iron mixture, it reduces the melting point from the normal 1130°C to 950°C. This technique was used in the early centuries, ceasing before the sixth century AD, when proper blast furnaces came into use which needed no such assistance.

Coal, which gave a high temperature, was used as a fuel from the fourth century AD, and probably earlier. One method was to put the iron ore in batteries of elongated, tube-like crucibles, and pack these round with a mass of coal which was then burnt. This had the extra advantage of excluding sulphur from the process. As late as the seventeenth century, unsuccessful attempts were being made in England to use coal for smelting iron.

Cast iron was at first the preserve of private speculators, who grew rich from it. But the Han Dynasty nationalized all cast iron manufacture in 119 BC so that the emperor could monopolize it. At about that time there were forty-six imperial Iron-Casting Bureaux throughout the country where government officials supervised the mass-production of cast iron goods.

The widespread availability of cast iron in ancient China had many side effects. It led to the innovation of the cast iron ploughshare in agriculture, along with iron hoes and other tools. Iron knives, axes, chisels, saws and awls all became available. Food could be cooked in cast iron pots, and even toys were made of cast iron. Cast iron statuettes of various animals have been found in Han Dynasty tombs dating between the second centuries BC and AD. Cast iron moulds for implements dating from the fourth century BC have also been discovered. Hoes and axes would have been cast in these, in either bronze or iron.

The expertise in cast iron enabled pots and pans to be made with very thin walls, impossible by other iron technology. One extremely important result was that salt could be mass-produced from evaporated brine, which can only be done in such thin pans. This in turn led the Chinese to exploit natural gas by deep drilling. This was in order to tap the energy from the burning gas to evaporate the vast quantities of brine required for the giant salt industry (which the Han Dynasty also nationalized along with the iron industry in 119 BC). The salt and gas industries could not have existed without the cast iron industry.

In the third century BC, the Chinese discovered how to make a malleable cast iron by annealing (that is, by holding it at high temperature for a week or so). It was then not brittle, and would therefore not shatter if subjected to a violent shock. This meant that objects like ploughshares could survive striking large stones with considerable force. Cast iron had something of the elasticity of wrought iron, but with the much greater strength and solidity that came from being cast. It was almost as good as steel.

Swords became 50 per cent longer when made in iron. In 218 BC a man named Chang Liang tried to assassinate the emperor using an iron mace which weighed 160 pounds. Perhaps it was too heavy, for he did not succeed. Some of the ancient Chinese feats of casting iron are so impressive as to be almost unbelievable, even when the results are before our eyes. For instance, there is a cast iron

ABOVE (21) The largest single piece of cast iron from ancient China, and still one of the largest such objects in the world: the Great Lion of Tsang-chou in Hopei Province, erected in the year 954 by the Emperor Shih Tsung in commemoration of his campaign against the Liao Tartars. It weighs about 40 tons, and stands 20 feet high and 16 feet long. It is not solid: its thickness varies from 8 inches to only 1½ inches in places. The sections were poured at the same time, but some of the joints between them are weak. This gigantic object was made more than four hundred years before any cast iron was to be available in Europe.

pagoda seen in Plate 20 (page 44). This fantastic structure can be precisely dated to the year 1105. It is 78 feet high, cast storey by storey. There are several other such iron pagodas.

But these were by no means the biggest cast iron buildings made by the Chinese. The largest seems to have been the temple built on the orders of the Empress Wu Tse in 688 AD. This building, which no longer survives, was a staggering 294 feet high, based on an area of about 300 square feet. It was in the form of a three-storey pagoda, and atop the structure was a 10-foot cast iron phoenix covered in gold plate.

Perhaps the grandest cast iron structure of all was not actually a building. The Empress Wu Tse had an octagonal cast iron column built, called the 'Celestial Axis Commemorating the Virtue of the Great Chou Dynasty with Its Myriad Regions'. It was built in 695 AD upon a base of cast iron 170 feet in circumference and 20 feet high. The column itself was 12 feet in diameter and rose 105 feet in the air; on top was a 'cloud canopy' 10 feet high and 30 feet in circumference. On top of this in turn stood four bronze dragons each 12 feet high supporting a gilded pearl. We have a record of the amount of metal used in this construction – 2,000,000 *catties*, which is about 1325 tons. The largest single cast iron object ever made (the pagodas were obviously not a single piece) was erected on the orders of the Emperor Shih Tsung of the Later Chou Dynasty in commemoration of his campaign against the Tartars in 954 AD. This extraordinary object, 20 feet tall, still stands and is known as the Great Lion of Tsang-chou. It is not solid, but its walls vary from 1½ to 8 inches in thickness. It is reproduced in Plate 21 (page 45).

THE DOUBLE-ACTING PISTON BELLOWS

FOURTH CENTURY BC

The double-acting piston bellows was a pump for air or fluids which enabled a continuous stream to be expelled. It was the capacity to provide continuous blasts of air which was a crucial factor in enabling the Chinese to achieve their superiority in metallurgy for so many hundreds of years. We do not know who invented the double-acting piston bellows, or exactly when it appeared. But it seems to have been widespread by the fourth century BC, so that we can probably safely place its invention in the fifth at least. It appears to be referred to in the great philosophical classic of Lao Tzu as follows:

> **Heaven and Earth and all that lies between,**
> **Is like a bellows with its tuyère [nozzle**
> **for the blast];**
> **Although it is empty it does not collapse,**
> **And the more it is worked the more**
> **it gives forth.**

The most conservative estimate dates Lao Tzu's book to about the fourth century BC, though the traditional dating is the sixth. Either could be correct. The third line is the crucial one, since it seems to refer to the continuous action of the double-acting piston bellows, which is a bellows that indeed does not collapse.

We see a double-acting piston bellows in Plate 22 (below), and in Figure 4 (opposite) we see a diagram of how it works. It is really a most simple but ingenious invention. A piston is pushed in and pulled out of a rectangular box, which acts as a cylinder. Feathers or folded pieces of soft paper are wedged round the piston to make sure that it is both airtight and lubricated in its passage. (These are the ancestors of modern piston-rings.) There is an inlet valve at each end of the box. When the piston is being pulled, air is sucked in from the far end. When

RIGHT (22) A late-eighteenth-century painting showing a portable double-acting piston bellows in use by a travelling tinker. (Victoria and Albert Museum, London.)

it is being pushed, air is sucked in from the near side. (See 'A' and 'B' in the diagram.) In both inward and outward strokes, air is sucked into the cylinder; and in both cases air in the portion being compressed (that on the other side of the piston) is pushed into a side chamber where it is then expelled through the tuyère, or nozzle. The device is so simple that it is difficult to understand why Westerners never invented it. It could provide blasts not only of air but of liquids, and an example of the latter is given in the account of the flame-thrower (page 254).

The earliest published pictures of double-acting piston bellows are found in a quaint book of 1280 entitled *Book of Physiognomical, Astrological and Ornithomantic Divination according to the Three Schools*.

In the West in ancient times, single-acting pumps were certainly known from at least the second century

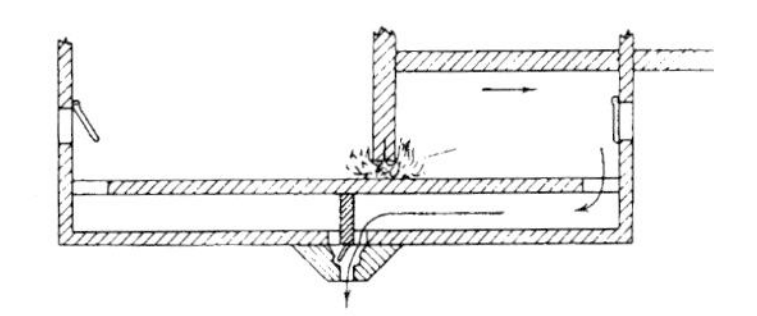

ABOVE (FIG. 4) A cut-away diagram of the double-acting piston bellows. A lengthwise section shows the piston being pulled out (to the right), compressing the air within the right-hand chamber and forcing it out of the nozzle at the bottom. When the piston is pushed in again, air is compressed in the chamber at the left, and is forced from the nozzle once more. The feathers shown are used to lubricate the movement of the piston while at the same time sealing off the air around its edges. The bellows is double-acting: whether the piston is being pushed or pulled it compresses air in alternate chambers, so that air rushes continuously from the nozzle. The nozzle has a valve like a swinging door, which here is seen pushed to the left; on the in-stroke of the piston it swings back to the right.

LEFT (23) Ulrich Hausmann, an expert on Chinese rural architecture, found this double-acting piston bellows in use in a farmhouse kitchen in Feng Huo village near Sian, and persuaded the housewife to step aside for a moment while he took the photograph. The bellows is in the enclosed box in the foreground, its wooden handle on the left. It provides a blast to keep the oven fire glowing, on this occasion for the preparation of traditional steamed bread.

BC; and syringes go back to indistinct antiquity, having been used for embalming in Egypt. But these were force-pumps which ejected air or liquid only in the outward stroke. The double-acting air bellows reached Europe, presumably from China, in about the sixteenth century. In 1716, J.N. de la Hire turned the principle to use with liquid for the first time in the West in a double-acting water pump.

THE CRANK HANDLE

SECOND CENTURY BC

If you want to turn a wheel which is mounted in place for some mechanical purpose, then it is silly just to push the wheel round. The obvious thing to do is to stick a rod into the side of the wheel, use that as a handle, and turn it. This is known as a crank handle. But no one ever thought of it until the Chinese adopted the idea in the second century BC.

The Chinese invention of the crank handle was for use on their rotary winnowing machine (page 24), which was crucial to agriculture. (A winnowing machine with a crank handle may be seen in Plate 8 (page 24).) It was only then that sticking a rod at right angles into the side of a wheel was seen to be useful as a handle to turn the wheel. Not for eleven hundred years would the same idea occur to Westerners.

The only other people who came close to the invention of the crank handle were the ancient Egyptians. They had a kind of oblique proto-crank as early as 2500 BC, used for a primitive hand drill. It featured a slanted handle at the top for turning, but was not attached to a wheel, nor was it at right angles.

The Chinese used the crank also on well-windlasses, querns, mills and the many machines used in the silk industry. The earliest published picture of a crank handle appeared in Wang Chen's *Treatise on Agriculture* in 1313. But the oldest surviving depiction of a crank handle is on a miniature pottery farmyard

ABOVE (24) A glazed pottery model of a farmyard scene from the first century BC, during the early Han Dynasty. A dog sits in the foreground. In the wall on the left there is an open chute, the opening for a rotary winnowing fan. Below the chute a crank handle used to operate the winnowing fan protrudes from the side of the wall. Crank handles were not known in Europe before the ninth century AD. (Nelson-Atkins Museum of Art, Kansas City, Missouri.)

model found in a Han Dynasty tomb, dating from about the first century BC (see Plate 24 (above)). The oldest European evidence for the crank handle is a picture of one on a rotary grindstone in the Utrecht Psalter, a manuscript of about 830 AD.

THE 'CARDAN SUSPENSION', OR GIMBALS

SECOND CENTURY BC

The 'Cardan suspension', or gimbals, takes its usual name from Jerome Cardan (Girolamo Cardano, 1501–76, who is also mentioned in the section on mathematics (pages 155-6). But Cardan neither invented the device nor claimed to have done so. He merely described it in his very popular book *De Subtilitate* (1550). The gimbals appeared in Europe as early as the ninth century AD) but it was invented in China by the second century BC at the latest.

This invention is the basis of the modern gyroscope, making possible the navigation and 'automatic pilots' taken for granted in modern aircraft. Anyone who has been fortunate to enter a nineteenth-century gypsy caravan will have noticed affixed to the walls the brass gimbals that hold lamps which remain upright no matter how violently the cart may jolt on the road. These interlocking brass rings can be moved around as much as you like, but the lamp suspended in the centre never turns over. This is the basic idea of the 'Cardan suspension'. A series of rings inside one another are each joined at two opposing points, enabling them to twist and turn freely. Consequently, if a heavy weight, such as a lamp, is positioned upright in the centre, it will remain upright. Whatever motions might occur to the rings around it will be taken up by the rings themselves, leaving the lamp unmoved. By the eighteenth century, Chinese mariners were using a gimbal-mounted compass. A ship's magnetic compass mounted in this way was free of disturbance by waves.

The earliest textual reference to the gimbals which has been found is in a poem called *Ode on Beautiful Women*, composed about 140 BC, by Ssuma

Hsiang-Ju. It describes a seduction scene, and among the bedclothes, hangings, furniture and so on, the poet mentions 'the metal rings containing the burning perfume'. The supposed inventor of the gimbals was Fang Feng, though his identity is uncertain. More than three centuries later, about 189 AD, the clever mechanic Ting Huan was given credit for inventing the gimbals a second time. A book called *Miscellaneous Records of the Western Capital* records:

> **Ting Huan ... also made a 'Perfume burner for use among Cushions', otherwise known as the 'Bedclothes Censer'. Originally such devices had been connected with Fang Feng but afterwards the method had been lost until Ting Huan again began to make them. He fashioned a contrivance of rings which could revolve [in all three dimensions], so that the body of the burner remained constantly level, and could be placed among bedclothes and cushions. So it was given the name of Bedclothes Censer.**

Ting Huan was connected also with the invention of the zoetrope, or 'magic lantern' (see page 98).

The gimbals turns up throughout Chinese literature after this point. In 692 AD the Empress Wu Hou was presented with 'wooden warming-stoves which, though rolled over and over with their iron cups filled with glowing fuel, could never be upset'. A book of 1734 called *Topography of the West Lake Region of Hangchow* mentions that 'interlocking pivots' were mounted inside paper lanterns which were then kicked and rolled along the streets without the lamps inside being put out. They were known as rolling lamps.

The various gimbals constructions had over the centuries such colourful names as perfume balls, globe-lamps, silver bags, rolling spheres and perfume baskets, and they were regularly used in annual processions as symbols of the moon carried in front of an undulating dragon (see Plate 26 (page 52)).

The gimbals reached Europe after eleven hundred years. And eight hundred years after that, the famous

LEFT (25) A Photograph of the interior of a brass Tibetan globe-lamp. The lamp is suspended by four separate rings, by which it is always held upright whichever way the lamp turns – even if it is upside-down. The outer cover is decorative and protective at the same time, and the lamp holds a candle-end in place of the wick it would originally have had. The use of interlocking rings to hold something upright in the centre is often called the 'Cardan suspension', after Jerome Cardan, a sixteenth-century Italian scientist, but the Chinese invention predates him by seventeen hundred years. (Collection of Dr Joseph Needham.)

OPPOSITE (26) A lacquer screen with a decorative inlay showing children taking part in a dragon procession. One child holds a globe-lamp, representing the Moon-pearl, in front of the dragon. Within the paper globe would have been a lamp, held upright by a gimbals made of bamboo rings, so that whichever way the boy danced the lamp would not be extinguished. (Collection of Dr Lu Gwei-Djen.)

scientist Robert Hooke and others adopted its principle in a new form, applying power from without rather than stabilizing a central element within, to formulate that Western invention, the universal joint. And it was this invention which resulted in the transmission of automotive power in contemporary motor cars. Perhaps today, as we drive, we should give a thought to how much we owe to that most unlikely of all sources of mechanical progress in mass transportation, the 'Perfume burner for use among cushions'.

MANUFACTURE OF STEEL FROM CAST IRON

SECOND CENTURY BC

Since the Chinese were the first to produce cast iron, they were also the first to make steel from cast iron. This was fully under way by the second century BC at the latest, and eventually led to the invention of the Bessemer steel process in the West in 1856. Henry Bessemer's work had been anticipated in 1852 by William Kelly, from a small town near Eddyville, Kentucky. Kelly had brought four Chinese steel experts to Kentucky in 1845, from whom he had learned the principles of steel production used in China for over two thousand years previously, and had made his own developments.

Iron, when melted and reformed into ingots, has a carbon content. This determines the nature of the metal as cast iron or steel, whichever the case may be. Cast iron is brittle because it contains a considerable quantity of carbon, perhaps as much as 4.5 per cent. 'Decarburization' is the removal of some or all of this carbon. Remove much of the carbon and you have steel; remove nearly all the carbon and you have wrought iron. The Chinese used wrought iron a great deal, most notably perhaps in building large bridges and aqueducts.

The Chinese invented the suspension bridge (see page 64), often constructing such bridges with chains whose links were of wrought iron instead of plaited bamboo. Cast iron was called 'raw iron', steel was called 'great iron', and wrought iron was called 'ripe iron' by the Chinese. In order to make iron 'ripe', they clearly understood that the iron was losing a key ingredient, and they described this as 'loss of vital juices'. But, without knowledge of modern chemistry, they could not identify the ingredient as carbon.

The Chinese were not the first to make steel. But they did invent two particular steel manufacturing processes, of which taking the carbon out of cast iron was the first (the second is described on page 76). The first, the process of 'decarburization', was accomplished by blowing oxygen onto the cast iron ('oxygenation'). We read of this in the classic *Huai Nan Tzu*, which dates from about 120 BC.

Making steel by this means was also called 'the hundred refinings method', since it was often done over and over again, the steel becoming stronger each time. Swords made by this method were highly prized. The back of the sword, not having an edge,

ABOVE (27) A traditional Chinese print showing artisans at work in the Imperial Workshops. Those on the left are forging a steel sword on an anvil.

LEFT (28) At the bottom right of this photograph is 'Sword Testing Rock'. The King of Wu is said to have tested his steel sword two thousand, five hundred years ago by slicing through the rock with it. The Chinese characters carved and painted on adjoining rocks tell of the legend. Nearby is a Sword Pond, presumably where steel was once tempered in water. This is near Suchow, at Tiger Hill, known in ancient times as Sea Surging Hill. When the King of Wu was buried at this place, three thousand steel swords are supposed to have been placed in his tomb for use in the afterlife.

would often be made of the more elastic wrought iron, and the harder steel would be welded onto it to bear the cutting edge for a sabre. The carbon content of the steel could be adjusted depending upon how much oxygen was applied to the molten iron.

Generally speaking, steel with a higher carbon content is stronger, but then strength is traded against brittleness. Steel can have a carbon content of between 0.1 per cent and 1.8 per cent. The Chinese could only make empirical judgements on the qualities of steel obtained from certain numbers of refinings. If very soft steel was desired, they could go on blowing more oxygen in, removing increasing amounts of the carbon. And they practised the world-wide technique of quenching, whereby steel that is cooled instantly in a liquid when still either red- or white-hot preserves its inner metallic micro-structure which it would lose if allowed to cool slowly. On the other hand, cooling steel slowly (tempering) has other advantages. The Chinese were great masters at manipulating their iron materials in countless different ways to obtain the exact type of metal they required. In iron and steel technology, they led the world until modern times. And they were the first to make steel by taking the carbon out of cast iron. But then, no one else could have done so at the time, since cast iron existed nowhere else but in China.

DEEP DRILLING FOR NATURAL GAS

FIRST CENTURY BC

The Chinese originated deep drilling by the first century BC and, with their traditional methods, were able to drill boreholes up to 4800 feet deep. The deep drilling for today's supplies of oil and natural gas is a development from these Chinese techniques.

The primary motive for deep drilling in China was the search for salt. Even as recently as 1965, 16.5 per cent of China's salt supplies came from brine pumped out of deep boreholes, making this source of supply second only to sea salt. In the beginning, the deep boreholes yielding brine were presumably artesian wells, with the brine spurting out of the wells under natural pressure. But, later, highly sophisticated methods of extracting the brine were developed for wells which had no artesian pressure. If the drilling went below the brine level, there would be vast supplies of natural gas, which was primarily methane. So the drilling for natural gas followed close on the heels of the drilling for brine and essentially developed at the same time.

The sheer size of Chinese drilling equipment was remarkable. Derricks could rise as much as 180 feet above ground. One huge derrick photographed at the turn of the century may be seen in Plate 30 (see page 58). Tubes for extracting brine could be as much as 130 feet long and were inserted down the boreholes. At the top of a borehole would be a shaft dug with spades, reaching down to the level of hard rock, whether this was one foot or dozens of feet down. Once the rock was reached, stones with holes through the middle were stacked one on top of another up to ground level, all perfectly centred so that a long hole 8 to 14 inches wide extended down through them all from ground level to rock level. Then, the drilling would begin. A drill would be suspended by bamboo cables from a derrick. Thanks to the advanced state of the iron industry in China (see the account of the invention of cast iron, page 44), cast iron drilling bits were available. These would be dropped onto the rock, and any depth from 1 inch to 3 feet a day might be drilled. The drilling of a truly deep borehole often took years. The only power used for drilling was manpower: men jumped onto a lever to raise the bit and jumped off to let it crash down again. They did this rhythmically for hours on end.

The bamboo cables were made of strips 40 feet long. A single-strength cable would be used down to 1500 feet, but at depths greater than that, the cable was of double thickness. The tensile strength of hemp rope is 750 pounds per square inch, whereas that of bamboo is nearly 4 tons per square inch, equal to some steel wire. The bamboo was also so flexible that it could easily be wound up round the borehead winding drum. Bamboo cables had the added advantage of becoming tougher when wet, the opposite of rope.

It has been remarked that because Europe had no material equivalent to bamboo, and because using hemp ropes as cables for drill bits would have been like drilling with a rubber ball on the end of a piece of elastic, Europeans could not have developed deep-cable-drilling techniques even if they had wanted to.

When drilling commenced for the day, the cable was lowered by hand very gently until a slight tremor indicated that the bottom had been reached. The cable could not be at all slack, so that even if the drill bit were thousands of feet down, the drilling foreman could adjust its height by inches using a device called

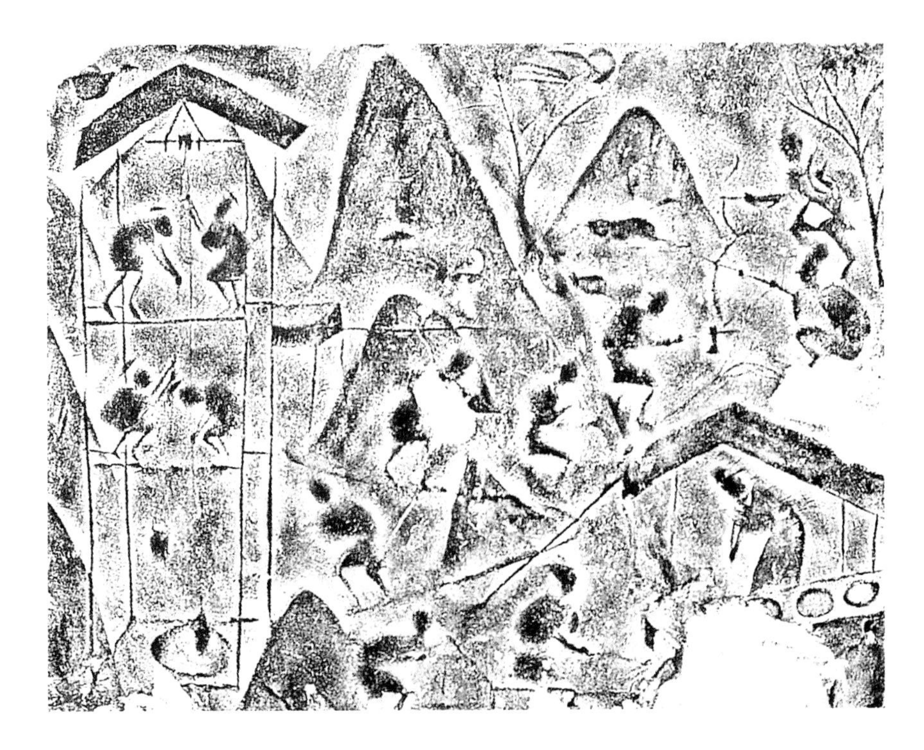

LEFT (29) A rubbing from a Chinese tomb of about the first century AD. On the left may be seen a derrick for deep drilling for brine deposits; four men are hoisting up a bucket full of brine within the three-storey derrick while another bucket descends. In the background is a hunting scene in the mountains.

a 'pounding regulator', which manipulates a loop made at the top of the drilling cable. Two kinds of bit were traditionally used. The larger was 10 feet long and weighed over 300 pounds. It was used to pound the rock and widen the hole made by the smaller bit. The smaller bit weighed only a few dozen pounds.

Large numbers of men were not needed to do the drilling. Shallow wells needed only two men to jump on the lever board, whereas four to six men sufficed for deep wells. With a typical 12-foot lever, the bit would be raised to a height of about 2 feet, and then let fall. But lest the hole not be even and round, a worker stationed at the mouth of the well twisted the cable at each drop so that the drill bit would strike at a slightly different angle each time. By rotating the bit in this way, an even hole was produced.

Extremely ingenious devices were developed for fishing out of boreholes anything that might get stuck or tangled. These things would be dealt with thousands of feet underground with relative ease. Other devices were invented called 'jars', which prevented the jarring of the derrick and platform every time the drill hit the rock.

The detritus left from several successive drill smashes would be removed by suction through hollow bamboo tubes with leather valves at the end; the pumps derived from the double-acting piston bellows described on page 46. The rock chips would either be mixed with water poured down the well (in the early stages of drilling) or would mix naturally with water which was seeping into the borehole, so that the rubble would be brought up as slush.

The boreholes were lined with bamboo tubing. These had male-female joints and were watertight. Water seepage was thus prevented when drilling, except at the bottom. And when brine was reached, the brine could thus be extracted without being diluted by seeping fresh water from higher levels. When extracted from the wells, the brine would be raised several feet above ground level and was distributed for miles all round in an elaborate network of bamboo piping. Huge cast iron evaporation pans were set up, into which the brine would be poured. Heat would be applied under the pans, and the brine vigorously boiled for long periods. So vast was the amount of salt produced by this enormous industry that the province of Szechuan, known anciently as Shu, enjoyed several periods of independence from central government on the basis of its self-sufficiency in deep-borehole salt supplies.

Although the salt was sometimes evaporated by fires burning either wood or coal, this was a more expensive way than using natural gas fires. It was cleverer to use the natural gas from the same borehole or an adjoining one to heat the pans, and as many as 5100 salt pans are reported to have been heated by the natural gas from a single well (the details of the use of natural gas as a fuel are recounted on page 89).

How many and how deep were the Chinese boreholes, whether for brine or for natural gas? Brine could not be found at less than 100 feet deep, and most of the boreholes had to be sunk to at least 600 feet. The majority of them were to depths averaging 3000 feet and the deepest recorded went to 4800 feet. There are more recent boreholes to 9600 feet, but we cannot be certain whether any of these deeper wells were drilled entirely by the traditional drilling technology or were assisted by imported Western technology, so the claims for the traditional drilling technology stop at a conservative 4800 feet.

As for the numbers of deep boreholes, there were hundreds of them. An imperial edict of 1089 limited the number of them in the prefecture and province of Ch'engtu alone to 160 wells. An earlier tally said that in six districts of Tzu-chou there were 157 wells in working order and 50 plugged by government officials as a penalty for tax evasion. The salt industry was supposed to have been nationalized in the second century BC, and entrepreneurs continually tried to conceal their wells by removing the derricks before government inspectors came along. And without signs on the surface, who is going to spot a hole a few inches across with a stone placed over it, even if it does extend downwards 3000 feet? But the government 'revenuers' were as assiduous in hunting down bootlegging brine drillers as their counterparts have been in modern times in looking for illicit stills of moonshine whisky in the Appalachian Mountains of Kentucky and Virginia.

One hotbed of tax evasion in the eleventh century was the district of P'ei-ch'eng, where, out of 55 boreholes, 45 were plugged by 'revenuers' who found them out. In 1095, a complaint to the emperor said: 'In the prefecture and province of Ch'engtu all the brine boreholes are plugged up, and because of their closure thousands of families are unemployed.'

The first information in Europe about Chinese drilling techniques seems to have arrived in garbled form in the seventeenth century through Dutch informants. But the first full description of the Chinese system sent to Europe was in the form of letters written in 1828 by a French missionary named Imbert. His account was discussed in 1829 by an incredulous French scientific society; Imbert then wrote another letter giving further details and insisting that it was indeed possible to drill deep holes with bamboo cables (Europeans had been attempting boreholes with linked rods, which were not working very well). Imbert said he had seen, with his own eyes, great wheels drawing up tubular buckets of brine; the wheels measured 40 feet in circumference and were turned fifty times, showing that depths of 2000 feet

RIGHT (30) A previously unpublished photograph taken in about 1900, showing what is apparently a huge derrick for deep drilling. The location is unrecorded.

LEFT (31) A photograph taken by Cecil Beaton during the Second World War, showing a traditional Chinese deep-drilling derrick over the well head of a brine and/or natural gas well.

could indeed be reached. A French engineer, Jobard, had tried the Chinese method as soon as Imbert's first letters had arrived. Various details of the Chinese system were also adapted for linked-rod drilling, bringing greater success.

By 1834 Chinese drilling techniques had become properly established in Europe for brine drilling, and by 1841 for oil drilling. In 1859, a well exclusively for oil was drilled at Oil Creek, Pennsylvania, by Colonel E.L. Drake, using the Chinese cable method. The oil was to be used for fuel, as had been the case in China for some time (see page 89). Drake and similar oil drillers in America may have obtained knowledge of the system not from France but from the hordes of Chinese indentured labourers used to build the railways in nineteenth-century America. In America, the method for oil drilling prior to the advent of steam power ('kicking her down', as it was called), was exactly the same as the Chinese technique of bowstring drilling which had derived from the men jumping on and off a lever. And even the modern rotary bits seem to have partial Chinese ancestry. In short, Western deep drilling was essentially an importation from China, and the modern oil industry is founded on Oriental techniques nineteen hundred years in advance of the West.

THE BELT-DRIVE (OR DRIVING-BELT)

FIRST CENTURY BC

The belt-drive or driving-belt transmits power from one wheel to another, and produces continuous rotary motion. It existed as early as the first century BC in China. It is attested by a passage in Yang Hsiung's book, *Dictionary of Local Expressions*, of 15 BC. It was developed for use in machines connected with silk manufacture, especially one called a quilling-machine, which wound the long silk fibres onto bobbins for the weavers' shuttles. These machines featured a large

ABOVE (32) The belt-drive (driving-belt) in use in a stone-crushing machine in modern Kuangsi Province.

wheel and a driving-belt and small pulley. The machines are mentioned again in the book *Enlargement of the Literary Expositor* compiled between 230 and 232 AD.

The driving-belt was essential for the invention of the spinning-wheel, which is described separately (page 134). The belts could run not only round normal wheels with rims, whether grooved or not, but also round rimless wheels. A rimless spinning-wheel may sound a contradiction in terms, and the use of a driving-belt with rimless wheels might at first seem an impossibility. But in fact a cat's cradle of fibres strung between wheel spokes which protrude slightly or exist in two sets placed in alternation can create an entirely adequate nexus for a belt. Needham photographed just such an archaic spinning-wheel in use in Shensi in 1942. It is extraordinary to think that the spinning-wheel of 1270 survives unchanged into the modern era. Yet another Chinese technique of using a driving-belt with a rimless wheel is to mount grooved blocks at the ends of the spokes, and run the belt through the successive grooves (see Plate 87 (page 135)).

A refinement of the driving-belt is the chain-drive, invented in China in 976 AD, and described on page 81. A chain-drive is essentially a driving-belt which instead of being solid is a chain into the links of which fit sprockets on the wheels around which it is wrapped.

The driving-belt was apparently imported to Europe as part of the technology of quilling-wheels and spinning-wheels introduced into Italy by travellers returning from China. The oldest actual representation of a driving-belt in Europe dates from 1430, on a rotary horizontal grindstone. Driving-belts remained extremely rare in Europe until the eighteenth and nineteenth centuries, indicating that Europeans did not appreciate the potential of this particular element of the Chinese textile machines for other purposes to any significant degree for more than three centuries. Flat belts and wire cables as driving-belts in Europe only began to be used in the nineteenth century.

WATER POWER

FIRST CENTURY AD

The harnessing of water power for the operation of blast furnace bellows commenced in 31 AD. The official history of that time recorded how Tu Shih, Prefect of Nanyang, invented a water-power reciprocator for the casting of iron agricultural implements: 'Thus the people got great benefit for little labour. They found the "water-powered bellows" convenient and adopted it widely.' Li Hsien commented on this passage in 670: 'Those who smelted and cast already had the push-bellows to blow up their charcoal fires, and now they were instructed to use the rushing of the water to operate it...'.

In the account of the double-acting piston bellows (page 46) we see that such bellows existed from the fourth century BC, so now the same machine was evidently being worked by water mills instead of by human muscle power. The water-powered bellows remained popular round Nanyang for two centuries. The *History of the Three Kingdoms* (290 AD) then continues the story:

> **Han Chi, when Prefect of Lo-ling, was made Superintendent of Metallurgical Production. The old method was to use horse-power for the blowing-engines, and each picul [about 130 pounds] of refined wrought iron took the work of a hundred horses. Man-power was also used, but that too was exceedingly expensive. So Han Chi adapted the furnace bellows to the use of ever-flowing water, and an efficiency three times greater than before was attained. During his seven years of office, iron implements became very abundant. Upon receiving his report, the emperor rewarded him and gave him the title of Commander of the Metal-Workers.**

This occurred around the year 238 AD. About twenty years after that, considerable improvements were introduced to the machines by the inventor Tu Yü, and water-powered bellows continued through each century to spread more and more widely throughout China. The use of a piston-rod and driving-belt by Wang Chen in his version, described in his *Treatise on Agriculture* of 1313, is discussed in the account of the essentials of the steam engine (page 72).

As for the use of water power by the metallurgical industry in Europe, this did not occur until the twelfth century when forge-hammers were thus powered. But the use of water power for bellows did not begin until the thirteenth century – a time-lag of twelve hundred years. This innovation by the Chinese in harnessing water power for industrial processes on a large scale was one of the most significant breakthroughs in energy supply before modern times. It was one of the major steps towards the Industrial Revolution.

ABOVE (33) A treadle-operated square-pallet chain pump being used in contemporary China for raising water for irrigation.

THE CHAIN PUMP

FIRST CENTURY AD

One of the inventions of greatest utility which has spread from China throughout the world, so that its origins are no longer realized, is the square-pallet chain pump. As may be seen in the accompanying illustrations, it consists of an endless circulating chain bearing square pallets which hold water, earth, or sand.

This pump can haul enormous quantities of water from lower to higher levels. The optimum angle of slope at which the chain of pallets can be laid out is about 24°. So, depending on how well the pallets were fitted to avoid leakage and on the sturdiness of the machine as a whole, the height that water can be raised by a single pump is about 15 feet.

By medieval times in China, the pumps had been adapted for use as conveyors of earth or sand rather than just water. They were thus the first conveyor belts.

We do not know who invented the chain pump, or exactly when. Although it may have existed some centuries earlier, we can take as its time of origin the first century AD. The philosopher Wang Ch'ung refers to its existence about 80 AD in his book *Discourses Weighed in the Balance*. Considerable improvements were made to the design during the next century. We know this from an account in the imperial history of the time, which discusses the lack of water in the capital, Loyang. The history tells us that the famous eunuch minister Chang Jang (died 189 AD) ordered various improvements for Loyang from the engineer Pi Lan:

> He further asked Pi Lan ... to construct square-pallet chain pumps and suction pumps, which were set up to the west of the bridge outside the Peace Gate to spray water along the north-south roads of the city, thus saving the expense incurred by the common people [in sprinkling water on these roads and carrying water to the people living along them]....

Chain pumps had achieved a standard form in China by 828. The imperial history for that year records:

> **In the second year of the T'ai-Ho reign-period, in the second month ... a standard model of the chain-pump was issued from the palace, and the people of Ching-chao Fu were ordered by the Emperor to make a considerable number of the machines, for distribution along the Cheng Pai Canal, for irrigation purposes.**

The pumps were used for civil engineering works and for draining all sorts of sites, as well as for irrigation and the supply of drinking water. The pumps were so spectacular in their results that visiting dignitaries and ambassadors from neighbouring lands eventually adopted them in their own countries. By medieval times, Korea and Annam (Vietnam) had transformed their agriculture and irrigation by this means. In 1221, when the pumps were introduced by some visiting Chinese to the inhabitants of Turkestan, the locals exclaimed with delight, 'You Chinese are so clever at everything!'

The first European square-pallet chain pumps were made in the sixteenth century, modelled directly on Chinese designs. By the end of the next century, the British Navy had copied them from Chinese junks for use as bilge-pumps aboard ship. The pumps were introduced to America by the Dutchman A.E. van Braam Houckgeest, where he said they had 'proved of great utility'. And the conveyor-belt form was adapted for use in the eighteenth century by Oliver Evans for flour-milling. This led to the development of the modern grain-elevator. A sixteenth-century European dredger using buckets instead of pallets is the ancestor of all the drainage devices on belt conveyors used in modern mining and excavation work. And as recently as 1938, the classic square-pallet chain pump was reintroduced to the United States from China for pumping crystallized brine from the Great Salt Lake in Utah.

ABOVE (34) A woodcut printed in *Exploitation of the Works of Nature* in 1637, showing two men working a square-pallet chain pump by treadles.

THE SUSPENSION BRIDGE

FIRST CENTURY AD

Few structures seem more typical of the modern world and its engineering achievements than the suspension bridge. And yet, the sophisticated form of the suspension bridge, with a flat roadway suspended from cables, was unquestionably invented in China. And it is highly likely that the two more primitive forms of suspension bridge also originated there, the simple rope bridge and the catenary bridge (where the walkway or roadway is not flat but follows the curve of the cables).

The simplest form of 'suspension' bridge – if we can even call it that – is simply a rope thrown across a gorge. Probably from the very beginning, the technique used for getting the rope across was that still used later for elaborate suspension bridges – shooting it across, tied to an arrow. After the Chinese invention of the crossbow (see page 244) greater power would have been available for heavier cables over longer distances.

Climbing or scrambling along a single rope above a gorge can be dangerous, and is hard on the hands. An ingenious solution is still in use in some areas, such as the Tibetan-Chinese border. The rope is threaded through a hollow piece of bamboo before being attached, and the person merely hugs the bamboo and slides along the rope without burning his hands or straining himself unduly. A more sophisticated method is by a cradle attached to the bamboo tube. Cable bridges of liana vines are known in the Andes mountains of Peru, dating back to at least 1290, and Needham suspects that this may be one of the many

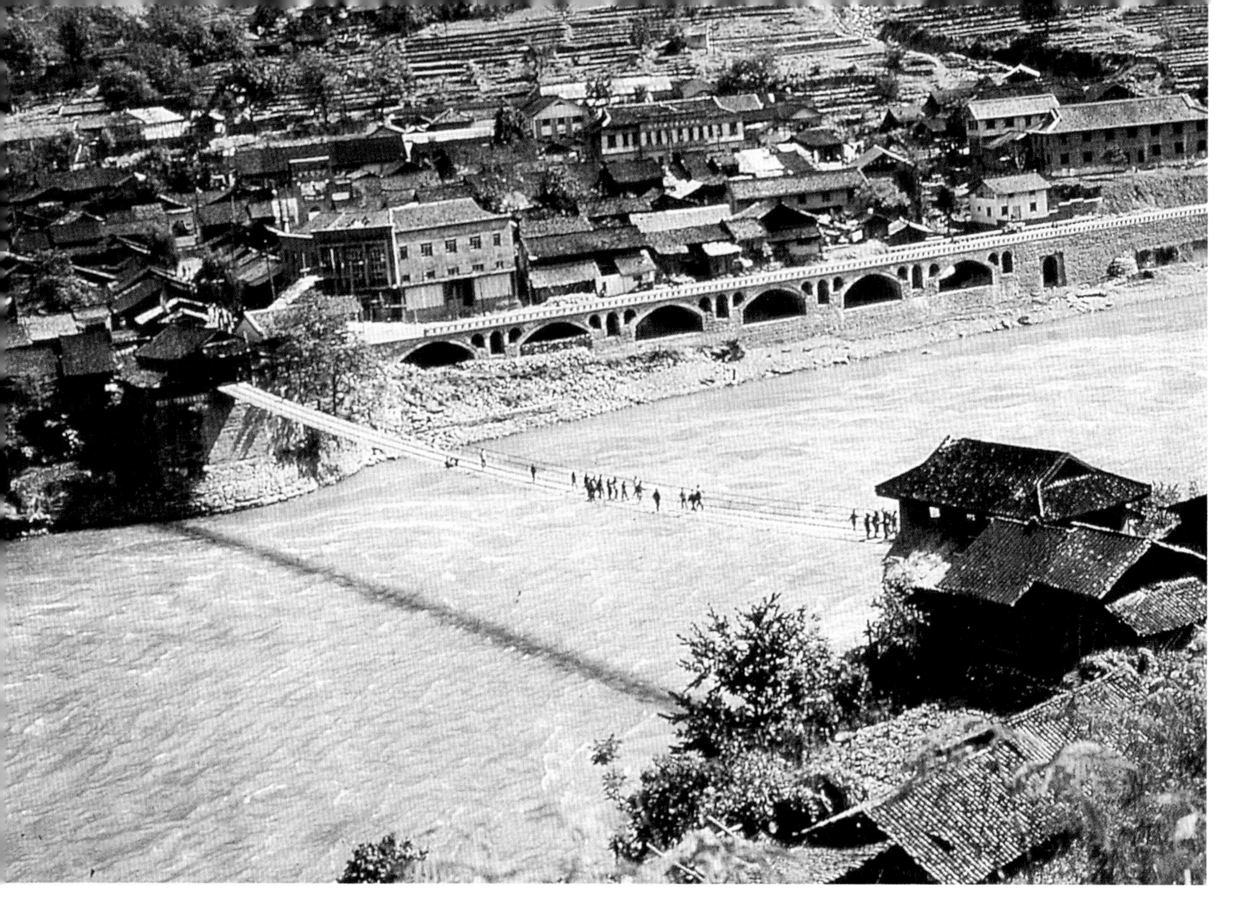

ABOVE (35) A well-known suspension bridge with a single span of 361 feet, at Lu-ting in Sikang. Although built as recently as 1705, it replaced an earlier bridge at the same spot. The suspension cables are wrought iron chains.

Chinese ideas to have spread to the New World across the Pacific.

Bridges of ropes and cables in China and Tibet evolved into multiple-cable bridges of various types. Sometimes three ropes or cables are stretched across together so that the person crossing can walk with his feet on two of them and hold a third above his head for balance. Or a woven walkway of matting is incorporated between the two bottom ropes or cables, to make the going easier. Another variation is to have a series of hanging straps by which the user pulls himself forward. All these and other variations occur in the area between China and Tibet, in the high mountains. A reference in the Chinese dynastic history for 90 AD appears to mention a suspension bridge which has planking and, hence, a proper platform upon which to cross:

> There the gorges and ravines allow of no connecting road, but ropes and cables are stretched across from side to side and by means of these a passage is effected.

This reference is rather vague. The same dynastic history for 25 BC describes a harrowing Himalayan suspension bridge:

> Then comes the road through the San-ch'ih-p'an gorge, thirty *li* long, where the path is only 16 or 17 inches wide, on the edge of unfathomable precipices. Travellers go step by step here, clasping each other for safety, and rope suspension bridges are stretched across the chasms from side to side. After 20 *li* one reaches the Hsien-tu mountain pass.... Verily the difficulties and dangers of the road are indescribable.

Fa-Hsien, the first Chinese Buddhist pilgrim to India, crossed this very bridge in 399 AD, and left this account of his experience:

> Keeping on through the valleys and passes of the Ts'ung-ling mountain range, we travelled south-westwards for fifteen days. The road is

ABOVE (36) This relatively primitive suspension bridge is still in use at Wolung in modern China.

difficult and broken, with steep crags and precipices in the way. The mountain-sides are simply stone walls standing straight up 8000 feet high. To look down makes one dizzy, and when one wants to move forward one is not sure of one's foothold. Below flows the Hsin-t'ou Ho. Men of former times bored through the rocks here to make a way, and fixed ladders at the sides of the cliffs, seven hundred of which one has to negotiate. Then one passes fearfully across a bridge of suspended cables to cross the river, the sides of which are here rather less than 80 paces [400 feet] apart.

Cable bridges in China were most efficient when made of bamboo. The cables were made with a centre formed of the core of the bamboo surrounded by plaited bamboo strips made of the outer layers of the wood. The plaiting was done so that the higher the tension, the more tightly the outer strips gripped the inner core. This led to the safety factor that it is the inner strands of a cable which snap first, rather than the outer strips which would otherwise unravel very fast. An ordinary 2-inch hemp rope can stand stresses of only about 8000 pounds per square inch, but bamboo cables can stand a stress of 26,000 pounds per square inch. Ordinary steel cables will only take twice as much stress (56,000 pounds), so bamboo is remarkably strong. (Modern steel alloys such as used in the Golden Gate Bridge at San Francisco can take stresses of 256,000 pounds per square inch.)

The most famous Chinese suspension bridge is a catenary bridge (which has a roadway following the curves of the cables rather than hanging flat): the An-Lan Bridge at Kuanhsien in Szechuan. It has a total length of 1050 feet, composed of eight successive spans, and there is not a single piece of metal in the entire structure. An account of a traveller crossing it in 1177 describes only five spans at that time. It has planking on which to walk, originally 12 feet wide but today only 9 feet wide, and it is believed to have

RIGHT (37) A previously unpublished photograph of the famous An-Lan Bridge at Kuanhsien in the southern province of Szechuan, taken by Ernst Boerschmann in about 1900. Composed of eight successive spans, this catenary suspension bridge has a total length of 1050 feet. The platform is wood planking and the cables are made of rope.

been built in the third century BC by Li Ping. It may be seen in Plate 37 (opposite).

The true suspension bridge became possible with the invention of the iron-chain suspension technique. As described elsewhere (page 53) the Chinese were in advance of the whole of the rest of the world in their iron and steel technology. Needham believes that they applied wrought-iron chains to suspension bridges by the first century AD. Massive stone abutments were built to contain the chain ends.

From these chains it became possible to suspend the planking gangway increasingly away from the catenary curve of the chains themselves so that it tended towards being a flat surface. The greatest span of which a Chinese iron-chain suspension bridge is known to have been capable is about 430 feet, at Lu-shan in Szechuan. The longest such bridge which still exists is 361 feet, at Lu-ting in Sikang. Its chains are embedded 40 feet deep into the stone pillars on both sides. The bridge may be seen in Plate 35 (page 65). In its present form, it was built in 1705, but it is presumed to have replaced earlier versions on the same site. It was the scene of a major incident during the Communist Army's Long March, when they successfully stormed the bridge under heavy fire despite the fact that its planking had largely been removed.

We have an interesting and vivid description written in 1638 by Hsü Hsia-K'o of a bridge in south-western Kweichow Province:

> **The P'an Chian bridge is held by iron chains which connect the cliffs on the eastern and western sides of the river, a distance of 150 feet. The warp so made has a weft of planks. The cliffs themselves are about 300 feet high and between them a swift raging stream of water, of unfathomed depth, rushed along. In earlier years ferry boats were often in grave danger of capsizing, whereupon people tried to span it by a stone structure, but they failed. Then in the fourth year of the Ch'ung-Chen reign-period the present Governor, then a judge, Chu Chia-Min, asked Major Li Fang-Hsien to build a suspension bridge. So now several tens of great iron chains are suspended from towers on each bank, and on them two layers of boards, about 8 inches thick and more than 8 feet long, are laid. The bridge looks flimsy and unsubstantial, but when people tread on it, it is as immovable as a mountain-peak; daily hundreds of oxen and horses with heavy loads pass over it. Each side of the bridge is protected by a high iron railing woven with smaller chains. On each bank there crouch two stone lions, 3 or 4 feet high, which clench these railing-chains tightly in their mouths.**

An old engraving of this bridge was published in 1665 in *Record of the Iron Suspension Bridge* by Chu Hsieh-Yüan. In the foreground is 'the stone of weeping', a memorial to all those who lost their lives trying to cross the river before the bridge was built in 1629. This bridge remained in service until 1939, when it was replaced with a steel suspension bridge, destroyed by the Japanese only a year later. In 1943, a new steel suspension bridge was erected half a mile downstream, which is still in use.

Chinese regions tend to have different types of bridge. Suspension bridges are found mostly in the south-west of China. Arch bridges (see page 77) are found mostly in the north. Only those who travel into the wilder and more remote parts of south-western China are likely to see the ancient suspension bridges which still exist there in profusion. Since the origins of suspension bridges are somewhat vague, with no precise date and no actual inventor, we may take the date of the first century AD, when iron chains were applied to bridges, as a highly conservative date for the earliest Chinese suspension bridges capable of carrying vehicles (though probably this happened at least four centuries earlier, using only bamboo cables, at Kuanhsien).

The first Western suspension bridge capable of carrying vehicles was built in 1809 across the Merrimac River in Massachusetts (span 244 feet). In the case of this invention, we can trace its transmission fairly well. The suspension bridges of Kweichow, such as the P'an Chian Bridge just described, came to the attention of the Jesuits and other Westerners who visited China in the seventeenth century. In 1655, Martin Martini described an iron-chain bridge over a river in Kweichow, which was incorporated in Blaeu's great *New Chinese Atlas* of that date.

An earlier mention of suspension bridges seems to have reached the European designer Faustus Verantius, for he proposed a sort of suspension bridge of linked rods in 1595. The Chinese also had such bridges of linked rods. But it was Martini's remarks which brought wide attention to suspension bridges in

Europe. The subject was picked up two years later, in 1667, by Athanasius Kircher, in his *China Illustrata.* Kircher wrote:

> **When several people cross the bridge at one time it moves and sways and oscillates up and down in such a way as to evoke in them no small fear of the danger of falling off; yet I find it impossible sufficiently to admire the skill of the Chinese engineers, who have executed so many and such arduous works for the greater convenience of wayfaring men.**

However, European engineers seem to have done nothing about this for nearly a century. Then the first Western iron-chain suspension bridge, the Winch Bridge, was built in 1741 over the Tees in England. It had only the cables, and no deck for vehicles. The same was true of some small bridges which were built about that time. But since the first Western suspension bridge capable of carrying traffic was not constructed until 1809, the Chinese are seen to have been at least 1800 years in advance of the West, if not indeed 2200 or more, in this particular field.

THE FIRST CYBERNETIC MACHINE

THIRD CENTURY AD

By the third century AD at the latest, the Chinese had a fully operational, navigational 'cybernetic machine', using the principles of feedback. It was called the 'south-pointing carriage', but had no connection with a magnetic compass. It was a large carriage, 11 feet long, 11 feet deep, and 9½ feet wide, surmounted by a jade statue of an 'immortal' – a sage who had achieved immortality. The figure's arm was raised, pointing ahead, and it always faced towards the south, no matter which way the carriage turned. Even if the road were circular, the jade figure would rotate, keeping the finger pointing in the same direction. How was this possible in the third century AD? This machine may have been invented even earlier, indeed, as much as twelve hundred years earlier. An official history for 500 AD describes how:

> **The south-pointing carriage was first constructed by the Duke of Chou [beginning of the first millennium BC] as a means of conducting homewards certain envoys who had arrived from a great distance beyond the frontiers. The country was a boundless plain in which people lost their bearings as to east and west, so the Duke caused this vehicle to be made in order that the ambassadors should be able to distinguish north and south.**

If this information is correct, the invention would date from about 1030 BC. But Needham suspects that the word 'carriage' was inserted in this account by scribes, and that what is being described is a 'south-pointer', that is, a compass, in which case the origin of the compass, described on page 162, must be pushed back even further.

The next person credited with building a south-pointing carriage is the astronomer and scientist Chang Heng, about 120 AD, although this is also regarded by Needham as doubtful. The only date which he is prepared to accept with certainty is the middle of the third century AD, with the famous engineer Ma Chün as the builder (and, thus, the inventor). The drawing of a pointing figure of jade, taken from the *Universal Encyclopedia* of 1601, was copied from a print of 1341, and is reproduced in Plate 39 (page 72).

If the machine did not use a magnetic compass, how did it work? The answer is that it had a train of differential gears, similar to those in a modern automobile. Perhaps the function of a differential gear should be explained as follows. When a wheeled vehicle is turning a corner the wheels on opposite sides of the vehicle are clearly going to need to turn at different rates since the near side is travelling a shorter distance than the far side. With a hand-cart or horse-drawn carriage, this may not pose such problems. But when a vehicle has power being applied to the axle to make the wheels turn, how is it possible for one wheel to be permitted to speed up a little, and the other slow down a little, on the same axle? This is made possible only by an ingenious combination of gear wheels and flywheels: the differential gear.

When Needham published his volume on mechanical engineering in 1965, he believed that the Chinese had invented the differential gear, and that it had made its first appearance in this south-pointing carriage. If the first south-pointing carriage were the one attributed to the Duke of Chou about 1000 BC, then the Chinese would indeed have been the inventors; but we must stay on the side of caution, and assume that the first south-pointing carriage was made in the second or third century AD. In that case, we must credit the Greeks with inventing the differential gear, a fact which became known only in 1975, when Professor Derek Price published his book *Gears from the Greeks*. In this work Price wrote the definitive account of a Greek differential gear dating from 80 BC, which Price said 'must surely rank as one of the greatest basic mechanical inventions of all time'. And although a transmission of this invention from Greece and Rome to China was possible, it is equally possible that the differential gear was independently re-invented in China for the south-pointing carriage.

The precision needed in the construction of the south-pointing carriage almost defies belief. For the outside road wheels alone, Needham points out, J. Coales, in *The Historical and Scientific Background of Automation*, 'has calculated that a difference of only 1 per cent between the wheel circumferences would lead to a change of direction of the pointing figure of as much as 90 degrees in a distance only fifty times that between the two wheels.' This was because the carriage would veer more and more to one side if one wheel were smaller (relative slip). So that for this south-pointing carriage, the size of the road wheels had to be accurate to a margin of error far less than one per cent, and a commensurate accuracy in size of gear wheels would have been necessary. This points to engineering of such a high order that we may well be justified in hesitating to apply the words 'ancient' and 'primitive' to it!

The south-pointing device was basically a reversal of the use of the differential gear in the modern

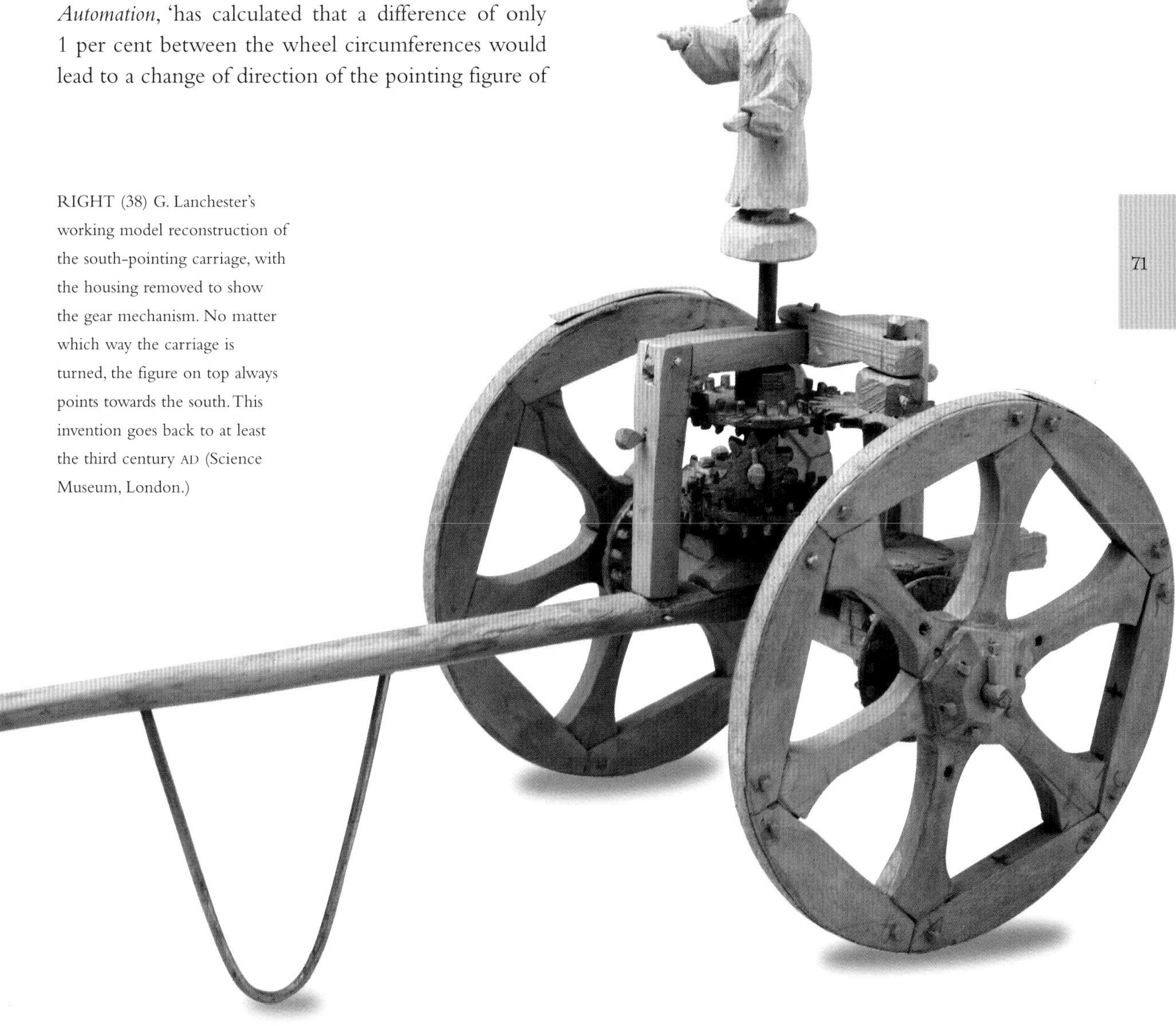

RIGHT (38) G. Lanchester's working model reconstruction of the south-pointing carriage, with the housing removed to show the gear mechanism. No matter which way the carriage is turned, the figure on top always points towards the south. This invention goes back to at least the third century AD (Science Museum, London.)

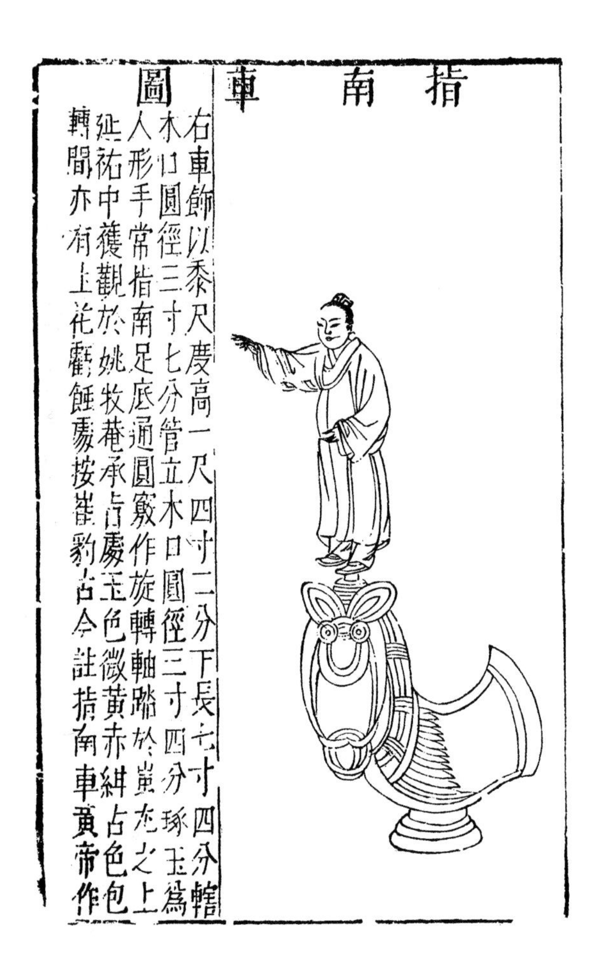

ABOVE (39) An old print of the jade figure on top of the south-pointing carriage, from the *Universal Encyclopedia* of 1601. It was redrawn from *Illustrated Record of Ancient Jade Objects* by Chu Te-Jun, published in 1341, which describes the south-pointing figure as 1.42 feet high and made of pale reddish-yellow jade.

automobile. Today, such gears are used to apply power to turn the wheels and make the vehicle move. But with the south-pointing carriage, which was pulled by animals, the power was transmitted from the wheels, and applied towards the continual adjusting of the position of the pointing figure. Thus it was the differential gear in the machine which turned the figure so that it always pointed to the south. And it did so just as the differential gear operates today, only in reverse.

There also remains evidence of a south-pointing boat, though few details of it survive. It was built during the Chin Dynasty (265-420 AD), and sailed round the palace gardens on the Ling-Chih lake. Needham speculates that it might simply have been navigated by means of a lodestone. But he adds: 'An alternative would be that the boat was a paddle-boat (paddle-wheel boats certainly existed in China at this time), and that someone attempted to apply to its two wheels the same kind of device as had already been applied to the carriage. Probably no one will ever find out.' Whether this boat was or was not a water-borne version of the south-pointing carriage, the principles had already been developed to make it possible.

Needham has called the south-pointing carriage 'the first homeostatic machine in human history, involving full negative feedback. Of course, the driver had to be included in the loop. But as Coales has acutely pointed out, an attractive carrot held by the pointing figure might have replaced the human driver and closed the loop more automatically.' Although Needham has occasionally spoken of the south-pointing carriage as the first actual cybernetic machine, he has qualified this by saying: 'The south-pointing carriage would have been the first cybernetic machine had the actual steering corrected itself, as we could easily make it do today'.

ESSENTIALS OF THE STEAM ENGINE

FIFTH CENTURY AD

The essential design of the steam engine, lacking only the crank-shaft, was invented in China before the steam engine existed. It was a water-powered flour sifting and shaking machine which operated in reverse mode to the later operation of the steam engine: instead of the piston of the steam engine working the wheels on a vehicle, the Chinese machine had wheels which were worked by rushing water in order to power pistons. The crank-shaft could not be incorporated in the Chinese machine because it is a Western invention which the Chinese never had. But it was not necessary.

The machines in question were south of the city of Ching Ming Ssu at the Buddhist monasteries of Loyang. They are mentioned in a book entitled *Description of the Buddhist Temples and Monasteries of Loyang*, dating from about 530 AD. These machines worked on the reciprocating principle of a piston being moved by a connecting rod attached to a crank (powered by a water wheel).

This principle came to be used more widely in connection with the metallurgical industry. A water-powered device of this kind was found to be the most efficient and labour-saving way to work the giant bellows of blast furnaces. The first published picture of such a machine appeared in Wang Chen's *Treatise on Agriculture* in 1313. Needham has published a mechanical diagram of how the machine worked (reconstructed from details of the text of Wang Chen's book and other sources), and gives a mechanically accurate description of it.

Here is Wang Chen's own fourteenth-century description:

> The design is as follows. A place beside a rushing torrent is selected, and a vertical shaft is set up in a framework with two horizontal wheels so that the lower one is rotated by the force of the water. The upper one is connected by a driving-belt to a smaller wheel in front of it, which bears an eccentric lug. Then all as one, following the turning of the driving-wheel, the connecting-rod attached to the eccentric lug pushes and pulls the rocking roller, and levers to left and right of which assure the transmission of the motion to the piston-rod. Thus this is pushed back and forth, operating the furnace bellows far more quickly than would be possible with man-power....
>
> When Metallurgical Bureaux are established, they often spend a great deal of money and hire much labour to work the bellows, which is very expensive indeed. But by these methods [using water-power] great savings can be made. Now it is a long time since the inventions were first devised, and some of them have been lost, so I travelled to many places to explore and recover the techniques involved. And I have drawn the accompanying diagrams according to what I found, for the enrichment of the country by the official metallurgists and the greater convenience of private smelters.

Wang Chen also mentions the flour sifting and shaking (bolting) machine.

A book called *Topography of Anyang District* preserves an account from a now lost work called *Old Manual of Metallurgical Water-Power Technology* which states that these machines were introduced in the sixth century by the eminent engineer, architect and city planner Kao Lung-Chih, who was at the time Director of the Ministries Department. He must have taken the idea of the Loyang flour machines and applied it to metallurgy. The lost work is even quoted as saying that the water wheels which he used were 1 foot broad and 7½ feet in diameter. He called the technique 'water power smelting'. This, then, was one of the secrets of Chinese metallurgical supremacy. For we know from another source that by driving the pistons from the smaller wheel on the machine, which spun fifteen times faster than the large driving-wheel, the pistons fanning the smelting fires could work at an enormous rate. The blast would have been continuous, since from the fourth century BC the Chinese had used the double-acting piston bellows for this purpose. Consequently, the Chinese had prodigious energies harnessed to provide fully automated continuous air blasts for their metallurgical operations by the sixth century AD. It is no wonder that this was the century which also saw the development of the steel co-fusion process. The excellent steel which the Chinese had been producing for several centuries previously provided the necessary bearings and parts to produce tolerance for such hard-working machinery.

It was seven hundred years before water power in Europe was to be harnessed for bellows, in the thirteenth century. In 1757, John Wilkinson patented a hydraulic blowing-engine which was essentially identical to that described by Wang Chen in 1313, except for the addition of a crank-shaft by way of refinement. In 1780, James Pickard patented a steam engine using essentially this apparatus in reverse, that is, power from the piston driving the wheel rather than vice versa. Pickard's patent forced James Watt (who was barred from this standard technique by not having the patent) to invent the sun-and-planet gear for his own steam engine. These European designs were all derived, through various intermediaries such as Agostino Ramelli (1588), from those of China. As for pistons driving wheels, rather than the other way round, Chinese stimulus was available separately there. Pistons driven by exploding gunpowder were tried in Europe on the idea, as Needham has put it, that 'the piston and piston-rod may be considered a tethered cannon-ball'. Since the Chinese invented both gunpowder and the gun (see pages 250 and 266), internal combustion as well as steam engines were partly inspired by the fact that a gun has a projectile which exactly fits the barrel and is expelled by force – further Chinese contributions to the ancestry of both engines.

'MAGIC MIRRORS'

FIFTH CENTURY AD

Those who like to think of China as a land of mystery are well served by the Chinese 'magic mirrors', which are some of the strangest objects in the world. They are known to go back to at least the fifth century AD, though their exact origins are unknown. About twelve hundred years ago, there still existed a book entitled *Record of Ancient Mirrors*, which apparently contained the secrets of these magic mirrors and their construction, but sadly it seems to have been lost for over a thousand years.

When magic mirrors came to the attention of the West in 1832, dozens of prominent scientists attempted to discover their secret. It was a hundred years before a satisfactory theory of magic mirrors was formulated (by the British crystallographer Sir William Bragg) in 1932. These strange objects had defied the best brains in Europe for a century.

What exactly, then, is a magic mirror? On its back it has cast bronze designs – pictures, or written characters, or both. The reflecting side is convex and is of bright, shiny polished bronze which serves as a mirror. In many conditions of lighting, when held in the hand, it appears to be a perfectly normal mirror.

However, when the mirror is held in bright sunshine, its reflecting surface

RIGHT & OPPOSITE (40) This is a typical example of a magic mirror. The microscopic irregularities on the apparently smooth reverse side of the mirror cannot be detected by the eye. However, when the mirror is held in the sun, so that a reflection of the sunlight is cast onto a dark wall by the seemingly smooth surface, the picture of cranes and trees and the two Chinese characters which have been cast in relief on the ornamented back of the mirror are miraculously visible as a projected image on the wall, as if the solid bronze had suddenly become transparent and the pattern had somehow 'passed through' the solid mirror and become visible in the wall reflection. The metallurgical process by which this eerie and seemingly supernatural phenomenon is achieved was not scientifically explained in the West until 1932, although the Chinese have been making magic mirrors and keeping the technique secret for 1600 years. This particular example in bronze belonged to Joseph Needham and is Japanese, rather than Chinese. The Chinese written characters on this mirror (many of which are regularly used by the Japanese) refer to a Japanese Noh play. (Needham Research Institute, Cambridge.)

can be 'seen through', making it possible to inspect from a reflection cast onto a dark wall the written characters or patterns on the back. Somehow, mysteriously, the solid bronze becomes transparent, leading to the Chinese name for the objects, 'light-penetration mirrors'.

But surely, the reader will protest, solid bronze cannot be transparent. This is true, and there was certainly a trick to it. But it was a sufficiently good trick to baffle Western scientists for a century, and even the earliest surviving Chinese discussion of magic mirrors consists of speculation on how they might work. This occurs in a fascinating work, *Dream Pool Essays*, by Shen Kua, published in 1086. Even at this date, Shen Kua thought of the mirrors as coming from some vague archaic period:

> There exist certain 'light-penetration mirrors' which have about twenty characters inscribed on them in an ancient style which cannot be interpreted. If such a mirror is exposed to the sunshine, although the characters are all on the back, they 'pass through' and are reflected on the wall of a house, where they can be read most distinctly.... I have three of these inscribed 'light-penetration mirrors' in my own family, and I have seen others treasured in other families, which are closely similar and very

ancient; all of them 'let the light through'. But I do not understand why other mirrors, though extremely thin, do not 'let light through'. The ancients must indeed have had some special art.... Those who discuss the reason ... say that at the time the mirror was cast, the thinner part became cold first, while the raised part of the design on the back, being thicker, became cold later, so that the bronze formed minute wrinkles. Thus although the characters are on the back, the face has faint lines too faint to be seen with the naked eye.

Although differences of cooling rate are not the explanation, Shen Kua was correct in suggesting that the shiny, polished mirror surfaces concealed minute variations which the eye alone could not detect. Needham says of the experiments carried out by European scientists: 'Careful and extended optical experimentation demonstrated that the surfaces of "magic" mirrors reproduced the designs on the backs because of very slight inequalities of curvature, the thicker portions being very slightly flatter than the thinner ones, and even sometimes actually concave.'

The basic mirror shape, with the design on the back, was cast flat, and the convexity of the surface produced afterwards by elaborate scraping and scratching. The surface was then polished to become shiny. The stresses set up by these processes caused the thinner parts of the surface to bulge outwards and become more convex than the thicker portions. Finally, a mercury amalgam was laid over the surface; this created further stresses and preferential buckling. The result was that imperfections of the mirror surface matched the patterns on the back, although they were too minute to be seen by the eye. But when the mirror reflected bright sunlight against a wall, with the resultant magnification of the whole image, the effect was to reproduce the patterns as if they were passing through the solid bronze by way of light beams. As Sir William Bragg said when he finally discovered this in 1932: 'Only the magnifying effect of reflection makes them plain.' Needham rightly calls all of this 'the first step on the road to knowledge about the minute structure of metal surfaces.'

What makes magic mirrors so mysterious is that, although they are made of solid bronze, it is nevertheless possible to 'see through' them. Their very existence was thus a great encouragement to those Taoist philosophers and mystics in China who tended to view normal matter as a grosser form of matter of a mundane sort, and felt that there must be other, far more rarefied forms of matter which were practically invisible. The Taoists even believed in a kind of immortality which meant that one continued to exist physically in a transparent and tenuous 'higher' state of matter. Such beliefs could have been justified and even demonstrated to the satisfaction of all onlookers by showing them a magic mirror and proving to them that under certain conditions, the most solid matter conceivable – cast bronze – could become 'transparent'.

THE 'SIEMENS' STEEL PROCESS

FIFTH CENTURY AD

As described in the previous steel process, cast iron has a high carbon content and wrought iron has practically no carbon content, whereas steel must be somewhere in between if it is to have all the desired characteristics. About the fifth century AD, the Chinese developed the 'co-fusion' process, in which cast and wrought iron were melted together to yield the 'something in between', which was steel. This is essentially the Martin and Siemens steel process of 1863, though carried out fourteen hundred years earlier.

This process was in full swing by the sixth century, from which time we have a Chinese description of it: 'Ch'iwu Huai-Wen also made sabres of "overnight iron". The method was to bake the purest cast iron, piling it up with the soft ingots of wrought iron, until after several days and nights, it was all turned to steel.'

Another account appears in the seventh century. In the *Newly Reorganized Pharmacopoeia* of 659 AD we read: 'Steel is made when the raw cast iron and the soft wrought iron are mixed and heated together. It is for making sabres and sickles.'

And in the eleventh century, Su Sung informs us that: 'By mixing the raw and the soft (cast iron and wrought iron), a metal is obtained for making the edges and points of sabres and swords; this is called steel....

We are given precise technical details by Sung Ying-Hsing in 1637:

The method of making steel is as follows. The wrought iron is beaten into thin plates or scales as wide as a finger and rather over an

LEFT (41) A Chinese blast furnace in operation in the year 1637. On the right, a manually-operated double-acting piston bellows fans the fire. One of the central labels is to 'lumps of steel dropping out', referring to lumps of viscous medium-carbon iron which will be turned into true steel later by the 'co-fusion' process invented in the fifth century. It was essentially the 'Martin and Siemens' steel process of 1863, fifteen hundred years earlier.

> inch and a half long. These are packed within wrought iron sheets and all tightly pressed down by cast iron pieces piled on top. The whole furnace is then covered over with mud (or clay) as well. Large furnace piston bellows are then set to work, and when the fire has risen to a sufficient heat, the cast iron comes to its transformation [i.e. melts] first, and dripping and soaking, penetrates into the wrought iron. When the two are united with each other, they are taken out and forged; afterwards they are again heated and hammered. This is many times repeated.

In our own time, experiments have been carried out at the steel works at Corby in England to reproduce the ancient Chinese steel-making techniques. The experiments were thoroughly successful. A very uniform steel was obtained, with the carbon from the cast iron spread evenly throughout, and a genuine blending of the cast and wrought iron. The original heating went up to 975°C, and the metal was taken out and forged with a hand hammer. It was then heated for eight hours at 900°C and came out beautifully.

It should be mentioned that another key ingredient to the Chinese success story in iron and steel technology was the invention of special bellows for blowing air (see page 46). It enabled a continuous stream to be applied, uninterrupted by the moving backwards of the piston. It would have been impossible to achieve such success with an unstable, huffing and puffing airstream in such a high temperature steel process. Furthermore, the application of hydrodynamic power to work the piston bellows by means of a horizontal water wheel made larger bellows possible. We must not forget, therefore, that the story of iron and steel in China is at all times surrounded by stories of other, interdependent inventions and techniques.

THE SEGMENTAL ARCH BRIDGE

SEVENTH CENTURY AD

A conceptual breakthrough occurred when a Chinese engineer was the first to realize that an arch did not have to be a semi-circle. A bridge could be built which was based not on the traditional semi-circular arch but on what is known as a segmental arch. The way to envisage this is to imagine a gigantic circle embedded in the ground, of which only the tip shows above ground level. This tip is a segment of a circle, and the arch it forms is a segmental arch. Such an arch forms the central arch of the bridge in Plate 42 (page 78). Bridges built in this way take less material and are stronger than ones built as semi-circular arches.

This advance took place in China in the seventh century AD. It was the concept of a genius, Li Ch'un, the founder of an entire school of constructional engineering whose influence lasted for many centuries. We are fortunate in that his first great bridge, built in

610, survives intact and is still very much in use today. Called the Great Stone Bridge, it spans the Chiao Shui river near Chao-hsien, at the foot of the Shansi Mountains on the edge of the North China Plain. It was featured on a postage stamp in 1962 and is one of the achievements of early Chinese engineers of which the modern Chinese have most knowledge, and of which they feel most proud. Many legends have attached themselves to the bridge over the centuries, and about the sixteenth century a poet spoke of it as 'looking like a moon rising above the clouds, or a long rainbow hanging on a mountain waterfall'.

It is difficult for us today to appreciate just how impressive a sight the Great Stone Bridge must have been to pre-modern eyes. For we are used to seeing pre-stressed concrete examples of segmental arch bridges everywhere, spanning all the expressways and motorways of the modern world. But in the seventh century, eyes were not so jaded, and the following inscription was carved on the Great Stone Bridge a little over a century after its construction, in 729:

> **This stone bridge over the Chiao River is the result of the work of the Sui engineer Li Ch'un. Its construction is indeed unusual, and no one knows on what principle he made it. But let us observe his marvellous use of stone-work. Its convexity is so smooth, and the voussoir-stones fit together so perfectly.... How lofty is the flying arch! How large is the opening, yet without piers! ... Precise indeed are the cross-bondings and joints between the stones, masonry blocks delicately interlocking like mill wheels, or like the walls of wells; a hundred forms organized into one. And besides the mortar in the crevices there are slender-waisted iron clamps to bind the stones together. The four small arches inserted, on either side, break the anger of the roaring floods, and protect the bridge mightily. Such a master-work could never have been achieved if this man had not applied his genius to the building of a work which would last for centuries to come.**

LEFT (42) The Great Stone Bridge spanning the River Chiao Shui is the world's first segmental arch bridge, built in stone by the architect Li Ch'un in the year 610 and renovated in the twentieth century. The bridge has a span of 123 feet. Not only was the great segmental arch in the centre an innovation (semi-circular arches had existed before), but the semi-circular arch spandrels to either side were also new. They let through additional flood waters and also allowed the structure to be lighter in weight. Such arches did not reach Europe for five hundred years.

The four small whole arches, 'breaking the anger of the roaring waters', were incorporated within the structure of the main bridge. They were an innovation of great consequence in bridge-building, for they were the world's first arched spandrels. Li Ch'un found that by punching these holes in the ends of the bridge he could accomplish several things at once: flood waters could rush through them, lessening the chance that the main bridge would be swept away at its supports in a sudden flood; the total weight of the bridge could be lessened, thereby diminishing the tendency to buckle by the ends sinking down into the river banks; and vast quantities of material could be saved, which would normally have gone to make solid ends for the bridge.

The Great Stone Bridge has a span of 123 feet. The largest surviving Roman whole arch bridge, the

LEFT (43) The so-called 'Marco Polo Bridge' just west of Peking (Beijing), spanning the Yung-ting river. It is China's greatest segmental arch bridge, built in 1189 and consisting of a succession of eleven segmental arches with a total span of 700 feet.

ABOVE (44) The chain-drive, the essential mechanism of the bicycle, was invented in China in 976 but did not appear in Europe until 1770.

Pont St Martin near Aosta, spans 117 feet. But the average whole arch Roman bridge spanned between 60 and 80 feet, whereas whole arches in Roman aqueducts had an average span of about 20 feet. The Great Stone Bridge was five hundred years ahead of Europe in one sense and six hundred years ahead in another. For the principle of the segmental arch in bridge construction, brought to Europe about Marco Polo's time, first began to be applied in the West at the end of the thirteenth century, as may be seen in the surviving Pont St Esprit across the Rhone and the small Abbot's Bridge at Bury St Edmunds in East Anglia. But it was only in the fourteenth century that the principle became widely and daringly applied. Examples are the famous Ponte Vecchio in Florence (1345), the covered bridge at Pavia (1351), and the Castelvecchio Bridge at Verona (1345). However, one has only to look at the last-named to see that arched spandrels had not yet been introduced. These came later in the same century to Europe.

The greatest segmental arch bridge in China is the famous 'Marco Polo Bridge', often so named because Polo described it at length. Just west of Peking (Beijing), it is often visited by tourists. It crosses the Yung-ting river at the small town of Lu-kou-ch'iao, and is 700 feet in length, consisting of a series of eleven segmental arches extending one after another across the river, each with an average span of 62 feet. It was built in 1189 and is still heavily used by modern truck and bus traffic. Marco Polo thought this bridge 'the finest in the world'. He was delighted by the elaborate carved balustrade, consisting of 283 marble lion heads, all different, and he enthused about how ten mounted men could ride abreast across it without the slightest inconvenience. This bridge may be seen in Plate 43 (page 78).

THE CHAIN-DRIVE

TENTH CENTURY AD

The chain-drive for transmission of power was invented in China by Chang Ssu-Hsün in 976 AD for use in his great mechanical clock (page 116). Ever since the first century AD, square-pallet chain pumps had existed in China (page 63). Hundreds of thousands of them were to be found all over the Chinese countryside by the tenth century, and the time of Chang Ssu-Hsün. When he came to the problem of transmitting power in his clock, Chang must have

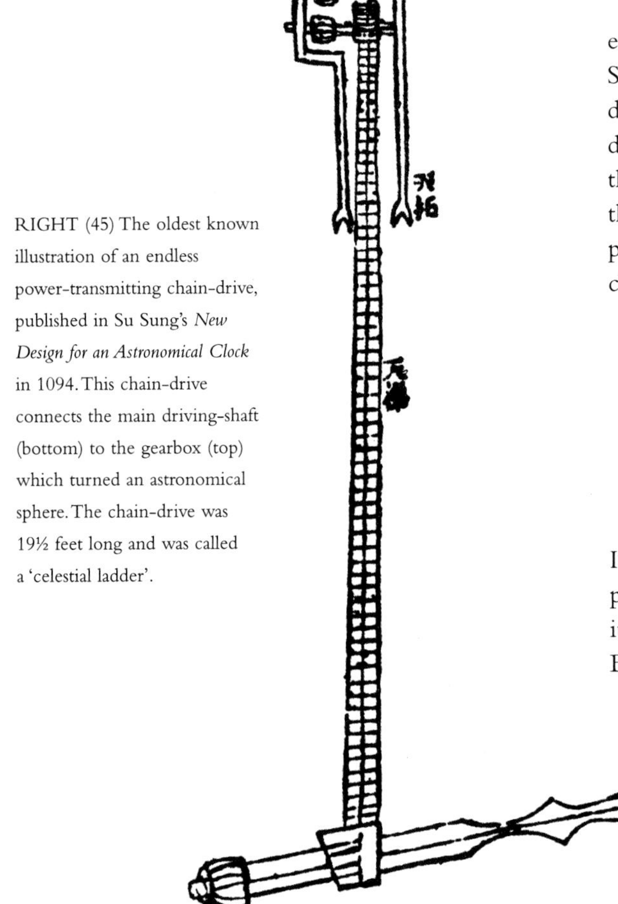

RIGHT (45) The oldest known illustration of an endless power-transmitting chain-drive, published in Su Sung's *New Design for an Astronomical Clock* in 1094. This chain-drive connects the main driving-shaft (bottom) to the gearbox (top) which turned an astronomical sphere. The chain-drive was 19½ feet long and was called a 'celestial ladder'.

been influenced by the common sight of the sprocketed chain of the familiar chain pumps. Though those chains did not transmit power, Chang evidently saw the possibilities of doing so with a similar design. And thus it was that he incorporated this important invention in his clock.

When, in 1090, Chang's more famous successor, the inventor Su Sung, built his enormous astronomical clock tower (see page 117), he tried at first to use a main vertical transmission-shaft. But this did not work. So he also adopted the chain-drive, which he called a 'celestial ladder'. We are fortunate that we actually have a drawing of this chain-drive made by Su Sung himself, and published in 1090 in his own book *New Design for an Astronomical Clock*. It may be seen in Plate 45 (above). Obviously, it is the oldest illustration of an endless chain-drive for power-transmission in the world. A shorter and tighter chain-drive was fitted to the clock as a further improvement, when it was discovered that the key to mechanical efficiency was to eliminate slack wherever possible in all mechanical connections.

Although we have little detail concerning Chang's earlier chain-drive of 976, we have some for that of Su Sung in 1090. The chain can be seen to have been double-linked. The chain-drive coupled the main driving-shaft of the clock apparatus with a series of three small pinions in a little gear-box, which operated the turning of the armillary sphere (an astronomical part of the clock). Su Sung himself described the chain-drive in his book by saying:

> **The 'celestial ladder' is 19.5 feet long – an iron chain with its links joined together to form an endless circuit, hanging down from an upper chain wheel ... and passing round a lower chain wheel on the driving-shaft.**

In the ancient Western world, loose chains and pentagonal sprocket-wheels were designed, though it is doubtful if they were built, by Philon of Byzantium about 200 BC. But no transmission of power was envisaged, and the intended use was to provide repeat-loading of a catapult. Even if this machine had been built, it would not have been a proper chain-drive.

The first genuine European chain-drive was made by Jacques de Vaucanson in 1770 for silk reeling and throwing mills. In 1869 J.F. Tretz used the chain-drive to make bicycles. Since bicycles are now the leading form of transport in China, it is ironical that only a tiny handful of the Chinese who use them have any idea that the bicycle chain-drive is a native Chinese invention eight hundred years in advance of a Western equivalent, and nine hundred years in advance of its application by a European for the bicycle.

UNDERWATER SALVAGE OPERATIONS

ELEVENTH CENTURY AD

In the eleventh century AD, at the instigation of a monk named Huai-Ping, the Chinese pioneered a technique of recovering heavy objects from a river or sea bed.

Between 1064 and 1067, the noted pontoon bridge at P'u-Chin near P'uchow over the great

Yellow River was destroyed by a sudden flood. This bridge, which had been constructed 350 years earlier, was a major crossing of the river. It consisted of floating boats securely linked together by a great wrought iron chain. The ends of the chain were attached to eight enormous cast iron figures in the shape of recumbent oxen set in the sandy beaches on the two banks.

When the flood washed away the bridge, the iron oxen were pulled into the river and sank deep under water. The dismayed local officials issued a public proclamation requesting ideas as to how they could be recovered, and it was to this plea that Huai-Ping responded.

On his instructions, workers filled two large boats with earth, and divers attached cables from them to the oxen in the river bed. Then earth was gradually removed from the boats, which caused them to float higher and higher in the water. To everyone's delight, the buoyancy thus created lifted the oxen from the river bed. They were then dragged into shallower water simply by sailing the boats towards the shore. From there, they were easily recovered.

This appears to be the first time in history that buoyancy techniques were used for an underwater salvage operation. These techniques are still employed today. When the giant ocean liner *Andrea Doria* sank in the Atlantic in the 1950s, the operation used to salvage the hulk involved attaching the ship – 225 feet below the surface – to ore ships full of water and then progressively emptying them, repeating the process as often as required: the very principle propounded by Huai-Ping nine hundred years earlier.

BELOW (46) An attempt at underwater salvage in the year 219 BC, shown in a bas-relief of the period. Nine bronze tripods had been lost in the River Ssu in 333 BC, and in 219 BC the emperor sent an expedition to recover them, but it was unsuccessful. However, by the eleventh century AD, the Chinese had developed sophisticated underwater salvage operations which were up to twentieth-century standards.

Part 4

DOMESTIC AND INDUSTRIAL TECHNOLOGY

LACQUER: THE FIRST PLASTIC

THIRTEENTH CENTURY BC

Lacquer was in use by at least the thirteenth century BC in China. Queen Fu Hao of that date was buried in a lacquered coffin, discovered when her intact tomb was excavated in 1976 at Anyang. Needham has said of lacquer: 'Lacquer may be said to have been the most ancient industrial plastic known to man.' It has been in use for well over three thousand years.

Lacquer is obtained rather like rubber, by tapping the sap of tree trunks. The lacquer tree (*Rhus vernicifera*, recently renamed *verniciflua* by botanists) is indigenous to China but not to Europe. It is particularly common in central China, growing at altitudes between 3000 and 7500 feet. The trees are tapped in summer and left to recover over a period of five to seven years, though in some cases they are cut down after tapping, and an inferior lacquer is obtained from their branches. The largest amount of lacquer a tree can produce is about 50 grams.

Lacquer is a plastic varnish which has remarkable powers of preservation, strength and durability. Strong acids and alkalis cannot damage it; it cannot be affected by heat less than 400°–500°F; it cannot be damaged by water or other liquids; it is insoluble to most solvents; and it is resistant to bacterial attack. Also, as an electrical insulator it is almost as good as mica. Its discovery was thus a major event. It is one of the toughest and most remarkable vegetable substances in the world, and is a natural plastic. The first artificial plastic to be discovered was celluloid, by John Wesley Hyatt, in 1869. But the modern plastics industry did not really commence until 1907 with the discovery of bakelite. Today many artificial lacquers are made, and in the plastics industry the word 'lacquer' is loosely applied to countless synthetic substances which have no connection with the lacquer tree. When a synthetic lacquer is pigmented, it is called an 'enamel'. These lacquers are therefore in every Western kitchen today.

But true lacquer was used for kitchen utensils thousands of years ago in China. Wood, bamboo or cloth utensils coated with many layers of thin lacquer formed the standard dinner service for rich Chinese in place of bronze vessels. They were able to withstand the heat of cooking and serving of food as well as metal. Chinese emperors gave lacquered articles to their officials as recognition of their services, and the monetary value of lacquerware actually exceeded that of bronzes.

Lacquer was used in China for furniture, screens, pillows and boxes of all sorts. It was worn as bonnets and shoes. Weapon accessories, such as sword scabbards, bows and shields were made of it. Lacquers were often inlaid with gold and silver or tortoiseshell. The fluidity of the lacquered surface when applied made possible a form of Chinese decoration which was as free and spontaneous as could be imagined, and this had a major impact on ancient Chinese art.

Lacquer came in many different grades, the best apparently being obtained from between the inner and outer barks of trees aged between 14 and 15 years. When lacquering an article, it is usual to commence with coats of inferior lacquers and save the best grade for the final coat. Each thin coat was applied and allowed to dry completely before the application of the next coat. In the best quality later carved work, over a hundred coats of lacquer were applied to obtain the necessary

OPPOSITE (47) A red lacquer throne of the Ch'ing Dynasty, dating from the period 1736-96. (Victoria and Albert Museum, London.)

depth for carving. Various pigments can be added to give colour, the traditional basic colours being black, red, brown, yellow, gold and green. A particularly intriguing colour was the 'pear-ground lacquer' made with gold dust and gamboge gum resin.

The lacquer industry in ancient China was highly organized, in the traditional Chinese bureaucratic way. There were both private and state lacquer manufacturing centres. There is a lacquered wood wine cup in the Musée Guimet, Paris, which can be dated precisely to the year 4 AD by an extraordinary inscription which, besides giving the date of manufacture, lists seven artisans involved in making the cup and five other officials of the company. This offers a fair idea of the Chinese approach to manufacturing as long as two thousand years ago. Twelve people to produce a single cup, nearly half of them functionaries who perhaps never even saw it! But on the other hand, it indicates also the use of something very like the modern industrial production-line.

As early as the second century BC, the Chinese had made important chemical discoveries about lacquer. They found a way to keep it from going hard by evaporation. What they did sounds like an old wives' tale – they threw crabs into lacquer to keep it liquid! The *Book of Master Huai Nan*, of 120 BC, however, says that crabs spoil lacquer so that it will not dry and cannot be used. Many subsequent authors mentioned this apparently ridiculous tradition. Li Shih in the twelfth century wrote that 'after coming in contact with crabs, lacquer will not concrete'.

Crustacean tissue does in fact contain powerful chemicals which inhibit certain enzymes, including the one which makes lacquer solidify! Needham comments on this bizarre business as follows:

What part, then, were the crab tissues playing? There can be no doubt that the ancient Chinese, before the 2nd century BC, had accidentally discovered a powerful laccase inhibitor. By preventing the action of the enzyme the darkening and polymerization were also prevented. So great an interference with the course of nature, analogous to the arrest of a spontaneously occurring rigidification

BELOW (48) As an inert plastic, lacquer survives for a long time without disintegrating. This lacquered toilet box dates from the Han Dynasty (207 BC–220 AD) – approximately two thousand years ago. The box is made of simple fabric, but its many layers of lacquer coating have ensured its preservation. (British Museum.)

LEFT (49) As a member of a production line making lacquer objects, this craftsman is depicted cutting decorative incisions in the lacquer. An inscription on a lacquer cup of 4 AD reads: 'Priming by Yi; lacquering by Li; outer coat by Tang; gilding of the handles by Ku; painting by Ting; inscribed by Peng; cleaning and polishing by Chung.' This worker, in a painting of the late eighteenth century, represents yet another stage in the process.

and ageing process, must have seemed highly significant to the alchemists, preoccupied as they were by the preservation of supple youth and the postponement or elimination of ankylosis and death. Moreover this action of crustacean tissues is not unique, for other researches have shown that they contain a powerful though somewhat enigmatic inhibitor for D-amino-acid oxidase.

Not only did the perpetual liquefaction of lacquer pose a model for immortality, in this proto-industrial biochemistry, but lacquer accompanied a Chinese from cradle to grave – he would be fed as a baby from lacquer vessels with lacquer ladles, and in death he would be buried in a beautifully ornamented lacquer coffin.

STRONG BEER (*SAKE*)

ELEVENTH CENTURY BC

Many Westerners are familiar with Japanese *sake*, but few will know that it represents a type of alcoholic drink which has only ever been made in China and Japan (where it was introduced from China). *Sake* is neither a wine nor a spirit. Some *sake* made today has been strengthened with spirit, but *sake* itself is a type of alcohol known in China as *chiu*, of which a very rough translation is 'strong beer'.

Ordinary beer, with an alcohol content reaching perhaps 4 or 5 per cent, has been made throughout the world for thousands of years. It was known to the Egyptians and Babylonians, and mention of it as an offering to the spirits in sacrifices occurs in early bone inscriptions in China dated about 1500 BC. There were many different kinds of this ordinary beer, with varying ingredients, from assorted regions.

By 1000 BC at the latest, and perhaps centuries earlier, the fermentation process was well established in China for the making of *chiu*. It was at least three times as strong as ordinary beers. Whereas in the West no beverage attained an alcohol content of more than about 11 per cent (achieved by some wines) before distilled alcohol appeared, this 'strong beer' in China is thought to have had a substantially higher alcohol content. Poems surviving from 800 BC or earlier describe people getting tipsy on this heady brew.

The strong beer fermentation process involved an entirely new concept in fermentation of grain. The major problem with ordinary beer is that the starch in grain cannot be fermented. Thousands of years ago, it was found that sprouting grain contains a substance (the enzyme now known as amylase) which degrades the starch of grain into sugars which can then be fermented. This was the basis of ancient beer around the world; sprouting grain would be dried as malt, which would 'digest' the starch of the normal grain for beer.

Although the Chinese made this sort of beer like everyone else, they found that a far better technique was to make something called *chü*. This consisted of ground, partially cooked wheat (or occasionally millet) grains which had been allowed to go mouldy. These moulds produce the starch-digestive enzyme amylase more efficiently than does sprouting grain. *Chü* therefore was a mixture of moulds plus yeast. The Chinese would mix it with cooked grain in water, which resulted in beer. The amylase broke the starch down into sugar and the yeast fermented this into alcohol.

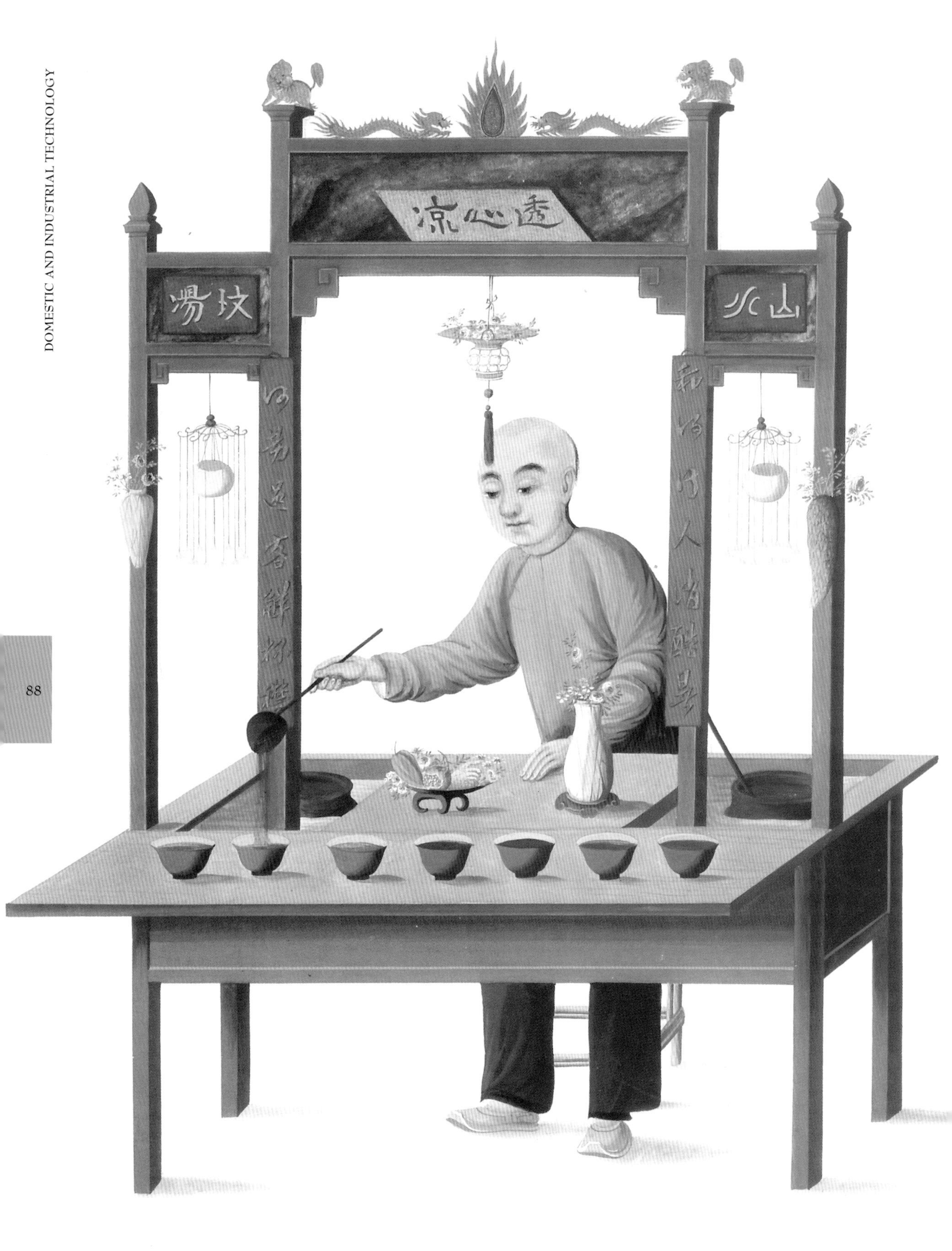

ABOVE (50) A late-eighteenth-century painting of a street seller of the Chinese drink known in Japan as *sake*. The drink is neither a wine nor a spirit, but is actually a 'strong beer' of which there is no Western equivalent. (Victoria and Albert Museum, London.)

It was also found that the alcohol content could be raised by continually adding more and more cooked grain in water to the brew as the fermentation proceeded. They call this 'killing' the grain. This 'topping-up' process eventually reached a peak above which it would not go, but it led to a very strong drink indeed. Apart from its spread to Japan centuries ago, 'strong beer' has never been adopted by other cultures, and indeed there remains no direct translation of its name into any Western language. This is one Chinese invention which has yet to be appreciated by the world at large.

PETROLEUM AND NATURAL GAS AS FUEL

FOURTH CENTURY BC

Petroleum and natural gas were used as fuel in China many centuries before the West and it is probably a conservative estimate to say that the Chinese were burning natural gas for fuel and light by the fourth century BC. We know that this occurred before the Han Chinese overmastered the local tribes in the southern province of Szechuan, starting in the third century BC. The deep boreholes drilled for brine also yielded natural gas from time to time. These methane gas deposits tended to occur under the brine, but many boreholes, including those intended for brine, yielded only natural gas and were known to the Chinese as 'fire wells'. These boreholes were being drilled systematically for brine by at least the first century BC, so that deep supplies of natural gas were tapped from that date by boreholes going down at least several hundred feet. And the systematic search for natural gas itself by deep drilling is recorded in the second century AD.

Ch'ang Ch'ü in 347 recorded in his book *Records of the Country South of Mount Hua*:

> **At the place where the river from Pu-p'u joins the Huo-ching River, there are fire wells; at night the glow is reflected all over the sky. The inhabitants wanted to have fire, and used to ignite the gas outlets with brands from household hearths; after a short time there would be a noise like the rumbling of thunder and the flames would shoot out so brilliantly as to light up the country for several dozen *li* around [several miles at least]. Moreover they use bamboo tubes to 'contain the light', conserving it so that it can be made to travel from one place to another, as much as a day's journey away from the well without its being extinguished. When it has burnt no ash is left, and it blazes brilliantly.**

The bamboo tubes here probably refer to pipelines, though we shall see that portable containers for the gas were later used. Bamboo pipelines did indeed

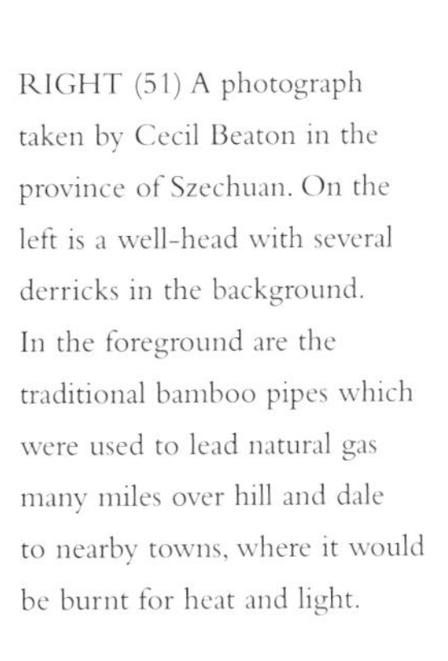

RIGHT (51) A photograph taken by Cecil Beaton in the province of Szechuan. On the left is a well-head with several derricks in the background. In the foreground are the traditional bamboo pipes which were used to lead natural gas many miles over hill and dale to nearby towns, where it would be burnt for heat and light.

RIGHT (52) A previously unpublished photograph taken in about 1900 by Ernst Boerschmann, showing the drilling derricks of the Szechuan natural gas fields.

carry both brine and natural gas for many miles, sometimes passing under roads and sometimes going overhead on trestles. If the brine needed lifting again to greater heights in order to flow downwards by gravity, pumps were employed. But this was not necessary for natural gas, which rose naturally.

The ignition and use of the natural gas for light and fuel posed problems which were successfully overcome by the ancient Chinese. The natural gas which comes from shallow boreholes is not very strong, and if piped directly to burners, can be lit safely. But the kind of pungent, powerful natural gas which the Chinese obtained from the majority of their boreholes, from depths below 2000 feet, had to be mixed with air before it could be safely burnt. There must have been many catastrophic explosions before the principles involved were learned.

Old texts describe in some detail the complicated arrangements which were eventually adopted to control the burning of natural gas. The gas from the 'fire wells' was fed first into a large wooden chamber about 10 feet below ground level over the mouth of the borehole. It was basically a cone-shaped barrel into which an underground pipe also conveyed air. The chamber therefore acted as a great carburettor, feeding into banks of pipes which led to other smaller conical chambers which rested on the surface of the earth. These too took in air, with a variety of entry pipes which could be opened or closed, so that a fine-tuning of the 'engine' was possible by a continuous manipulation of the fuel/air mixture. If the pressure of the mixture were to flag, dangerous flash-backs and explosions could occur, so the main chamber would be opened up further. But fires could result if the mixture were too rich, so surplus gas was allowed to escape through what was called a 'sky thrusting pipe' exhaust system.

Although uncontrolled emissions of natural gas, when ignited, could either flame up 100 feet in the air, or even explode, the wells with 'carburettors' had their gas piped along tubes to the various outlets where flames 1½ feet high were commonly produced. These would be used to heat evaporation pans of brine. The cast iron pans were gigantic and weighed as much as 1000 pounds, though there were outlying pans which were smaller. The brine was gently warmed in the smaller pans, and then transferred to the great pans for fierce boiling. Natural gas from very shallow wells could easily heat between ten and forty pans. But from the more common deep wells of about 3000 feet depth, the gas commonly serviced 600 to 700 burners, and one case at Fu-ch'ang is recorded where as many as 5100 evaporation pans were heated by a single 'fire well', though this figure must include small as well as large pans.

Flames smaller than 1½ feet high were used for other purposes, such as providing lights in certain Szechuan towns. These early gas lamps anticipated Victorian England by many centuries. The gas was also available for heating in these towns, though details of how it was employed are lacking. It seems doubtful that proper gas stoves existed, and it is more likely that the heating applications were generally for cooking and boiling uses. More research is necessary in the old gazetteers of Szechuan in order to try to find further information about the uses to which natural gas and petroleum were put there over the millennia. The thousands of regional Chinese gazetteers, which go back many centuries, have never been properly studied, and priceless information relating to Chinese science and technology has been found in those few which have been examined.

It is one such gazetteer, dating from before the tenth century, and surviving only in quotations in a later work, which tells us how the Szechuanese anticipated modern portable butane gas cylinders for carrying supplies of natural gas:

> Lin-ch'iung has a fire well which is over 600 feet deep and which burns brightly at the opening. People use tubes to carry it around, and they can go over a hundred *li* [dozens of miles] and still be able to light it.

Thus, a man could travel for a couple of days and then open up his gas cylinder, probably through a tap, and cook himself a meal. Other portable heat sources included petroleum products, which often went under the name of 'stone lacquer' because they looked like lacquer but seeped from the stones. A book of 980 records:

> Some fire wells produce a liquid too. The people of the commandery use bamboo tubes to carry it to light their way in the same manner as ordinary people carry torches. This is because the liquid is inflammable.

Another book by Yang Shen in the sixteenth century recorded of the 'fire wells' at K'ang-ting and Chien-wei:

> The streams produce oil which when heated bursts into flame. People take it and use it in lamps like candles. This oil is just an accumulation of solar essence. It is nothing strange.

The domestic uses of petroleum products seem to have been confined to modest applications such as oil lamps and oil-fired torches, but they were used on a large scale for breaking up rocks by fire. Since burning oil could burn in water, boulders in harbours were sometimes broken apart by having burning oil poured over them.

The bamboo tube 'butane cylinders' have apparently been seen in use in the hinterland of China by modern scholars, though no photograph appears to exist of one in operation. Leather bags were also employed. A seventeenth- or eighteenth-century account survives in one of the Szechuan gazetteers, entitled *Historical Geography of Fu-shun in Szechuan*:

> There is a fire well 90 *li* west of the prefectural seat of Fu-shun which is 40 to 50 feet deep and has an opening 5 or 6 inches in diameter. Gas rises up from it. Bamboo tubes with their partitions removed lead it off from the well and when fire is applied the gas ignites. The flames may be small or they may leap up several feet into the air making a noise like thunder as they burn; the gas is led through underground conduits to heat the brine, never ceasing day and night. When not wanted for use, it is doused with water and extinguished. Furthermore, bamboo tubes which have been bored through are joined together to lead the gas away. It can be used as a

LEFT (53) A photograph taken by Cecil Beaton during the Second World War, showing a workman at the Ts'iu-ch'ing Salt Wells in Szechuan Province. He is stirring brine in a vat which is being heated by the jet of a natural gas flame, in the manner practised since the fourth century BC in China. Natural gas is found below brine deposits in sufficiently deep wells. Using their traditional drilling techniques, the Chinese could reach depths of 4800 feet.

> **substitute for firewood and torches. Travellers going to a distance carry the gas with them in leather bags. After travelling even as far as several thousand *li*, a hole is punctured, touched with fire, and light and heat come instantaneously.**

But because these phenomena were rather regional in nature, the Chinese of other regions did not necessarily have much familiarity with them. About 1600, Sung Ying-Hsing wrote:

> **The 'fire wells' of Western Szechuan are a highly amazing phenomenon; there is cold water in them but not the slightest evidence of the essence of fire. Yet men take long bamboo stems and split them, then put them together again so as to form a pipe, wrapping it securely with varnished cloth; this tube they insert into the well ... it will be seen that hot dry flames are bursting forth from the pipe.... Yet if the bamboos are opened and examined, no sign of charring or burning can be seen. To use the spirit of fire without seeing the form of fire – this is indeed one of the strangest things in the world.**

PAPER

SECOND CENTURY BC

Although the word 'paper' is derived from the word 'papyrus', paper and papyrus have nothing whatever to do with one another. Papyrus, which existed in Egypt as early as the third millennium BC, is made from the inner bark of the papyrus plant (*Cyperus papyrus*). Apart from the fact that it gives a sheet on which one can write, it is completely and totally different from paper. The Chinese invented paper, by the second century BC at the latest.

Paper in the modern world is mostly made of wood pulp. But, just to confuse the issue even more, paper in ancient times was never made of wood pulp. So, what then is paper?

Paper is the sheet of sediment which results from the settling of a layer of disintegrated fibres from a watery solution onto a flat mould, the water being drained away, and the deposited layer removed and dried. The fibres can be of any material whatever, though plant fibres are by far the most commonly used ones, and as remarked above, fibres from trees are the mainstay of the paper industries today. The earliest European paper was not made of wood pulp, but of disintegrated and pounded rags of linen. Anyone who owns or has handled a book printed in the seventeenth century in Europe will be aware of how durable and springy the paper is; this paper is made of linen, and it will still be youthful and fresh when most of the books printed in the twentieth century have disintegrated to dust.

The oldest surviving piece of paper in the world was discovered by archeologists in 1957 in a tomb near Sian in Shensi Province, China. It is about 10 cm square and can be dated precisely between the years 140 and 87 BC. This paper and similar bits of paper surviving from the next century are thick, coarse, and uneven in their texture. They are all made of pounded and disintegrated hemp fibres. From the drying marks on them, it is evident that they were dried primitively on mats woven as pieces of fabric, not on what we know as paper moulds. In these early days, the water just drained slowly through the underlying mat of fabric, leaving the paper layer on top. This was then peeled off and dried thoroughly. But so thick and coarse was the result that it could not have been very satisfactory for writing.

However, paper does not appear to have been used as a medium for writing until a considerable time after its invention. The oldest surviving piece of paper in the world with writing on it was discovered under the ruins of an ancient watchtower in Tsakhortei near Chü-yen in 1942. The watchtower was abandoned by Chinese troops during the rebellion of the Hsi-ch'iang tribe, and the paper can thus be dated to 110 AD. It contains about two dozen readable characters.

It is probable that paper was in use for a century or more in China before its possibilities as a medium for writing were noticed. Its earliest uses were in connection with clothing, wrapping, lacquerware and personal hygiene. A text of 93 BC records an imperial guard recommending to a prince that he cover his nose with a piece of paper – the first Kleenex! A record of a murder case from 12 BC notes that the poison used had been wrapped in red paper. By the time of the Emperor Kuang-Wu (reigned 25-26 AD), an official of the imperial secretariat was already responsible for 'the seals and cords of office, and for paper, brush, and ink.'

The use of paper for clothing may at first seem strange; we think of paper as being thin today, and hardly the proper material to keep out the cold. But the use of paper as protective clothing against the cold was practised by the Chinese from the second century BC onwards. We are not certain when the Chinese in the south began making paper of the bark of the paper-mulberry tree (*Brousonetia papyrifera*), but the pounded bark of this remarkable tree was found to be serviceable for clothing from an early date. And it would seem that not long after paper proper was invented, the disintegrated fibres of paper-mulberry bark were employed, as well as the more common hemp, to make real paper.

However, the earliest uses of paper were derived from the simple pounded bark of this tree. We know that in the sixth century BC, a disciple of Confucius named Yüan Hsien from the State of Lu wore a hat made from paper-mulberry bark. The historian Ssuma Ch'ien records that in the second century BC, huge quantities of this substance were in commercial circulation. A paper hat, a paper belt, and a paper shoe dating from 418 AD were discovered in an excavation at Turfan, and reported in 1980.

But surely articles of clothing made from paper were too flimsy? Perhaps the inferior paper of today, made from wood pulp, would be. But the paper of those days, made from much stronger and tougher fibres, was not. So tough was paper then that it was frequently used as a shoe liner. And paper clothing was so very warm and impenetrable by cold winds that people complained that it allowed no circulation of air round the body and was too hot to wear! Beds in winter were kept warm with paper curtains, and thin curtains, also of paper, were used as mosquito nets.

The poet Lu Yu wrote a letter to the philosopher Chu Hsi about the year 1200, thanking him for the gift of a paper blanket. The letter survives, and we are able to enjoy the amusement of reading Lu Yu's rapturous account: 'I passed the day of snow by covering me with a paper blanket. It is whiter than fox fur and softer than cotton.'

Not only was paper used for clothing, it was used for military armour! In the ninth century a provincial governor named Hsü Shang is recorded as keeping an army of a thousand soldiers ready at all times clothed in pleated paper armour which could not be pierced by strong arrows. Paper armour became common on land and at sea. When two pirate ships surrendered in an amnesty in the twelfth century, 110 suits of paper armour were handed over by them. And in the twelfth century, 'Chen Te-Hsiu is recorded as saying that he had sufficient weapons at his fort for defence, but of his hundred sets of iron armour, he had kept half of

BELOW (54) A late-eighteenth-century painting showing the manufacture of paper. The large vat contains a watery solution of disintegrated fibres. A mould (with mesh screen, similar to the one resting on the side, at left) is being lifted from the vat and drained. The remaining layer of sediment will be dried and peeled away as a sheet of paper. The piles of paper sheets resulting from this process are at the right, resting on planks over the vat. (Victoria and Albert Museum, London.)

ABOVE (55) A modern demonstration of the traditional Chinese method of paper manufacture at the Ontario Science Centre. Here, the mesh screen mould is being lifted from the vat with its watery solution of disintegrated fibres. Although most paper in the modern world is made from wood pulp, the Chinese never used it, preferring the stronger fibres of linen and a variety of other plant materials, most of which would be too expensive for mass manufacture today.

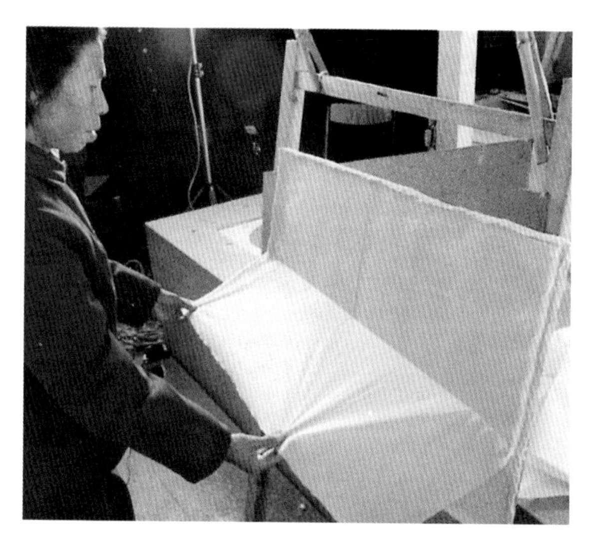

ABOVE (56) A later stage of the modern demonstration at the Ontario Science Centre. Here, the layer of sediment accumulated on the mould from the vat has dried and is being delicately peeled off – a sheet of paper. Paper (which is completely different from papyrus) was invented in China by the second century BC, but did not reach Europe until a thousand years later. Its secret remained undiscovered for another four hundred years after that, and paper was not made in Europe until the twelfth century.

them and traded in the other fifty for sets of the better paper armour. Even bullets from guns were said to be unable to pierce good paper armour, as we read in this account in 1629 by Mao Yüan-I:

> **Armour is the basic equipment of soldiers, with which they are able to endure without suffering defeat before sharp weapons. The terrain in the south is dangerous and low, and where foot soldiers are generally employed they cannot take heavy loads on their backs when travelling swiftly. If the ground is wet or there is rain, iron armour easily rusts and becomes useless. Japanese pirates and local bandits frequently employ guns and firearms, and even though armour made of rattan or of horn may be used, the bullets can nevertheless pierce it. Moreover, it is heavy and cannot be worn for too long. The best choice for foot soldiers is paper armour, mixed with a variety of silk and cloth. If both paper and cloth are thin, even arrows can pierce them, not to say bullets; the armour should, therefore, be lined with cotton, one inch thick, fully pleated, at knee length. It would be inconvenient to use in muddy fields if too long and cannot cover the body if too short. Heavy armour can only be used on ships, since there soldiers do not walk on muddy fields. But since the enemy can reach the object with bullets, it could not be defended without the use of heavy armour.**

The Chinese also invented wallpaper, apparently as a result of hanging up large printed paper sheets which it was found more convenient to glue to the walls. Wallpaper was brought to Europe from China in the fifteenth century by French missionaries.

As for the sanitary uses of paper, there are some staggering statistics for this from China. In 1393, the Bureau of Imperial Supplies manufactured 720,000 sheets of toilet paper, measuring 2 feet by 3 feet each, for the use of the imperial court for one year. In addition, 15,000 special sheets, 3 inches square, 'thick but soft, and perfumed' were prepared for the exclusive use of the imperial family for the same year. Toilet paper was generally made from rice straw fibres, which were cheap and easy to process. Untold millions of sheets of toilet paper were in use in the Middle Ages. In and around the year 1900 in the province of Chekiang alone, the annual production of toilet paper amounted to ten million packages of between 1000 and 10,000 sheets each. This means that Chekiang was producing between ten and one hundred billion sheets per year at a time when barely any at all was presumably being used in the West. If we multiply the Chekiang statistics to take in the whole of China, it means that many thousands of billions of sheets of toilet paper were being used there a century ago. How far back did the use of toilet paper go in China? We can trace it in texts from as far back as the sixth century, when the scholar official Yen Chih-T'ui wrote in 589: 'Paper on which there are quotations or commentaries from the *Five Classics* or the names of sages, I dare not use for toilet purposes.' And an Arab of 851 wrote: 'They [the Chinese] are not careful about cleanliness, and they do not wash themselves with water when they have done their necessities; but they only wipe themselves with paper.'

ABOVE (57) A wheelbarrow for carrying particularly heavy loads, as used in southern China in the seventeenth century, and still in use today by farmers and workmen. This engraving was published in 1637 in the book *The Creations of Nature and Man* (*T'ien-kung k'ai-wu*).

The other uses of paper in China were so many that it would take too long to enumerate them in this volume. Obviously paper was important for the making of kites, another Chinese invention (page 188) and the Chinese were also the world's leading paper-folders, as well as paper-cutters for decorative designs. Some ancient paper flowers survive to this day, and the art of *origami* (fancy paper-folding), which originated in China, is now popular around the world. Paper umbrellas and paper money both originated in China and are discussed elsewhere (pages 108 and 131). The large number of substances whose fibres were used for Chinese paper-making make a study in themselves, and include bamboo, straw of rice and wheat, sandalwood, hibiscus, seaweed, floss silk from silk cocoons, rattan, jute, flax, and ramie. Chronologically, hemp was the main material in earliest times, followed by paper-mulberry fibres, then rattan, then bamboo, and later straw. All sorts of fancy, perfumed, glossy and other special papers were prepared in China, some of which probably excel any which have ever been produced in the West.

Paper reached India in the seventh century and West Asia in the eighth century. For five centuries the Arabs jealously guarded the secret of paper-making and would not reveal it to the Europeans, but sold them paper instead – at great profit. The Arabs had learned the techniques of paper manufacture from some Chinese prisoners of war captured after a battle at Samarkand. Europe obtained its first paper through the Arabs at around the end of the eighth century. However, the next signs of paper being used in Europe date no earlier than the eleventh century, and paper seems to have been slow to replace papyrus in the West. The first manufacture of paper in Europe dates from the twelfth century, and it was not until the thirteenth century that an Italian paper industry could be said to be in full swing. This is fifteen hundred years after its invention in China.

THE WHEELBARROW

FIRST CENTURY BC

It may seem difficult to believe, but wheelbarrows did not exist in Europe before the eleventh or twelfth century. The earliest known Western illustration of a wheelbarrow is in a window at Chartres Cathedral, dated about 1220. Considering that the use of wheelbarrows could cut the number of labourers required for any building project by half, the lack of them before this must have been as appalling as the welcome of them must have been ecstatic.

The wheelbarrow was apparently invented in south-western China in the first century BC by a semi-legendary personage called Ko Yu. We say 'apparently' because, first, it may have existed before that and, secondly, Ko Yu may either have been an actual individual of that name, or otherwise may be a sort of artisan's deity for wheelbarrow-makers. In both Kan Pao's *Reports on Spiritual Manifestations* (348 AD) and Liu Hsiang's *Lives of Famous Immortals* (compiled between the first century BC and fourth century AD) we are told that Ko Yu, a noted Taoist from Szechuan Province in the south-west, constructed a wooden goat or sheep and rode away into the mountains on it. This was a conventional early way of speaking of wheelbarrows, for in the third century AD, wheelbarrows constructed by Chuko Liang were called a 'wooden ox' and a 'gliding horse'. The former was said to be pulled by shafts in front, while the latter was said to be pushed by shafts behind.

The oldest surviving picture of a wheelbarrow dates from about 100 AD. It is a frieze relief from a tomb-shrine excavated near Hsüchow, which very clearly shows a wheelbarrow with a man sitting on it. In another contemporary illustration, this time from a tomb of about 118 AD at Ch'engtu in Szechuan, a man is seen pushing a wheelbarrow. There are several other illustrations from this period, the Han Dynasty, indicating that wheelbarrows were increasingly commonplace.

The earliest descriptions of the construction of wheelbarrows are couched in coy and obscure language. For the first few centuries, wheelbarrows were of great military importance, and specifics of their construction were closely guarded secrets. Various sorts were produced which could carry hundreds of pounds each. Some carried men on seats, and others carried supplies. Huge numbers of them were used to supply armies fighting in difficult, hilly terrain, in which China abounds. Many battles could never have been fought and won without wheelbarrow supply brigades. Another use of wheelbarrows was to form protective movable barriers against cavalry charges, which could be arranged in any shape at a moment's notice. The ingenuity of the Chinese at exploiting the wheelbarrow was limitless, and they were even given sails, with which they could achieve speeds over land or ice of 40 miles per hour (see account on page 216).

Scenes from medieval China showing wheelbarrows abound. We are fortunate that a great painting by Chang Tse-Tuan of the city K'aifeng (then the capital in the year 1125 survives, in which various types of wheelbarrow may be seen. An empty one stands before a draper's shop, while a loaded one passes a dyeing establishment. Another is seen being loaded with sack-like objects outside the best hotel, while beside it in the street passes one so heavily laden that a mule pulls it, with one man pushing and another man pulling. Wheelbarrows were thus ubiquitous features of Chinese life – and still by this time they had not yet reached Europe.

A large variety of designs existed, some with wheels in the dead centre, with the weight resting entirely on the axle, and others with wheels forward as they are today in the West. Some had tiny wheels, some had huge ones. Additional small wheels were sometimes fitted in front to ease the passage over potholes and other obstacles. Practically any shape and size of wheelbarrow existed – and still exists – in China. Many of these designs have still not passed beyond the confines of China and made their way West, despite the fact that, for particular uses, many of them are far superior to the kind Westerners generally use. One can honestly say that the wheelbarrow in all its forms is still an invention which the West has yet to discover!

OPPOSITE (58) Barbed wire being transported by a traditional Chinese wheelbarrow near Shanghai in 1938. The Chinese invented the wheelbarrow in the first century BC, but it did not reach Europe before the twelfth or thirteenth century AD, and many of the best wheelbarrow designs have never been adopted in the West. This type, with the large central wheel, can be pulled from the front as well as pushed from behind, and can carry more than 2 tons.

SLIDING CALLIPERS

FIRST CENTURY BC

A measuring tool very much like the modern adjustable wrench (spanner) was used in ancient China at the time of Christ. Plate 59 (page 98) shows a diagram of one.

The tool, which is made of bronze, is an adjustable sliding calliper gauge with slot and pin. The only difference between it and a modern adjustable wrench is that it does not have the small revolving worm (screw). The side which was used for measurement is decimally graduated in 6 inches and in tenths of an inch. On the other side there is an ancient inscription which, translated, reads: 'Made on a *kuei-yu* day at new moon of the first month of the first year of the Shih-Chien-Kuo reign-period.'

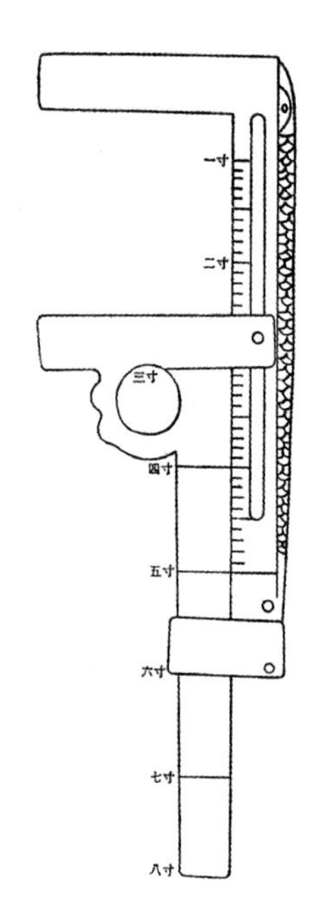

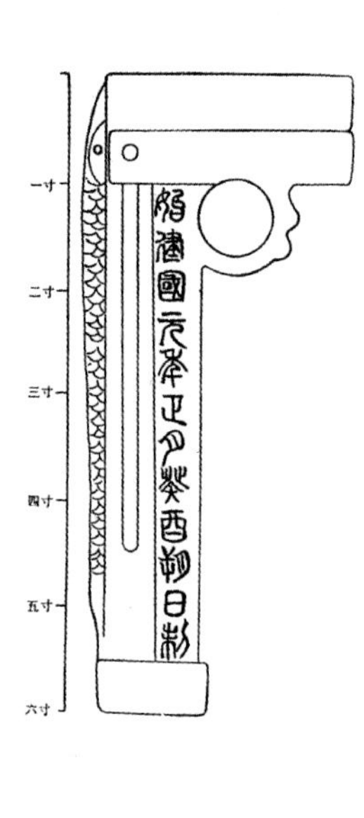

ABOVE (59) These sliding callipers are an adjustable spanner, without the worm. Unknown in Europe before Leonardo da Vinci sketched a set, they first came to be used in France in 1631, but the Chinese invented them in the first century BC. This illustration, published in China in 1925, depicts a surviving example of the measuring tool in bronze, with slot and pin for the sliding adjustment outside the calliper gauge, which dates from 9 AD, and bears an inscription on the right side which says it was 'Made on a kuei-yu day at new moon of the first month of the first year of the Shih-Chien-Kuo reign-period'. The left side of the tool is graduated in 6 inches and tenths of an inch.

This dates the implement to the first year of the reign of the Emperor Wang Mang, or 9 AD. This is the most impressive measuring instrument surviving from any ancient culture. These sliding callipers graduated in inches and tenths of inches must have been developed by the preceding century, the first century BC.

In Europe, sliding-scale callipers were introduced by Pierre Vernier in 1631, and the screw micrometer by William Gascoigne in 1638 for use in astronomy. The date of the earliest European calliper gauge is not known for certain before this, though the first such idea seems to have occurred in sketches made by Leonardo da Vinci over a century earlier. For the full sliding-scale calliper, however, the Chinese were in advance of Europe by approximately 1700 years.

RIGHT (60) The magic lantern, also known as the 'pacing horse lantern', had been invented by the second century AD in China. This traditional example shows horsemen which move round a central light. Magic lanterns seem to have reached Europe in the seventeenth century.

THE MAGIC LANTERN

SECOND CENTURY AD

The magic lantern, or zoetrope, is a device which held audiences spellbound before the modern era, and has always had a particular fascination for children. It can exist in various forms, but essentially it consists of a series of pictures through which light shines, and which move in succession to give the illusion that the figures in the pictures are themselves moving. It is thus the earliest ancestor of the cinema. The pictures can be turned by hand or automatically (as by vanes turned by hot air currents rising from a lantern). The pictures are generally seen projected on a wall or a screen, though street shows in portable boxes have instead a peep-hole through which one peers at the moving pictures inside the box. When projected, it is better that lenses be used, though they are not essential.

In 1868, W.B. Carpenter, the Vice-President of the Royal Society, wrote that the zoetrope, or magic lantern, had been invented by Michael Faraday in 1836, only thirty-two years earlier. This was wrong, since John Bate had described the same thing in his book *Mysteryes of Nature and Art* in 1634. But the real truth is that the zoetrope was invented in China.

The projection of moving images on a screen is recorded as having been practised in 121 BC, when a magician named Shao Ong staged a kind of seance for an emperor in this way. But another early form of magic lantern was in the possession of an emperor who died in 207 BC; after the lamp was lit, one could see the sparkling of scales of turning dragons. The same emperor had a similar object called 'the pipe which makes fantasies appear'. It seems to have had a small windmill or air turbine connected to it, for we are told in the book *Miscellaneous Records of the Western Capital* written in the sixth century AD: 'There was a jade tube two feet three inches long, with twenty-six holes in it. If air was blown through it, one saw chariots, horses, mountains, and forests appear in front of a screen, one after another, with a rumbling noise. When the blast stopped, all disappeared.'

The next record we find of a magic lantern occurs in about 180 AD. At that time, the inventor Ting Huan had perfected a 'nine-storied hill censer', which seems to have been a vastly complicated multiple magic lantern. Attached to it were strange birds and mysterious animals, which moved around when a lamp was lit. A similar device is described by T'ao Ku in his book *Records of the Unworldly and Strange*, published about 950 AD: 'Moving shapes were seen and tinkling noises heard, after the lighting of a candle or lamp'.

By the twelfth century, magic lanterns were called 'horse-riding' or 'horse-pacing' lamps, since after the lamp was lit, a succession of prancing horses was projected round about on the walls, moving as if on their own. Europeans discovered these toys when they began visiting China, and a Jesuit missionary named Father Gabriel de Magalhaens has left this description of one from the middle of the seventeenth century:

The Lamps and Candles, of which there are an infinite number in every Lanthorn [lantern], are intermix'd and plac'd withinside, so artificially and agreeably, that the Light adds beauty to the Painting; and the smoak gives life and spirit to the Figures in the Lanthorn, which Art has so contriv'd, that they seem to

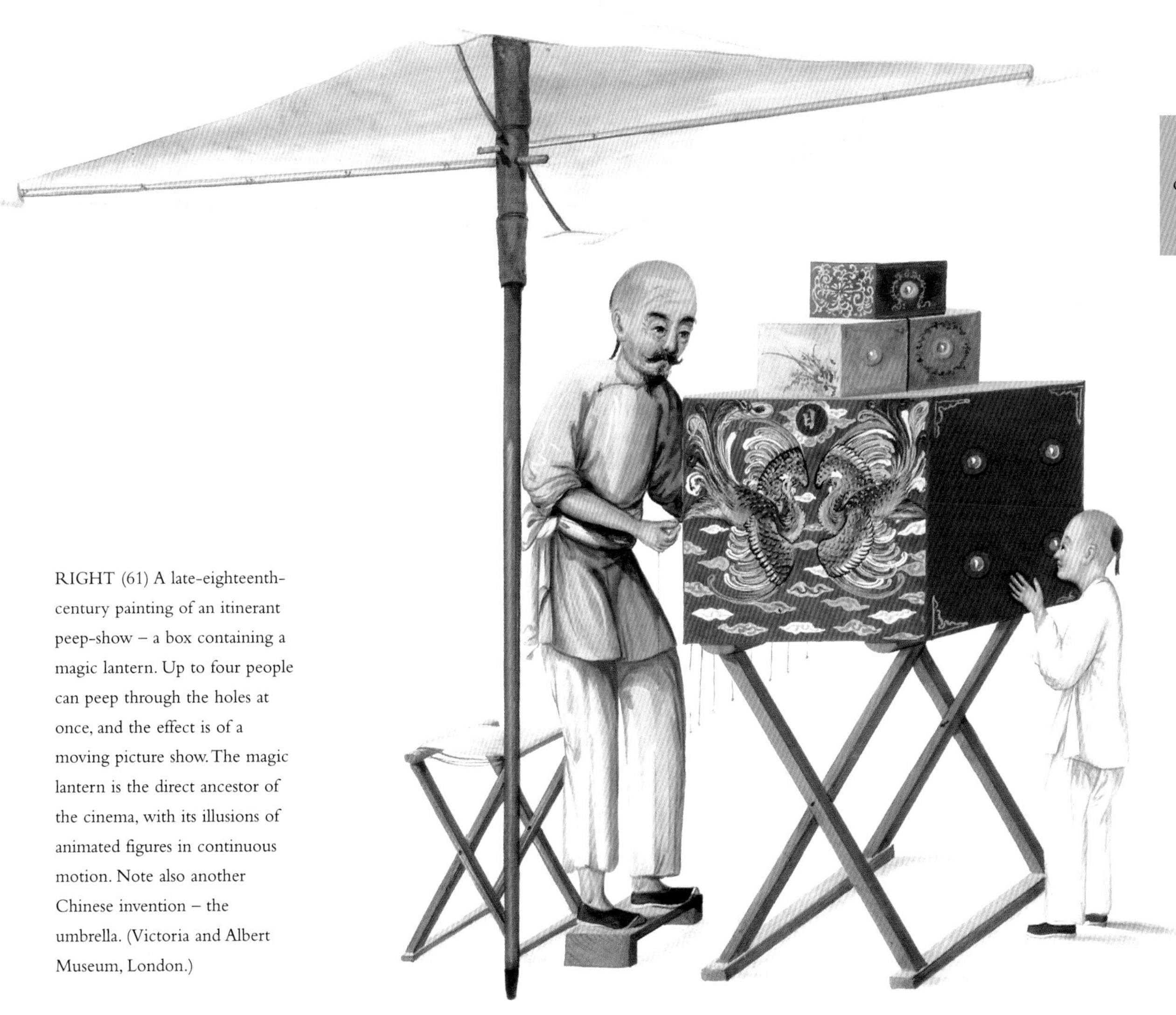

RIGHT (61) A late-eighteenth-century painting of an itinerant peep-show – a box containing a magic lantern. Up to four people can peep through the holes at once, and the effect is of a moving picture show. The magic lantern is the direct ancestor of the cinema, with its illusions of animated figures in continuous motion. Note also another Chinese invention – the umbrella. (Victoria and Albert Museum, London.)

walk, turn about, ascend and descend. You shall see Horses run, draw Chariots, and till the Earth; Vessels Sailing; Kings and Princes go in and out with large Trains; and great numbers of People both a-Foot and a-Horseback, Armies Marching, Comedies, Dances, and a thousand other Divertisements and Motions represented....

The projected slides which lecturers use, which some people still call 'lantern slides', derived from these Chinese magic lanterns. Needham says: 'It is not generally known that the first lecturer to use lantern slides was a Jesuit of the China mission, Martin Martini (16141).... His lectures illustrated by the new technique were given at Louvain in 1654.'

The magic lantern itself was also transported to Europe from China. We do not have all the details. Did the Chinese use lenses with their magic lanterns? They certainly used them a great deal for a number of other purposes, and Needham thought it possible 'that someone had the idea of placing one or more lenses at the pinhole of a closed chamber.' He goes on to suggest that this occurred during the T'ang Dynasty (618–906 AD). He also stresses that the 'persistence of vision' of the moving successions of images is the basis of cinematography, and that therefore the Chinese may perhaps be considered the earliest pioneers of the cinema.

Their images may only have been painted on pieces of paper or mica, and have gone round and round on a closed loop, but the effect for ancient times was surely profound, as the Jesuit father so vividly testified. We may look upon it as primitive but the ancient Chinese magic lantern had, in its day, a kind of cinematic glory in its turning picture show.

THE FISHING REEL

THIRD CENTURY AD

The Chinese invented the crank handle in the second century AD. It will therefore not be so surprising that they had invented the fishing reel, a small windlass, by the third century. A painting by the Chinese artist Ma Yuan, dating from about 1195 AD, is perhaps the oldest surviving picture of a fishing reel. Called 'Angler on a Wintry Lake', it clearly shows the use of the reel in the twelfth century. No representation of the fishing reel in the West is known before the year 1651.

The ancient Chinese name for the fishing reel is *tiao ch'e*. The earliest reference to it found by Needham occurs in *Lives of Famous Immortals*, which dates from the third or fourth century AD. However, parts of this book have been reliably dated to 35 BC and 167 AD, so it may be that the passage about the fishing reel is older than the book itself.

LEFT (62) This illustration from the *Universal Encyclopedia* of 1601 shows a fisherman catching a river turtle. The fishing reel can be plainly seen.

A predecessor of the fishing reel is mentioned in the *Mo Tzu (Book of Master Mo)* which dates from about 320 BC. This book was the canon of a group of warrior-philosophers and proto-scientists, the Mohists, who made many innovations in military technology. One of their machines for warfare was an *arcuballista*, an early form of artillery which fired groups of javelins at the enemy. Javelins were too valuable to waste, so they were attached to cords and could be retrieved for re-use by means of a reel and windlass – though probably if a javelin was sticking through somebody's chest, it was left there. It is certainly ironical that this military device led to the development of that most peaceable of inventions, the fishing reel.

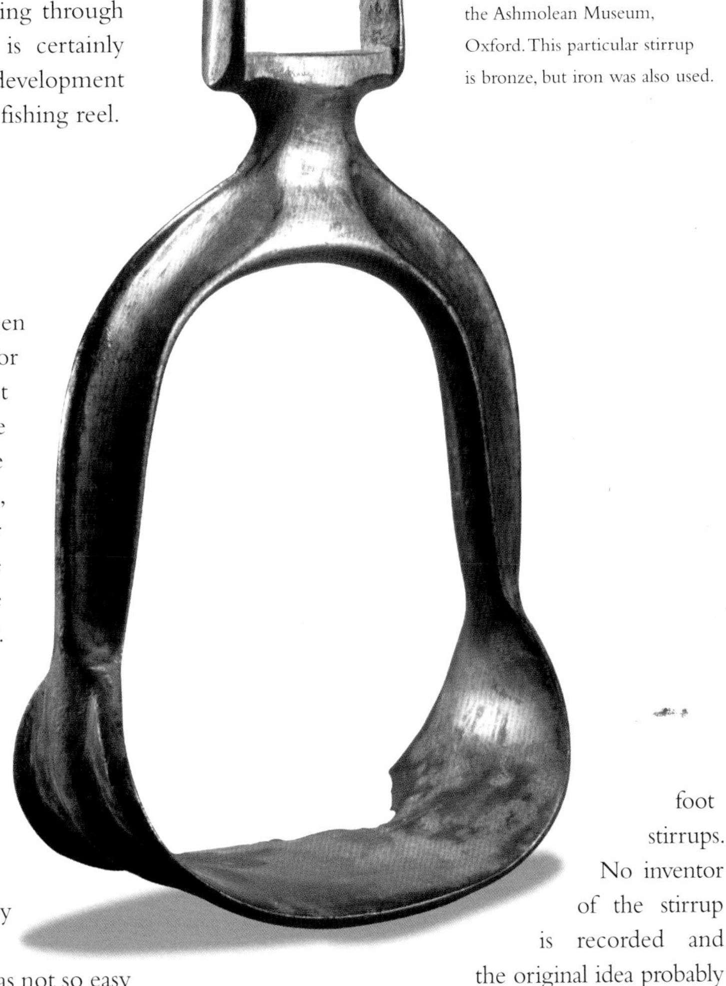

LEFT (63) One of the earliest surviving stirrups. It dates from the sixth or seventh century, and is preserved in the Ashmolean Museum, Oxford. This particular stirrup is bronze, but iron was also used.

THE STIRRUP

THIRD CENTURY AD

For most of the time that man has been riding horses, he has had no supports for his feet. Stirrups were unknown to most of the great armies of ancient times – the Persians and Medes, the Romans, the Assyrians, the Egyptians, the Babylonians, the Greeks. The horsemen of Alexander the Great made their way across the whole of Central Asia without being able to rest their feet while in the saddle. When galloping or jumping, horsemen had to hold the horse's mane tightly to avoid falling off. The Romans devised a kind of hand-hold on the front of the saddle which gave them something of a grip when the going got rough; but their legs just dangled whenever they were not pressed tightly against the horse.

Mounting a horse without stirrups was not so easy either. Fierce warriors took pride in their flying leaps, gripping the mane with the left hand and swinging themselves up; and some bareback riders still do this today. Cavalry-men of ancient times used their spears to help them up, either by hoisting themselves aloft as in pole-vaulting, or by using a peg sticking out of the spear as a foot-rest. Otherwise it was necessary to rely on a groom for a leg-up.

By about the third century AD, the Chinese had remedied this situation. With their advanced metallurgical expertise they began to produce cast bronze or iron foot stirrups. No inventor of the stirrup is recorded and the original idea probably came from the occasional use of a loop of rope of leather to assist in mounting. Of course, such loops could not be used for riding, because if one fell off, one would be dragged along and come to a sticky end. Such loops may have been first used by the Chinese, the Indians, or the nomads of Central Asia bordering on China. The essentials of the stirrup may thus have originated in the steppes, the product of ingenious men whose lives were lived on horseback. Apparently from the third century, the Chinese were casting perfect metal stirrups. The earliest surviving depiction of a stirrup is on a pottery

dated to 302 AD.

An excellent depiction of a stirrup may also be seen in a relief from the mausoleum of the Chinese Emperor T'ai Tsung (627–49). We may see in Plate 63 (page 101) an actual stirrup of the sixth or seventh century. Stirrups very quickly made their way to Korea, where they are seen in fifth-century tomb paintings. It has also been pointed out that only after stirrups came into use could the game of polo be properly played.

The transmission of stirrups westwards took place with the migrations of a fierce tribe called the Juan-Juan, who came to be known as the Avars (a Turkish word apparently meaning 'exiles'). Their cavalry was devastatingly effective because they had the use of cast-iron stirrups. About the middle of the sixth century, they were driven westwards and moved across south Russia to settle between the Danube and the Theiss. By 560, the Avars were a serious threat to the Byzantine Empire, and the Byzantine cavalry was entirely reorganized in order to counter them. The Emperor Maurice Tiberius prepared a military manual, the *Strategikon* (which still survives, untranslated) in 580, specifying the cavalry techniques to be adopted. He mentions the need to use iron stirrups – the earliest mention in European literature.

Stirrups then spread to the rest of Europe by means of the Vikings and possibly the Lombards. One Avar-style child's stirrup has even been excavated in London, brought by a Viking. But the use of stirrups in Europe (other than by the Byzantines and the Vikings) was long delayed, for reasons which are not entirely clear. Conventional armies of Europe do not seem to have adopted them until the early Middle Ages. Perhaps the lack of metallurgical expertise was a handicap, with stirrups having to be of wrought rather than cast metal for a long time. Mass production of stirrups was only possible with cast metal.

When we think of medieval Europe we think of knights in armour carrying heavy lances and riding on horseback, but that would have been impossible without stirrups, for such heavily weighted riders would have fallen off too easily. It was the Chinese invention of the stirrup which made Western medieval knights possible, and gave us the age of chivalry.

OPPOSITE (64) Heavily armoured men on horseback were only able to stay on their horses through the adoption of the stirrup, invented in China in the third century AD. This T'ang Dynasty ceramic figure dates from between the seventh and ninth centuries. (British Museum, London.)

PORCELAIN

THIRD CENTURY AD

Ordinary pottery is made from clay baked in a kiln at temperatures ranging from 500°C to 1150°C, and is called earthenware. Porcelain is something quite different: it consists of a body of fused clay covered by a glaze, a glassy substance, and is fired at a high temperature – about 1280°C. The secret of making porcelain lies in the use of a pure clay, kaolin or China clay, which when fired at a sufficiently high temperature changes its physical composition, a process known as vitrification, and becomes translucent and totally impervious to water. The reason why China was able to 'invent' porcelain at a very early age compared to the rest of the world was that the Chinese potters both found the clay and were able to produce the high temperature necessary to fuse it.

In practice, pure kaolin is mixed with 'porcelain stone', known in China in recent centuries as *petuntse* or *baidunzi*, a substance which is the intermediate product when igneous rocks decompose into clay over many millennia. Porcelain stone contains a high percentage of feldspar, which supplies the alkali as a flux to lower the vitrification temperature of the clay and the additional silica to enhance the porcelain's translucency after vitrification. Porcelain stone is also used for preparing high temperature glazes which also contain a strong flux such as wood ash or lime to enable the glaze mixture to turn

LEFT (65) This 1637 engraving from the book *The Creations of Nature and Man* (*T'ien-kung K'ai-wu*) shows a small porcelain kiln in operation, the main fire visible on the right. This produces heat rising from the bottom upwards for the first 24 hours. The men are inserting lighted wood through holes or 'skylights' at the top. This wood will burn for a further four hours, to produce the heat which will travel from the top downwards. A clay cup passed through a minimum of 72 separate processes before it became a porcelain cup. Further bundles of sticks lie on the ground at the bottom of the picture.

LEFT (66) Traditional Chinese kilns and firing of porcelain, from a late-eighteenth-century painting.

into a glass when fired at a temperature for porcelain production. The result is porcelain, which can often be seen through. Since the clay is opaque while the glass is transparent, the resulting fusion of the two gives a substance which is partially transparent, or translucent.

Just how astounding this was when first seen by foreigners may be judged from the remarks made by the Arabic merchant Suleiman, who wrote in 851 AD in his *Chain of Chronicles* of porcelain which he had seen in China: 'There is in China a very fine clay from which are made vases having the transparency of glass bottles; water in these vases is visible through them, and yet they are made of clay.'

It was incomprehensible to foreigners that pottery could be translucent and let the light through, since clay was well known to be opaque. This seemed to be a miracle: how could one see through something which everyone knows cannot be seen through? Pots and vases through which the light could pass? Incredible! Absurd!

Porcelain was not a sudden invention which took place at a particular time. It was arrived at gradually in China. Textual evidence is hopelessly confused and vague, owing to problems of terminology, so that archeological discoveries are the only means of determining when porcelain first came into being. Fresh discoveries keep pushing back the dates further into the past. 'Proto-porcelain' or 'primitive porcelain' made of kaolin clay, of compact texture, and surprisingly lustrous, apparently goes back to the eleventh century BC. But these wares do not have the fusion of the clay with feldspar and quartz, so that they are not true porcelains.

Historians of porcelain and ceramics are not always in agreement about when the line was crossed from 'proto-porcelain' to proper porcelain. However, it now seems that archeological finds push back the date of true porcelain to the first century AD. By the third century AD, in any case, true porcelain was undeniably in use.

By the Sung Dynasty (960–1279), porcelain had reached heights of artistry which some believe were never surpassed, and have not been equalled since. Porcelain manufacture by this time was a highly organized trade employing hundreds of thousands of people. There were teams of men who specialized in washing the clay, others who concerned themselves only with glazes, others who maintained the kilns, and so on. One kiln of this period which has been excavated could accommodate twenty-five thousand pieces of porcelain at a single firing. It was built on the slope of a hill, the gentle incline of about 15° reducing the speed of the flames through the kiln. The sophistication of the kilns was most impressive. Some were fired by burning of wood, while others were down-draught burners of charcoal. Flues were naturally employed, along with sophisticated layers of insulation, buttresses and clay linings. Control of the firing process was of the utmost precision. Porcelain could be fired either in oxidizing or de-oxidizing (reducing) flames. In the Ming Dynasty, when the famous blue-and-white ware was largely produced, the best lustrous quality of the cobalt blue pigment could only be obtained at certain specific temperatures, and in a reducing flame. Various metals used as pigments spread themselves chemically throughout the bodies of the porcelain objects in quite different ways depending on whether oxygen is being taken in or given off. A reducing flame forces porcelain to give off oxygen, leading to some of the most beautiful effects. The achieving of certain colours and effects in porcelain is therefore the result of intricate and subtle control of firing conditions of the kilns.

The secrets of porcelain manufacture were jealously guarded, and visitors from Europe such as Marco Polo could but gape and wonder. Porcelain objects were still a very great rarity in Europe by the fifteenth century. They were gifts for kings and potentates. Not until 1520 did the first sample of kaolin clay reach Europe, brought by the Portuguese. Europeans then thought that if only they could find deposits of this white clay, they would be able to make porcelain. Frantic efforts were made to locate deposits, in ignorance of the fact that kaolin clay alone was far from sufficient for the making of porcelain.

The countless experiments carried on with various earths and solid substances in furnaces eventually had a most unpredictable result. Scientists and craftsmen began to notice that upon cooling down again, molten minerals could crystallize. Until this began to be observed, Western scientists had been convinced that crystals could only be formed from liquids. About the middle of the eighteenth century in Europe, the idea began to gain ground that perhaps the Earth's rocks could have been formed from the cooling of molten masses of lava. Such an idea was heresy, but a heresy increasingly tolerated. It was the year 1776, famous for the American Declaration of Independence, that saw the first publication of a paper (by James Keir) suggesting a kind of declaration of independence in geology:

> **Does not this discovery of a property in glass to crystallize reflect a high degree of probability on the opinion that the great native crystals of basaltes, such as those which form the Giant's Causeway, or the Pillars of Staffa, have been produced by the crystallization of a vitreous lava, rendered fluid by the fire of volcanoes?**

The pottery manufacturer Josiah Wedgwood (who was trying to make porcelain) even became involved in scientific experiments to determine whether the Earth's crust might have been formed from lava flows. And by 1785, James Hutton the geologist presented his revolutionary new theory of the Earth based on these ideas, which we remark elsewhere (page 182) had been

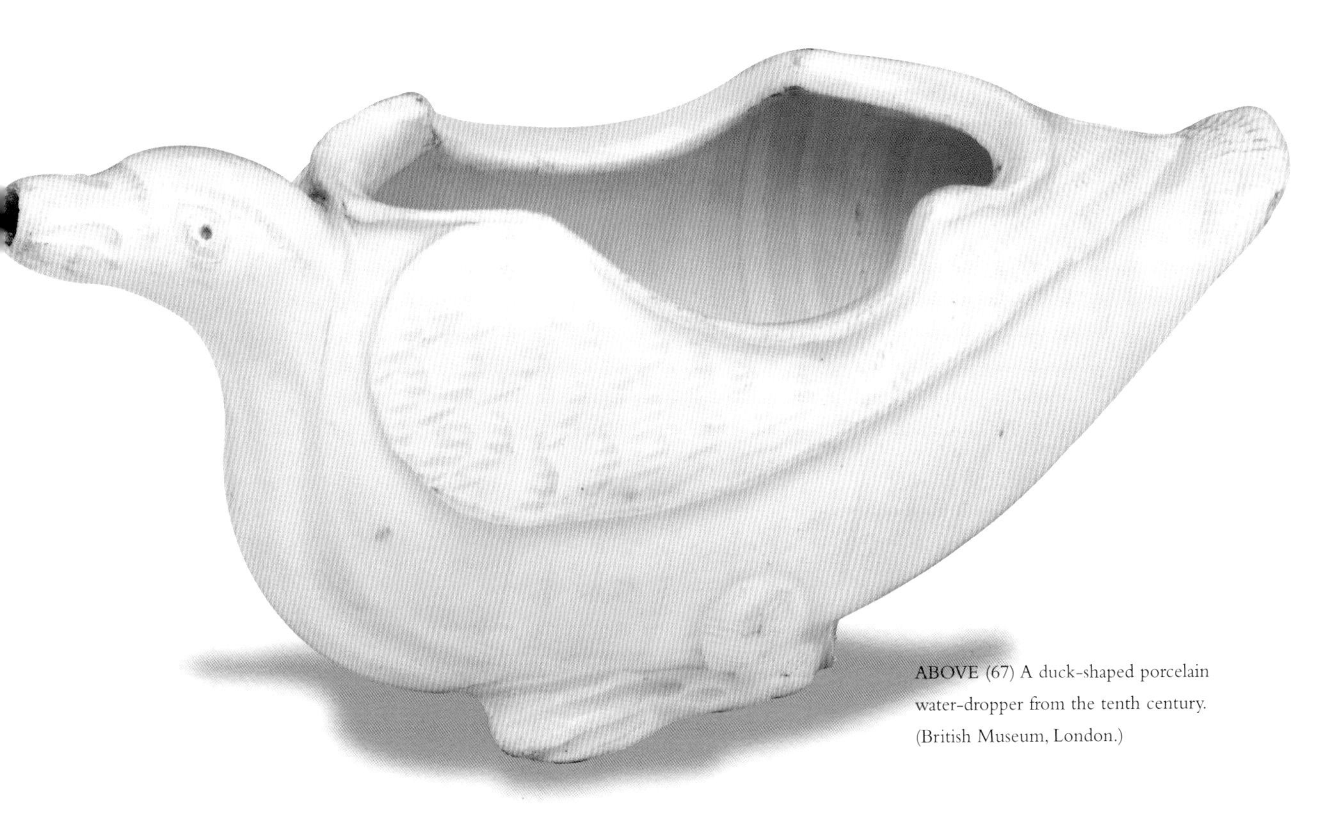

ABOVE (67) A duck-shaped porcelain water-dropper from the tenth century. (British Museum, London.)

RIGHT (68)
This magnificent porcelain plate dates from the reign of the Emperor Yung-cheng (1723–1735) of the Ch'ing Dynasty. The man kneeling with a banner is a soldier announcing a military victory, which has just taken place, to the two men who are sitting playing the game of *wei ch'i*, more commonly known by its Japanese name of *go*. A delicate rose is the predominating colour in this exquisite work, and is even used for the soldier's banner, which was not a very martial choice! The pale lustrous green used for the grass, and the sky as limpid as a daydream, combine to make this plate a triumph of the most refined subtlety. (Nanking Museum.)

arrived at by the Chinese centuries before. And so, one of the great scientific advances in the Western world took place as a direct consequence of the attempts by Europeans to find the secret of porcelain manufacture. It is highly ironical that this great advance itself was merely a late duplication of an indigenous Chinese idea (albeit unknown to Europeans, and soon surpassed in the West).

In Europe, porcelain was finally developed in the eighteenth century, some 1700 years after the Chinese. Previous European 'porcelains' were not true ones: they would melt if placed in ovens at the temperatures used to fire Chinese porcelain. At last the goal had been achieved, and one of the greatest secrets of China fell to a sustained Western onslaught which had lasted for centuries.

BIOLOGICAL PEST CONTROL

THIRD CENTURY AD

Mandarin oranges are ferociously attacked by black ants, caterpillars and other predators, and if this is not somehow controlled, no orange will be left intact on a tree. For 1700 years, the Chinese have controlled insect pests by biological means, using one insect to kill another. This has only really been practised in the West in the twentieth century. Perhaps the most striking and important Chinese use of biological pest control was in the use of yellow citrus killer-ants to protect the mandarin trees. Here is how a text of 304 AD, *Records of the Plants and Trees of the Southern Regions*, by Hsi Han, describes the use of the carnivorous yellow ants:

> **The mandarin orange is a kind of orange with an exceptionally sweet and delicious taste.... The people of Chiao-Chih sell in their markets [carnivorous] ants in bags of rush matting. The nests are like silk. The bags are all attached to twigs and leaves which, with the ants inside the nests, are for sale. The ants are reddish-yellow in colour, bigger than ordinary ants. These ants do not eat the oranges, but attack and kill the insects which do. In the south, if the mandarin orange trees do not have this kind of ant, the fruits will be damaged by many harmful insects, and not a single fruit will be perfect.**

The killer-ants employed for this purpose are still used. They have been identified by Western entomologists as *Oecophylla smaragdina* (see Plate 69 (right)). Originally, whole nests of citrus ants were sold, as we have seen, but later they were caught by being trapped in bladders. This is described by Chuang Chi-Yü in his *Miscellaneous Random Notes* of 1130:

> **In Kuangchow there is a shortage of arable land so people often plant mandarin oranges and *chü* oranges for income; but they suffer considerable losses caused by small insects feeding on the fruits. However, if there are many ants on the trees then the injurious insects cannot survive. Fruit-growing families buy these ants from vendors who make a business of collecting and selling such creatures. They trap them by filling hogs' or sheep's bladders with fat and placing them with the cavities open next to the ants' nests. They wait until the ants have migrated into the bladders and then take them away. This is known as 'rearing orange ants'.**

But how could the ants spread easily through an entire orange grove? Bridges of bamboo have traditionally been stretched between the orange trees to allow the citrus ants to move from one tree to another throughout the grove. This practice continues today. They were in use at least four hundred years ago, as we learn from Wu Chen-Fang's *Miscellanies from the Southern Regions* of 1600:

> **In Li-chih village, west of Kaochow, oranges and pomelos are important secondary crops. Trees planted on several *mou* (each 6.6 acres) of land are connected to each other by bamboo strips to facilitate the movement of the large ants which ward off insect pests. The ants build**

ABOVE (69) Carnivorous citrus ants (*Oecophylla smaragdina*) used by the Chinese for seventeen hundred years as a means of biological pest control to protect their mandarin orange trees from predators. Some of the ants here hold leaves for a nest, while others sew them together with silk from larvae.

nests among the leaves and branches in the hundreds and thousands. A nest may reach the size of a *tou* (1.6 gallons).

This biological pest control first came to Western attention when a paper on the subject was published in the *North China Herald* on 4 April 1882, by H.C. McCook; but few people noticed it. It was not until a serious outbreak of citrus canker occurred in the Florida citrus groves in the 1910s that a plant physiologist was sent to China by the US Department of Agriculture in 1915 to search for canker-resistant oranges, and discovered the citrus ants. The study of this subject by Westerners was interrupted by the Japanese invasion of China in the 1930s. Then, in 1958, a Chinese scientist, Ch'en Shou-Chien, recommended a renewed study of the ants. Their use in Chinese orange groves continues to this day.

THE UMBRELLA

FOURTH CENTURY AD

The umbrella as we know it was invented in China towards the end of the fourth century AD. An earlier type, made of silk, had been used as a chariot rain-cover in the rain or many centuries; the umbrella proper appeared during the Wei Dynasty (386–532 AD). Instead of silk, it used a special kind of oiled heavy paper made from the bark of the mulberry tree (see the account of paper, page 92), and was used as protection from both rain and sun.

The Wei emperor used an red and yellow umbrella, whereas ordinary people used blue ones. Umbrellas became common, and in 1086 we find the author Shen Kua using the umbrella as a descriptive analogy, referring to the astronomical 'lunar mansions' in the sky as radiating from the celestial pole 'like the spokes of an umbrella'.

ABOVE (70) Photograph taken by Cecil Beaton during the Second World War, showing disabled Chinese soldiers making oiled paper umbrellas.

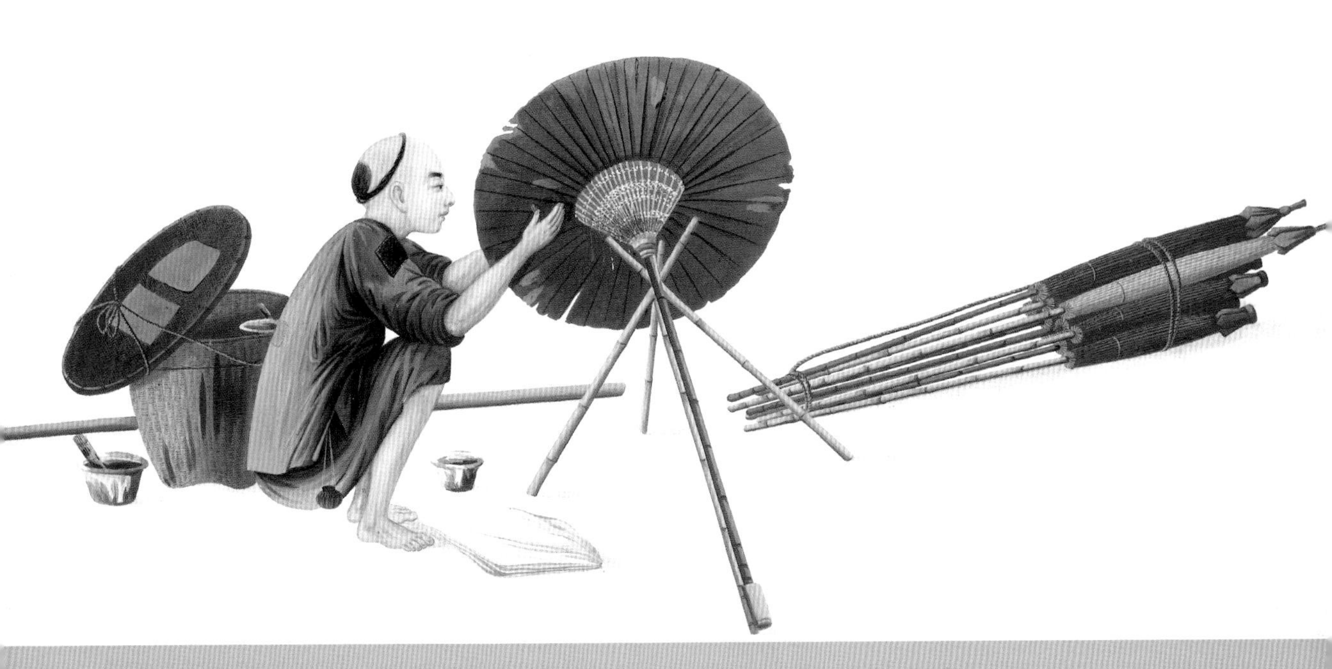

ABOVE (71) A late-eighteenth-century painting of an umbrella-mender, with a pile of rolled umbrellas on the right. The umbrella was invented in China towards the end of the fourth century AD, probably inspired by the silk canopies placed over chariots during the Han Dynasty (207 BC–220 AD). (Victoria and Albert Museum, London.)

BELOW (72) A man holding a fifteenth-century umbrella in a storm. This painting is by the Ming Dynasty Emperor Chu Kao-Chih, and appears in his unpublished 1425 treatise *Essay on Astronomical and Meteorological Presages*, the manuscript of which survives (see another reproduction from it in Plate 11 (page 29)). (Cambridge University Library.)

By the fourteenth century silk umbrellas must have been available as well as the oiled-paper ones, for in 1368 an imperial decree announced that silk umbrellas were to be reserved for the exclusive use of the royal family. This law does not say a lot for the Ming Dynasty, which promulgated it; but perhaps the ruling cliques thought that people with silk umbrellas were getting 'above themselves'. The umbrella seems at this time to have taken on quite a symbolic significance. It was used in ceremonies, and the emperor would give special signed umbrellas to his most trusted officials.

How and when the umbrellas came to Europe is apparently not known. Perhaps paper umbrellas sold in China made their way to Europe, where the design was copied and its origin soon forgotten.

MATCHES

SIXTH CENTURY AD

Every time we strike a match, we are using a Chinese invention. The first version of a match was invented in the year 577 AD by impoverished court ladies during a military siege, in the short-lived Chinese kingdom of the Northern Ch'i. (This kingdom was presided over by the psychopathic ruler described at greater length in the account of man-flying kites on page 191). Hard-pressed during the siege, they must have been so short of tinder that they could otherwise not start fires for cooking, heating, etc. The neighbouring kingdoms of the Northern Chou and the Ch'en agreed to attack the Northern Ch'i – which consisted of the entire North China plain – from both sides at once. The attack was so successful that the Ch'i were annihilated. Later, the two conquering forces warred against one another and were in turn absorbed in the next unification of China, under the Sui Dynasty (581–617 AD).

Early matches were made with sulphur. A description is found in a book entitled *Records of the Unworldly and the Strange* written about 950 by T'ao Ku:

> **If there occurs an emergency at night it may take some time to make a light to light a lamp. But an ingenious man devised the system of impregnating little sticks of pinewood with sulphur and storing them ready for use. At the slightest touch of fire they burst into flame. One gets a little flame like an ear of corn. This marvellous thing was formerly called a 'light-bringing slave', but afterwards when it became an article of commerce its name was changed to 'fire inch-stick'.**

There is no evidence of matches in Europe before 1530. Therefore, the Chinese were using them for just short of a thousand years before they arrived in Europe. Matches could easily have been brought to Europe by one of the

RIGHT (73) A late-eighteenth-century painting of a boy selling joss-sticks and matches. (Victoria and Albert Museum, London.)

Europeans travelling to China at the time of Marco Polo, since we know for certain that they were being sold in the street markets of Hangchow in the year 1270 or thereabouts. This is recorded in one of the six-and-a-half thousand ancient topographical books which survive about regions and cities of China, *Institutions and Customs of the Old Capital (Hangchow)*, which dates from 1270 and recounts events from 1165 onwards. Since the earliest European matches used sulphur, the invention seems to have been directly transmitted by a European traveller. Sulphur is still used in matches today, though it was in 1830 that Sauria in France and Kammerer in Germany made the breakthrough which gave us the modern match, by using a compounded mixture of yellow phosphorus, sulphur and potassium chlorate.

It is easy for us to take for granted such small, daily necessities as matches. After all, they are cheap, used once and thrown away. But where would we be without them? Many adults in today's developed world carry with them, all unknown as they go about their daily business, small tokens of the inventiveness of a group of anonymous Chinese women of the sixth century – the products of their desperation in the face of starvation and eventual violent death.

ABOVE (74) A set of seventeenth-century Chinese chessmen, made of ivory pieces set into dark 'heartwood' ebony, kept in an ebony box with a sliding lid. Chinese chessmen were not carved as figures, but were always discs identified by the Chinese characters on top – one player's painted red, the other's painted black. No Chinese chessboard of this age appears to survive, and it is believed that only one other set of chessmen similar to this exists (represented in A. E. J. Mackett-Beeson, *Chessmen*, New York, 1968, – see Plate 89.) (Collection of Robert Temple.)

CHESS

SIXTH CENTURY AD

Although most historians of chess believed that the game was invented in India, Needham has been able to establish that it originated in China. Chess took its present form as a militaristic combat game in India, but its origins were connected with astrology, magnetism and divination. According to Needham: 'The battle element of chess seems to have developed from a technique of divination in which it was desired to ascertain the balance of ever-contending *yin* and *yang* forces in the Universe (sixth-century AD China, whence it passed to seventh-century AD India, generating there the recreational game).'

The surviving form of 'Chinese chess' played today is not the same either as the ancient Chinese forms or as modern Western chess. This has led various scholars to overlook the game's Chinese origins. For instance, A.E.J. Mackett-Beeson, in discussing the history of chess pieces in 1968, remarks: 'Although the Chinese claim to have invented chess some 2072 years ago during the Hansing campaign, *choke-choo-hong-ki* [the name applied to the Chinese game] is merely a variation of *shatranj*, the medieval game of India.' This modern Chinese chess was a reintroduction to China, and was indeed influenced by the Indian game, which had originally come from Chinese sources and been altered in India.

In modern Chinese chess the board has the usual sixty-four squares. Over the middle of the board, however, is a river, which originally symbolized the Milky Way, across which some of the pieces cannot move. For instance, each player has two 'men' called 'horses'. These move just as our knights do, but they cannot cross the river. On the other hand, each player has two chariots which are equivalent to our rooks, and these can cross the river. The king moves one space at a time, as our king does, but is confined to a fortress consisting of four squares, king one and two and queen one and two. Each player has only five pawns; they can cross the river but do not queen. There is the further amusing curiosity that each player has two pieces called 'rocket boys' which do not capture the piece directly ahead but instead capture the one directly beyond it, as if they had fired artillery over the head of the first piece.

Needham believes that the opposing kings of chess were originally the Sun and Moon. The planets were represented by pieces divided between the two sides. The pawns, at fourteen a side, were the twenty-eight equatorial constellations. The 'rocket boys', Needham thinks, 'may well have been comets'. He quotes an 'Essay on Chess' by the historian Pan Ku of the first century AD:

> Northerners call chess (*ch'i*) by the name of *i*. It has a deep significance. The board has to be square, for it signifies the Earth, and its right angles signify uprightness. The pieces of the two opposing sides are yellow and black; this difference signifies the *yin* and the *yang* – scattered in groups all over the board, they represent the heavenly bodies. These significances being manifest it is up to the players to make the moves, and this is connected with kingship. Following what the rules permit, both opponents are subject to them – this is the rigour of the Tao.

BELOW (75) An earthenware tomb model of the second century AD, showing two men playing the board game *liu-po* ('Six Learned Scholars'), which goes back to the third century BC at least. Needham believes this game to be one of several astrological board games which together led to the development of chess, and which were 'militarized' in India to provide the combat elements recognizable to us today. (British Museum, London.)

In ancient China, there was a wide variety of chess games. Needham stresses that: 'The Chinese language makes no distinction between these forms and the war-game of chess itself, both being expressed by the character *ch'i*.' So, although the forms of chess vary enormously, not only do the Chinese designate them all by the same name, but they may clearly be seen to be related. It was not the specific rules, number of men, and modes of play that were the main point to the game, but the magical purpose of foretelling future events.

As for a date for the Chinese origins, Needham says: 'The oldest Chinese name for a chess-like game played on a board is *i*, to which there are two references in Mencius (fourth century BC)...'. An improved version known as 'image-chess' seems to have been invented in the sixth century AD. At that time, a cavalry general named Yü Hsin wrote an essay entitled 'The Image-Chess Game'. An emperor, or one of his advisers, actually seems to have invented this later form of chess. As a book entitled *Red Lead Record* tells us:

> **Tradition handed down says that image-chess was invented by the Emperor Wu of the Northern Chou Dynasty [561–78 AD]. According to the [official history] it was in 569 AD that the Emperor finished writing his *Image-Chess Manual*. He assembled all his officials in a palace hall and gave lectures to them about it ... in image-chess images of the Sun, Moon, stars, and constellations were used ... it was necessary to have scholarly commentaries on it, and lectures to the hundred officials.**

Just as modern Chinese chessmen are often flat wooden discs, so many ancient Chinese chessmen are simply bronze discs bearing written characters saying what they represent. Several examples are in the British Museum, where they have been mislaid and forgotten in recent years. Some have pictures of constellations. Many have been found with the picture of the Great Bear (the Plough or Big Dipper), which, as we see in the account of the compass (page 162), was represented on the lodestone pointer in early Chinese compasses. Although the Chinese origin was forgotten, the astrological importance of chess was realized by the Indians and Arabs. Indeed, there is a thirteenth-century Latin poem which gives the astrological symbolism of each chess piece, showing that this knowledge still existed at that time even in Europe.

The earliest European references to chess date from about 1010 in the Pyrenees. The game seems to have entered Spain through contact with the Arabs, who obtained it from India, where it was known as *chaturanga* from the early seventh century AD.

In summarizing the unique factors in China which enabled a game such as chess to develop, Needham says that 'the important thing to notice is that in China, and in China alone, on account of the dominance of the *yin–yang* theory of the macrocosm, could a divination technique or "pre-game" have been devised which was both astrological and yet had a sufficient combat element to enable it to be vulgarized into a purely military symbolism.'

BRANDY AND WHISKY

SEVENTH CENTURY AD

Readers will be doubly surprised to learn that the Chinese invented brandy, because though the fact in itself is impressive enough, it is not widely realized that the Chinese drank wine made from grapes at all, much less distilled it into brandy. Grape wine was being drunk by the second century BC at the latest in China, since we begin to have textual evidence of it by then. The envoy and traveller Chang Ch'ien brought good wine grapes (*Vitis vinifera*) back with him from Bactria about 126 BC. Before the importation of those grapes, however, there were wild vine species, or 'mountain grapes', which were already being used for wine, namely *Vitis thunbergii* and *Vitis filifolia*. Wine made from them is mentioned before Chang Ch'ien in the book called *Classical Pharmacopoeia of the Heavenly Husbandman*.

The fact that an even stronger drink could be obtained from wine first came to attention through the production of 'frozen-out wine' among the Central Asian tribes. No doubt because of the extremely cold conditions in which these people lived, it was often noticed that wine and other fermented beverages (such as fermented mare's milk) when frozen would have small amounts of unfrozen liquid in the middle. This was the alcohol, which had remained liquid while the water in the drink had frozen. It was presumably this to which Chang Hua was referring in 290 AD in his book *Records of the*

常
豊
常

Investigation of Things, when he said: 'The Western regions have a wine made from grapes which will keep good for years, as much as ten years, it is commonly said; and if one drinks of it, one will not get over one's drunkenness for days.'

The tribal peoples of Kao-Ch'ang (Turfan) presented 'frozen-out wine' as tribute to Chinese emperors more than once, commencing in 520 AD. The freezing-out technique for obtaining spirits eventually became a test applied to distilled spirits. Yeh Tzu-Ch'i tells us in his book of 1378, entitled *The Book of the Fading-like-Grass Master*, that people would test their spirits by leaving them outside in the winter to freeze. If they did not freeze, they knew the distilled spirits were pure and unadulterated, but if they partly froze, they knew they were watered-down or otherwise impure. 'Frozen-out wine' is not mentioned in Europe until Paracelsus, in the *Archidoxis*, written in 1527 but not published until 1570. Paracelsus's remarks caused something of a sensation among Europeans. So unfamiliar was the phenomenon he described that Francis Bacon in 1620 wrote, half-incredulous: 'Paracelsus reporteth, that if a glass of wine be set upon a terras [terrace] in bitter frost, it will leave some liquor unfrozen in the centre of the glass, which excelleth *spiritus vini* drawn by fire.' (For by then, distillation of alcohol was known.)

Distilled wine, or brandy, was known in China as 'burnt wine'. The English word 'brandy' itself comes from the Dutch *brandewijn* ('burnt wine'). And the German word for brandy or spirits is *Branntwein* ('burnt wine'), while a distiller is a *Branntweinbrenner* ('burnt wine burner'). It is possible that all of these words result from a direct translation of the Chinese

LEFT (76) A late-eighteenth-century painting of a small still in operation. The tube carrying the distilled substance leads to a receptacle at bottom left. (Victoria and Albert Museum, London.)

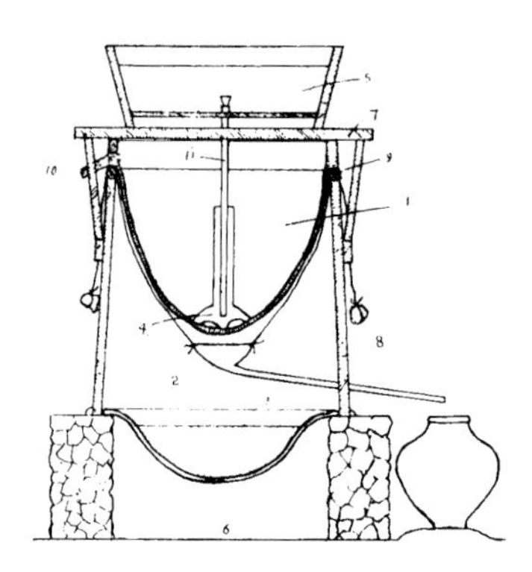

ABOVE (FIG. 5) A diagram of a Chinese still from *China at Work* by R.P. Hommel. The distilled substance, which can be any of a number of substances, including brandy or whisky, is drawn off by the side-tube (2) at the bottom, into a cool vessel. The fire and grate (6) are underneath a cast iron bowl (3) which forms the bottom of the still and is filled with the mash to be distilled. At the top is a wooden reservoir of cold water (5) supported by a wooden framework (7) and in the middle is a pewter cooling reservoir (1). A central pipe (11) lets cold water flow down into the cooling reservoir or condenser at the still head.

shao chiu ('burnt wine') by Dutch sailors. If not, they are a curious coincidence indeed.

We have a description of the making of 'burnt-wine' brandy by the author Li Shih-Chen, in his book *The Great Pharmacopoeia*:

> **Strong wine is mixed with the fermentation residues and put inside a still. On heating, the vapour is made to rise, and a vessel is used to collect the condensing drops. All sorts of wine that have turned sour can be used for distilling. Nowadays in general glutinous rice or ordinary rice or glutinous millet or the other variety of glutinous millet or barley are first cooked by steaming, then mixed with ferment and allowed to brew in vats for seven days before being distilled. The product is as clear as water and its taste is extremely strong. This is distilled spirits (*chiu lu*).**

Here we have a description not only of brandy, but of various kinds of whisky. This passage was published in 1596, but is merely one of the clearest passages describing distilled spirits which had been made in China from the seventh century AD. Many passages are rather obscure and coy because of the problems over the excise duty on spirits which was levied by the government. The Chinese were the world's first large-scale bootleggers. The Emperor Wang Mang (who reigned 9–23 AD) nationalized the fermentation and brewing industries. So strict were the taxes and the prohibitions against private manufacture of wine or spirits during the Northern Wei Dynasty (386–535 AD) that the penalty for private brewing was death. People had to evolve a series of 'cover-names' for brandy and whisky just as in modern times the terms 'hooch', 'white lightning', and 'moonshine' have developed. One type of drink was called 'the Sage', another was called 'the Worthy'. And in the eleventh century, if you wanted to offer your guests a drink of your own moonshine, you would say, 'Have a drop of "wisdom soup" '.

The distillation of alcohol in the West was discovered in Italy in the twelfth century. Spirits came to be known as *aqua ardens* ('the water that burns') or *aqua vitae* ('the water of life'). By the thirteenth century, several writers mention it, and talk about brandy ten times distilled, until it reached what must have been about 90 per cent alcohol. The word 'alcohol' (which derives from Arabic) was introduced by Paracelsus in the sixteenth century. By 1559, *aqua vitae* was being praised by Conrad Gesner in terms which we can all find familiar: 'Yea, it changeth the affections of the minde, it taketh away sadnes and pensivenes, it maketh men meri, witty, and encreaseth audacitie...'. And in this increased audacity, wit and merriment, the Chinese were in advance of Europeans by about five hundred years.

THE MECHANICAL CLOCK

EIGHTH CENTURY AD

The difficulty in inventing a mechanical clock was to figure out a way in which a wheel no bigger than a room could turn at the same speed as the Earth, but still be turning more or less continuously. If this could be accomplished, then the wheel became a mini-Earth and could tell the time. For, after all, the time is nothing more nor less than how far the Earth has turned today.

Accomplishing this mechanical feat was one of the greatest steps forward of the human race. Where would we be today without clocks? The mechanical clock was invented in China in the eighth century AD. But still in 1271, Robertus Anglicus in his commentary on the *Sphere* of Sacrobosco tells us that in Europe 'artificers are trying to make a wheel which will pass through one complete revolution for every one of the [Earth's], but they cannot quite perfect their work. If they could, it would be a really accurate clock, and worth more than any astrolabe or other astronomical instrument for reckoning the hours...'.

By 1310, this had finally been achieved in Europe. And the stimulus for it seems to have been some garbled accounts of Chinese mechanical clocks which came to the West by way of traders. This was the same century that brought to Europe the

BELOW (77) A model of the 'Cosmic Engine', Su Sung's great astronomical clock of 1092. The framework has been left uncovered to reveal the mechanisms. The original clock tower was 30 feet high. At the top is the power-driven armillary sphere for observing the positions of the stars. In the original, this was bronze, and the power for turning it was transmitted by a chain-drive. Mid-right (B) may be seen a celestial globe which was inside the tower and turned in synchronization with the sphere above. The central element in the reconstruction (D) is the water wheel escapement, which, though turned by water power, was a mechanical escapement. This was a mechanical clock rather than a water clock, even though its power came from falling water or mercury. (Science Museum, London.)

Chinese inventions of gunpowder, segmental arch bridges, cast iron, and printing.

Apart from the fact that the Chinese are obviously an inventive people, what other factors can account for the fact that they were the first to invent mechanical clocks? Was there some special reason why they urgently needed to know the hours of the day and the days of the year with a precision not required in Europe? The answer is yes, but few could possibly imagine why.

The Chinese emperor was a cosmic figure, the equivalent on Earth of the Pole Star. His every move was regulated in conjunction with astrology. His heir was not necessarily his eldest son. Many examples in Chinese history exist of fourth sons, or other lesser offspring, being selected as the next emperor. How, then, was it determined who should be the heir? Part of the process of selection involved the astrological computation of the moment of the child's conception (since in China horoscopes commence at the estimated time of conception rather than at birth). And the moments when conception might take place were carefully set aside for the highest-ranking wives and concubines of the emperor to sleep with him. Access to the emperor's person had to be precisely timed in order for this to work properly. From the *Record of Institutions of the Chou Dynasty* compiled about the second century BC, we find the following astonishing passage about the emperor's sex life:

> **The lower-ranking women come first, the higher-ranking come last. The assistant concubines, eighty-one in number, share the imperial couch nine nights in groups of nine. The concubines, twenty-seven in number, are allotted three nights in groups of nine. The nine spouses and the three consorts are allotted one night to each group, and the empress also alone one night.**

RIGHT (78) John Combridge's model of the water wheel escapement of Su Sung's great astronomical clock. The water or mercury poured from the tank on the right, regulated by means of a water clock, and turned the wheel clockwise. Each spoke bears a bucket which, when filled, turns the wheel a fraction. The motion was jerky, but it sufficed. (Science Museum, London.)

On the fifteenth day of every month the sequence is complete, after which it repeats in the reverse order.

The emperor, known as the Son of Heaven, was full of a powerful *yang* force, which was the essence of masculinity. But he needed to be fed with a matching *yin* force, the essence of femininity, to achieve a balance. At the time of the full moon, when the *yin* force was at its peak, the empress would sleep with the Son of Heaven, feeding him with her powerful *yin*. This would be the most propitious time for a conception to take place. The lesser women, during the time of the Moon's waning, had to sleep with the emperor in groups in order to pool their respective *yin* forces to overcome the lack of the Moon's strength. For most of the nights of his life, therefore, the emperor slept with nine women at a time.

If a likely lad were to be chosen to be the next emperor, the astrologers would go back to the precise time of his conception, plot the stars which were culminating and consider any comets or novae, or other astronomical phenomena. If the astrological configurations were indicative of a strong leader, a valiant warrior, or whatever, this would weigh in the young prince's favour. But the eldest son might well have been born under the influences of stars concerning death or disaster. So he would be ruled out quite early on.

But let us see what the situation was at about the time when the mechanical clock was invented in China. Was the succession principle operating very well? In the ninth century, Pai Hsing-Chien bewails the chaos of the system. We read in his Poetical Essay of the Supreme joy:

Nine ordinary companions every night, and the empress for two nights at the time of the full moon – that was the ancient rule, and the secretarial ladies kept a record of everything with their vermilion brushes.... But alas, nowadays, all the three thousand palace women compete in confusion.

Clearly, the succession to the throne was in peril. Bad princes might be chosen. The timing of their conceptions was not being noted properly. Time for a clock to be invented! And in 725 AD, this was done.

The Chinese did not invent the first clock of any kind, merely the first mechanical one. Water clocks had existed since Babylonian times, and the earliest Chinese got them indirectly from that earlier civilization of the Middle East, just as they got much of their earliest forms of astronomy from them. The Chinese certainly did invent improved water clocks of various kinds, including a 'stop-watch' portable one which used mercury rather than water and measured small periods of time. It used weighted balances, or steelyards, rather than just a rising indicator in a bucket as water flowed in and buoyed it up. But these were improvements of an invention which was not originally Chinese. Nor did the Chinese invent the clock dial. That was an invention of either the Greeks or the Romans, and is mentioned by the architectural writer Vitruvius in the first century BC.

The world's first mechanical clock was built by the Chinese Tantric Buddhist monk and mathematician I-Hsing. This was actually an astronomical instrument which served as a clock, rather than simply a clock. A contemporary text describes it:

[It] was made in the image of the round heavens and on it were shown the lunar mansions in their order, the equator and the degrees of the heavenly circumference. Water, flowing into scoops, turned a wheel automatically, rotating it one complete revolution in one day and night [24 hours]. Besides this, there were two rings fitted around the celestial sphere outside, having the sun and moon threaded on them, and these were made to move in circling orbit. Each day as the celestial sphere turned one revolution westwards, the sun made its way one degree eastwards, and the moon 13⁷⁄₁₉ degrees eastwards. After twenty-nine rotations and a fraction of a rotation of the celestial sphere the sun and moon met. After it made 365 rotations the sun accomplished its complete circuit. And they made a wooden casing the surface of which represented the horizon, since the instrument was half sunk in it. It permitted the exact determinations of the time of dawns and dusks, full and new moons, tarrying and hurrying. Moreover, there were two wooden jacks standing on the horizon surface, having one a bell and the other a drum in front of it, the bell being struck automatically to indicate the hours, and the drum being beaten automatically to indicate the quarters.

All these motions were brought about by machinery within the casing, each depending on wheels and shafts, hooks, pins and interlocking rods, stopping devices and locks checking mutually [i.e. the escapement].

Since the clock showed good agreement with the Tao of Heaven, everyone at that time praised its ingenuity. When it was all completed in 725 AD it was called the 'Water-driven Spherical Birds'-Eye-View Map of the Heavens' and set up in front of the Wu Ch'eng Hall of the Palace to be seen by the multitude of officials. In 730 AD candidates in the imperial examinations were asked to write an essay on the new astronomical clock. The text continues:

But not very long afterwards the mechanism of bronze and iron began to corrode and rust, so that the instrument could no longer rotate automatically. It was therefore relegated to the museum of the College of All Sages and went out of use.

From this description we may see that the first mechanical clock was a transition from the water clock to the purely mechanical clock of Europe which used no water power anywhere. However, this does not mean the clock of I-Hsing was a water clock by any means. What it means is that the first mechanical escapement for the clock was worked not by a falling weight or by springs, but by water power. This is quite a different thing from having an actual water clock whose time indicator rises with the level of water (or mercury) in a tank. It is not unreasonable that in developing a mechanical clock, the Chinese turned to water as a power source, since all previous clocks had been water clocks, and the association of ideas was natural to them. The perpetual flow of water was likened to the perpetual turning of the heavens. As Su Sung wrote in 1092 in speaking of his own improved clock:

The principle of the use of water power for the driving mechanism has always been the same. The heavens move without ceasing and so also does water flow and fall. Thus if the water is made to pour with perfect evenness then the comparison of the rotary movements of the heavens and the machine will show no discrepancy or contradiction; for the unresting follows the unceasing.

The actual workings of the clock of I-Hsing consisted of a vertical water wheel which instead of paddles (such as are turned by a rushing stream) had cups at the ends of the blades. These cups were filled by water dripping from a water clock. I-Hsing's clock was therefore a mechanical clock driven by a water clock. When one of the cups was sufficiently full, it would weigh enough to turn the great wheel by one notch, overcoming the resistance of a restraining tooth which had held the wheel still while the cup filled. Various gear arrangements would then transmit the movement to the time indicators, etc. As the ancient text tells us, shafts, hooks, pins, and interlocking rods were all part of the apparatus. Some idea of what these may have been like may be gathered from the fuller description of Su Sung's clock which we shall see opposite. (See Figure 6.)

This clock of I-Hsing would have been a poor time-keeper, moving jerkily and representing more the first realization of a wonderful idea than a superb piece of machinery. It was because of the enormous promise that this clock offered for future improvements that contemporaries would have had every reason to be excited. And its promise was more than borne out in succeeding centuries. I-Hsing himself died only two years later, and was unable to construct a 'second generation' model.

I-Hsing's clock was, like water clocks, subject to the vicissitudes of the weather. In order to keep the water in them from freezing, torches generally burnt beside them. This was of course not necessary for the small mercury clocks, as mercury does not freeze at any temperature likely to occur in the Earth's weather. Therefore, in the next great clock of which we have accounts in China, mercury was substituted for water because of the freezing problem. This clock was built by Chang Ssu-Hsün in 976 AD). It represented a considerable elaboration and improvement over I-Hsing's machine, though the latter no longer existed and had been lost even before the end of the T'ang Dynasty in 906.

Chang Ssu-Hsün's clock was apparently much larger than I-Hsing's. It was certainly far more complex. The dynastic history of the time describes it:

...a tower of three storeys each over ten feet in height, within which was concealed all the machinery. It was round at the top to symbolize the heavens and square at the bottom to symbolize the earth. Below there was set up the

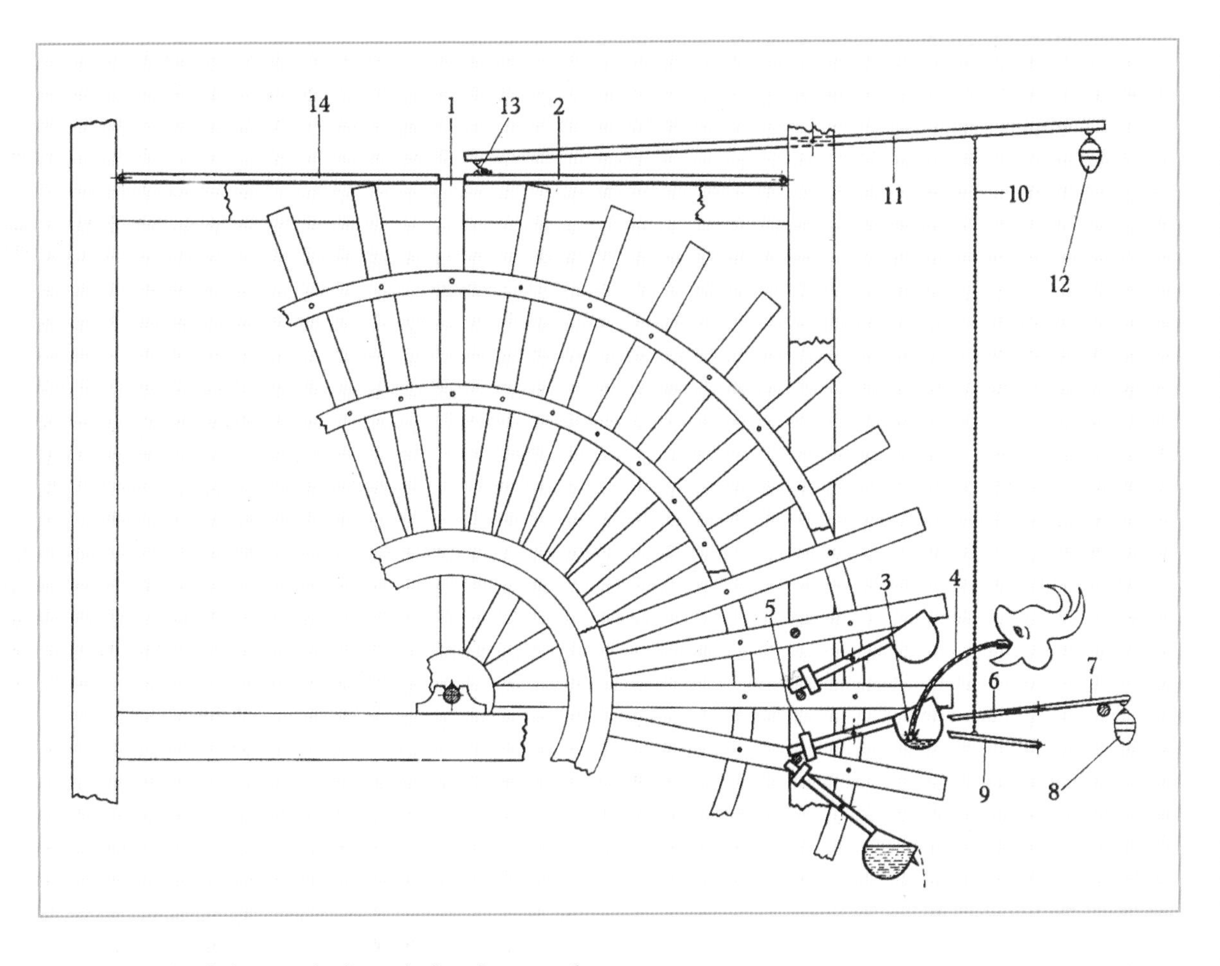

lower wheel, lower shaft, and the framework base. There were also horizontal wheels, vertical wheels fixed sideways, and slanting wheels; bearings for fixing them in place; a central stopping device and a smaller stopping device [i.e. the escapement] with a main transmission shaft. Seven jacks rang bells on the left, struck a large bell on the right, and beat a drum in the middle to indicate clearly the passing of the quarter-hours. Each day and night [i.e. each 24 hours] the machinery made one complete revolution, and the seven luminaries moved their positions around the ecliptic. Twelve other wooden jacks were also made to come out at each of the double-hours, one after the other, bearing tablets indicating the time. The lengths of the days and nights were determined by the varying numbers of the quarters passing in light or darkness. At the upper part of the machinery there were the top piece, upper gear-wheels, upper stopping device [escapement], upper anti-recoil ratchet pin, celestial ladder gear case [possibly the first chain drive in history, or otherwise this was invented shortly afterwards by Su Sung for his clock], upper beam of the framework, and the upper connecting-rod. There were also on a celestial globe the 365 degrees to show the movements of the sun, moon, and five planets; as well as the Purple Palace [north polar region], the lunar mansions in their ranks and the Great Bear; together with the equator and ecliptic which indicated how the changes of the advance and regression

ABOVE (FIG. 6) John Combridge's diagram of the water wheel escapement of Su Sung's clock. The buckets on the ends of the spokes are shown for the three lowest spokes, the first empty, the second filling from the spout of the tank, and the third full (upon turning further, it would empty into a trough). When the bucket was filled, it depressed a lever which pulled the long chain (10), which then pulled the upper balancing lever (11), which was counterweighted (12). When this happened, the upper lock (2) was jerked away, enabling the wheel to turn. There was another upper lock (14) to prevent recoil of the wheel. Each bucket had its own counterweight (5) to regulate the exact amount of water necessary for tipping the bucket downwards to operate this series of movements.

> of heat and cold depend upon the measured motions of the sun. The motive power of the clock was water, according to the method which had come down from Chang Hen in the Han Dynasty through I-Hsing ... but ... as during the winter the water partly froze and its flow was greatly reduced, the machinery lost its exactness, and there was no constancy between the hot and the cold weather. Now, therefore, mercury was employed as a substitute, and there were no more errors....

All of these efforts were preparatory for the greatest of all Chinese medieval clocks, the 'Cosmic Engine' of Su Sung, built in the year 1092. Su Sung's clock is known in considerable detail due to the miraculous preservation, over nine centuries, of his book *New Design for a Mechanized Armillary Sphere and Celestial Globe*. This work describes the design and construction of the great clock in full detail. Some drawings in the book have recently been discovered to be those of Chang Ssu-Hsün in 976, to whose earlier clock Su Sung's must have borne a closer resemblance than had been thought until now. Modern working models of the clock have been constructed, and in Plates 77 and 78 (pages 117–118) may be seen one at the Science Museum in London. These models are, of course, based on the full descriptions and drawings in Su Sung's book.

Su Sung's clock was actually an astronomical clock tower more than 30 feet high, like the previous one of Chang. But on top of Su Sung's tower was additionally a huge bronze power-driven astronomical instrument called an armillary sphere (see page 38), with which one could observe the positions of the stars. A celestial globe inside the tower turned in synchronization with this sphere above, so that the two could constantly be compared. We are told that the observations made on the demonstrational globe inside and by the observational sphere above 'agreed like the two halves of a tally'.

On the front of the tower was a pagoda structure of five storeys, each having a door through which mannikins and jacks appeared ringing bells and gongs and holding tablets to indicate the hours and other special times of the day and night. All of these time-indicators were operated by the same giant clock machinery which simultaneously turned the sphere and the globe.

This machinery consisted, as usual, of a huge vertical water wheel with scoops at the end of each blade, into which water dripped from a water clock. Every time the wheel turned one notch upon the filling of a scoop, there was a ratchet-pin which came down to prevent the wheel recoiling backwards. As for the forward motion of the wheel, it went forward one scoop every quarter of an hour. Needham describes the machine as follows:

> The wheel was checked by an escapement consisting of a sort of weigh-bridge which prevented the fall of a scoop until full, and a trip-lever and parallel linkage system which arrested the forward motion of the wheel at a further point and allowed it to settle back and bring the next scoop into position on the weigh-bridge. One must imagine this giant structure going off at full-cock every quarter of an hour with a great sound of creaking and splashing, clanging and ringing; it must have been very impressive, and we know that it was actually built and made to work for many years before being carried away into exile.

In fact, Su Sung's clock ran from 1092 until 1126 when the capital, K'aifeng, was lost by the Sung Dynasty. The clock was then dismantled, moved to Peking, and reassembled there, where it ran for some years further. The clock had previous vicissitudes which shed light upon the spirit of the times. Members of political factions opposed to the one of which Su Sung had been a member (he was a conservative) wanted to destroy his clock for political reasons. We are told this by Chu Pien in his book *Talks about Bygone Things beside the Winding Wei* (river in Honan) of 1140:

> But at the beginning of the Shao-Sheng reign-period [1094 AD] Ts'ai Pien, Minister of State, suggested that the armillary clock of Su Sung ought to be destroyed as something which belonged to the previous Yuan-Yu reign period [of only two years before]. At that time Ch'ao Mei-Shu was Assistant Director of the Imperial Library, and as he greatly admired the accuracy and beautiful construction of Su Sung's instruments, he struggled to argue against Ts'ai Pien, but at first his efforts proved unsuccessful. However, he sought the help of Lin Tzu-Chung who talked to Chang Tun, the Prime Minister, and thus the destruction of the clock

was averted. However, after Ts'ai Ching and his brother came into power nobody dared to say anything to prevent Su Sung's machinery being torn down. How shameful!

Thus do we see how personal envy and political disagreement between a liberal and a conservative could result in the destruction of one of the greatest mechanical contrivances in the history of mankind, within only a few years of its construction. But by the strenuous efforts of generations of scholars, Su Sung's book with its diagrams and detailed text survived intact from 1094 until the present time, when Joseph Needham has translated and published it in his book *Heavenly Clockwork*.

Su Sung's clock was possibly the greatest mechanical achievement of the Middle Ages anywhere on the globe (for further details of its working see pictures and captions accompanying this text); knowledge of its principles spread to Europe leading to the development of mechanical clocks in the West two centuries later.

PRINTING

EIGHTH AND ELEVENTH CENTURIES AD

Woodblock printing on paper and silk arose in China around the seventh century AD, and actual specimens survive from the eighth century. But the origins of printing go even further back into the distant past: there were many related techniques which preceded printing.

First, seals were used to stamp impressions of names, and even as many as a hundred Chinese characters at once, onto various surfaces. The Chinese got the idea of seals from the Middle East, where the Babylonians and Sumerians used them in profusion long before Chinese civilization arose. Cutting a seal is rather like cutting a woodblock in printing, and it is easy to see how the latter technique derived its inspiration from the former. But that is quite different from saying that the idea of using seals *led to* printing. This did not happen with the Sumerians or Babylonians. It was the need to make enormous numbers of copies of certain writings which led the Chinese to invent ways of mass-producing written material on paper, a substance which already existed in China (see page 92).

Another proto-printing technique in China was connected with bronze casting. Chinese characters were often incorporated in the designs on Chinese bronzes. But a close study of these bronzes shows that the characters were set into the moulds either separately or in small groups. (This was especially the case when the 'lost-wax' method of bronze casting was used.) We therefore find bronze inscriptions using the forerunner of movable type as early as the seventh century BC. Individual typecasting was therefore in progress nearly two millennia before it was adopted for printing.

Other means of mass-producing written scriptures and sacred texts were by ink rubbings from stone carvings of the original writings. Stone inscriptions of promulgated decrees occur from at least the third century BC. By the second century AD, stone was used to preserve permanent canonical versions of Buddhist, Taoist, and Confucian texts. Between the years 175 and 183, the complete texts of the seven main Confucian classics, amounting to 200,000 characters, were engraved on forty-six stone tablets. These do not survive, but a similar set made between 833 and 837 still exists near Sian, and is known as the Forest of

ABOVE (79) A traditional Chinese engraved wood block, the print made from it, and graving tools. (Victoria and Albert Museum, London.)

Stone Tablets. Also still surviving are 7000 stone tablets of Buddhist scriptures carved between the sixth and eleventh centuries.

The Chinese were the world's leading experts at stone-rubbing. The devout of all Chinese religions came with their ink and sheets of paper and made as many copies of the standardized sacred texts as they wanted. And indeed this was done constantly. But just as supply sometimes generates demand, so the availability of rubbings and scriptures seems to have acted as a tempting teaser: many people could now have the scriptures in their homes. But many was not most; for every person who had a rubbing there must have been a dozen who were encouraged to want their own. This new demand must have been an important stimulus leading to the development of actual printing.

The Chinese also used stencils and composite inked squeezes, the former being used particularly by the Buddhists. Paper had a pattern made in it consisting of rows of tiny dots. It was then pressed down on top of a blank sheet, ink was applied to the back, and a stencilled design in ink was the result. This was indeed very close to printing, for it enabled the cheap reproduction of a quantity of clear images – such as the Buddha sitting in meditation.

The composite inked squeezes were remarkable in that in a curious sort of way they anticipated photography. Using these sophisticated techniques, three-dimensional objects could be represented on flat paper with perspective, but with no problems of focus (i.e., no 'depth of field' problem existed). Rubbings would be taken of round bronze vessels whereby the more distant parts would be inked more lightly, and the nearer parts more heavily – giving the perspective effect to an eerie and uncanny degree. Inscriptions and decorations in three dimensions represented on paper could be carried around and kept in multiple copies in the home. This technique was known as 'whole shape rubbing' and it necessitated very careful study and preparation before being attempted. It ranks as one of the most highly skilled crafts ever practised anywhere. A good 'whole shape rubbing' looks so startlingly like a photograph in perfect focus that it seems nothing short of an inventor's miracle.

All of these techniques were still not sufficient for the needs of the Chinese. Foremost in pushing back the frontiers of printing technology were the Buddhists; they simply had to have many more copies of their sacred texts than hand-copiers and proto-printing could produce. Therefore it was no accident that the earliest printed text in the world was a Buddhist charm scroll printed in China and preserved in the Pulguk-sa Temple in Kyongju, south-east Korea, where it was discovered in 1966. It was printed some time between the years 704 and 751.

The print runs of the eighth century were quite literally fantastic – almost unimaginable, even by modern standards. The Buddhists took the new technology to Japan where the same Buddhist *sutra* of the Korean scroll was used as the source of a printed charm, and produced in a print run of *one million copies*! Many of these copies still survive today, even though they were printed in 764 AD.

The first complete printed book is thought to be the Buddhist *Diamond Sutra*, printed in the year 868 and discovered by Sir Aurel Stein in 1907. It is preserved in perfect condition in the British Museum. It consists of a scroll 17½ feet long and 10½ inches wide and contains the complete text of a Sanskrit work translated into Chinese, with a very elaborate and impressive frontispiece showing the Buddha in discourse with his disciple Subhuti, surrounded by attendants and divine beings. It bears a colophon at the end which says: 'On the fifteenth day of the fourth moon of the ninth year of Hsien-t'ung [868 AD], Wang Chieh reverently made this for blessings to his parents, for universal distribution.'

There were also large print runs for ordinary books. For instance we know that between 847 and 851 several thousand copies were printed of a biography of the alchemist Liu Hung. Calendars were also very popular in printed form, and were even personalized. One calendar surviving from the year 882 is headed: 'Family calendar of Fan Shang of Ch'engtu-fu in Hsi-ch'uan, province of Ch'ien-nan'. So many of these privately printed calendars were circulated that as early as 835, a regional official in the southern province of Szechuan suggested that they be banned – because they were being sold in marketplaces before the Board of Astronomers could approve and issue them, thus pre-empting and anticipating the government's important prerogative. (Chinese emperors looked upon it as a sacred and politically essential matter to revise and promulgate calendars; hundreds were produced during Chinese history.)

By the tenth century, *belles lettres* and philosophy were fully represented in print. Collections of the works of individual poets were being printed for them and circulated to friends. The scholar Ho Ning

LEFT (80) The frontispiece of the world's oldest surviving book printed on paper, the *Diamond Sutra*, discovered at Tunhuang by Sir Aurel Stein in 1907. The book is in the form of a roll with a total length of 17½ feet, and it was printed in the year 868. The frontispiece depicts the Buddha discoursing with his disciple Subhuti, and surrounded by divine beings, monks and officials in Chinese dress. This is the earliest woodcut illustration in a printed book. (British Library, London.)

(898–955) collected his poems and songs and had printed several hundred copies as gifts for his friends; in 913, the Taoist monk Tu Kuang P'ing printed his own commentary upon the classic of Lao Tzu, the chief Taoist sage.

During the tenth century, the prime minister Feng Tao decided to print the eleven classics of Confucianism, together with two supplementary books. Feng Tao managed to survive ten reigns of five different dynasties – a miracle of political survival in a troubled time – which alone enabled him to complete his gigantic task. Finally, after twenty-two years, the Confucian classics were printed in 953. Filling 130 volumes, they were the world's first official printed publications, sold to the public by the Chinese National Academy.

By this time, printing had come of age. Vast quantities of certain works continued to be issued, ranging into many millions of copies. Of one Buddhist collection of the tenth century, over 400,000 copies still survive. So we can imagine what the initial print run must have been! Twenty thousand copies of a printed picture of a goddess on silk still survive from that century, and 140,000 of a picture of a pagoda from the same period. Three other works of that time survive in quantities of 84,000 copies each. Also, during the tenth century, paged books were produced in the modern style, replacing the earlier printed scrolls. This, then, was the establishment of a woodblock printing industry which in the quantities produced rivalled the most modern efforts of our own times.

The blocks used by the Chinese for their printing tended to be made of fruit woods. Coniferous woods were found unsuitable, because they were impregnated with resin which affected the evenness of the ink coating. For cutting delicate lines in illustrations, a favourite material was the extremely hard wood of the Chinese honey locust tree. For regular text, the soft and easily worked boxwood was often used. But the best all-round wood for block printing was pear: this has a smooth and even texture with a medium hardness, and can be carved in any direction with no grain problems. The slightly harder wood of the jujube date tree was probably the second most commonly used. Naturally, there had to be no knots or spots in the wood, and after cutting the blocks it was customary to soak them in water for a month. But if there was insufficient time, they were boiled, then left to dry in a shaded place and planed on both sides, for it was common to print pages simultaneously on each side of a block. Vegetable oil would be spread over the planed surface, which would be polished with the stems of polishing grass (*Achnatherum*).

Printing blocks were normally kept for long periods of time – sometimes centuries – in the same family. Large print runs of books tended not to be the rule: a few tens of copies would be run off, then the blocks would be stored away until a few dozen

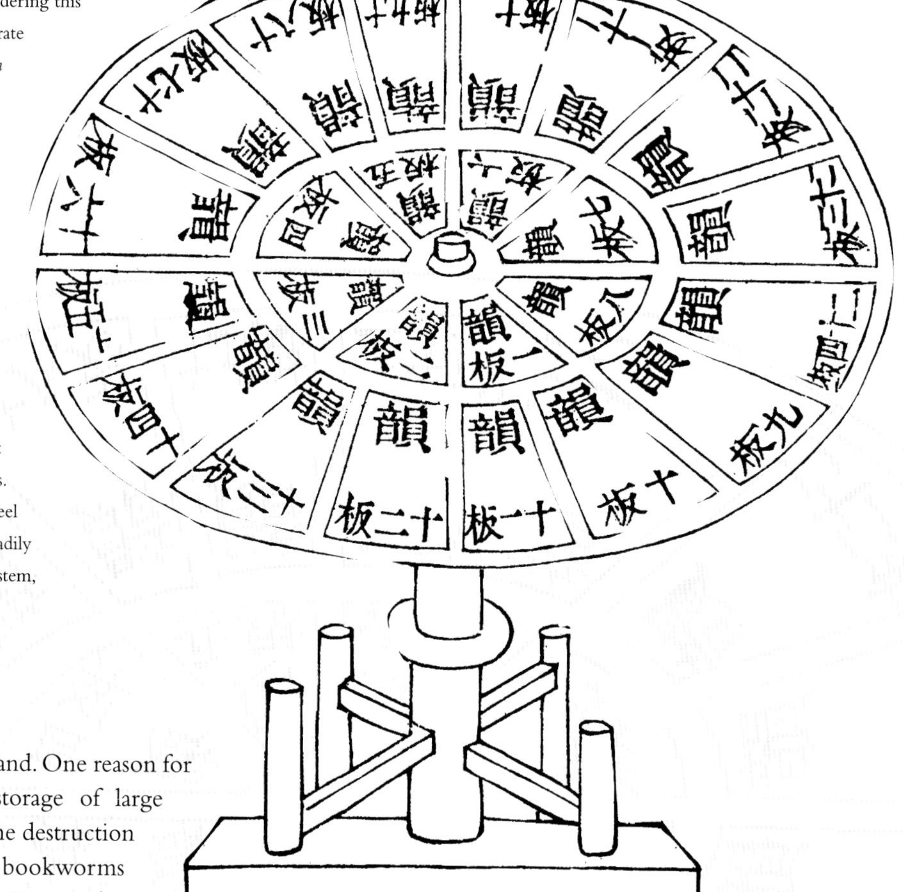

RIGHT (81) This drawing is from a book printed in 1313 called (surprisingly, considering this subject: the reason being to illustrate the rotating wheel) the *Treatise on Agriculture* (*Nung Shu*). It shows a wheel-shaped rotating case for the classified storage of fonts of movable type, which were invented circa 1045 by Pi Sheng. All the characters were either carved into wooden fonts or cast into metal fonts and stored according to their rhyming spoken sounds, in twenty-four compartments, consisting of eight inner ones and sixteen outer ones. By spinning the compositor's wheel round, the pieces of type were readily accessed according to this easy system, making for speedy typesetting.

more copies were in demand. One reason for cutting down on the storage of large quantities of copies was the destruction regularly wrought by bookworms in some areas of China. It was also economically advantageous to wait for orders before buying the paper.

Korea was the first country to which printing spread from China, around the year 700. Many old printing blocks survive: at the Haein-sa Temple high on Mount Kaya in southern Korea, 81,258 blocks of magnolia carved on both sides are preserved intact today, which were used to print a Buddhist classic between the years 1237 and 1251.

Such a large number of printing blocks would not have been exceptional in China. Chinese printing was on a large enough scale to cope with the country and its population, and one individual printer in Chiangsu in the seventeenth century named Mao Chin is known to have printed 600 titles on a variety of subjects. For the Thirteen Confucian Classics he used 11,846 blocks, for the Seventeen Standard Histories he used 22,293 blocks, and for another collection he used 16,637 blocks. Even during the early stages of his career, this single private printer employed twenty block cutters and had a store of over 100,000 printing blocks. But these quantities are small compared to those of the imperial printers, who produced several editions of the great imperial encyclopedias – over five thousand *printed volumes* each. Many of these sets still survive today.

The Chinese were also the inventors of multi-colour printing. Paper money produced in the year 1107 was printed in three colours as a precaution against counterfeiting, though in this case it is possible that the two additional colours may have been *stamped* onto the paper. The money in question had legends in black, a circle design in vermilion, and a 'blue face' in indigo. Two-colour printing of texts seems to have begun during the next century, and an edition of the *Diamond Sutra* of 1340 survives, which uses black for the text, and red for prayers and pictures. It became common for commentaries to appear in red beside black texts. The earliest surviving four-colour printing dates from the early twelfth century, and is in black, grey, green and red on a large single sheet discovered in 1973 in the cavity of an old pillar. It shows a legendary figure called Tungfang Shuo, and would have been used as a wall decoration.

Chinese colour-printing techniques used watery rather than oily inks, providing subtle effects. The inks in fact exactly matched the ones used in the original artwork, and were made from the same earth pigments. When it was necessary to overlay colours, it was done in a variety of different ways. Sometimes inks were allowed to dry before having others printed over them: others were left wet. Subtle variations in pressure on different portions of the blocks resulted in gradations of printing strength, reproducing the expression and texture of brush strokes. Often the same colour would be printed over and over from the same block, with varying portions inked so that deeper tones could be obtained in selected portions of the print. Ink was also allowed to run on the block, or it was wiped away in certain places. One expert on the subject, Dr J. Tschichold, has said: 'There is hardly another graphic art in the world that depends so entirely on the artistic sympathy and understanding of the printer as does the Chinese colour print.'

A traditional Chinese form of printing which required special expertise on the part of the printer and his staff was printing by movable type. Typesetters had to be linguistic experts and scholars knowledgeable in the history of language and literature. This is because of the thousands of characters in the Chinese language, many of which are obscure and rarely used, therefore only known to learned scholars. It is common in traditional Chinese typesetting to have at least twenty different pieces of type for each of the commonest characters. So the imperial printing works had to make 200,000 bronze characters in 1725, and in 1733 250,000 wooden characters were produced for another project. In the early nineteenth century, one private printer is known to have had a stock of no less than 400,000 different bronze characters. It is obvious that organizing this vast mass of material for printing purposes was a major problem in contrast to the ease with which movable type was used to print in the alphabets of Western languages. This was a major reason for the rare use of movable type in China, despite the fact that it was invented there four centuries before its 'invention' in Europe by Johannes Gutenberg.

Effective movable type was invented between the years 1041 and 1048 by an obscure commoner named Pi Sheng, who lived from about 990 to 1051. In his *Dream Pool Essays* of 1086, the famous scientist Shen Kua recorded the invention as follows:

During the reign of Ch'ing-li, Pi Sheng, a man of unofficial position, made movable type. His method was as follows: he took sticky clay and cut in it characters as thin as the edge of a coin. Each character formed, as it were, a single type. He baked them in the fire to make them hard. He had previously prepared an iron plate and he had covered his plate with a

LEFT (82) A modern reproduction of the movable type invented by Pi Sheng between 1041 and 1048, and a page printed from it. Movable type was not invented by Johannes Gutenberg, as is universally believed in the West. The reproduction was made from the detailed description by Shen Kua which survives from 1086.

BELOW (83) It is hard to believe that this magnificent example of multi-colour Chinese printing dates from the seventeenth century, but it does. A bird is seen on a snow-covered blossoming branch, in early spring, indicating a northern Chinese location (where it would be cold at that time). Multi-colour printing was invented in China no later than the eleventh century, as three-coloured paper money was printed (a precaution against counterfeiters) in a mass print-run in 1107 and still survives, as does a two-coloured book of 1340. This print comes from the famous multi-coloured book in several volumes, *Ten Bamboo Studio*, of Hu Cheng-yen, from the 1663 edition. Hu named his art and printing studio at Nanking 'Ten Bamboo' after the trees outside his window. The prints in this book heighten their effects by subtle variations in hand pressure on the blocks during the act of printing, as may be seen dramatically in the gradations of pink in the blossoms, themselves made possible only because inks of a watery rather than oily nature were used. Although the pages of this book are flat, the decorated letter paper produced at the same studio in 1644 by Hu and his colleague Tsao Sung-hsüeh gave yet another dimension to the colour printing by combining it with special hand-embossing that enhanced the colour pressure variations even further, creating a subtle varying depth in three-dimensions. The Ten Bamboo Studio team achieved such spectacular printing effects that they have still not been rivalled even today, as the subtle skills of variable hand-pressure on wooden printing blocks are now a completely lost art. (Chinese Department, British Library.)

mixture of pine resin, wax, and paper ashes. When he wished to print, he took an iron frame and set it on the iron plate. In this he placed the types, set close together. When the frame was full, the whole made one solid block of type....

If one were to print only two or three copies, this method would be neither simple nor easy. But for printing hundreds of thousands of copies, it was marvellously quick.... For each character there were several types, and for certain common characters there were twenty or more types each, in order to be prepared for the repetition of characters on the same page. When the characters were not in use, he had them arranged with paper labels, one label for words of each rhyme-group, and kept them in wooden cases. If any rare character appeared that had not been prepared in advance, it was cut as needed and baked with a fire of straw. In a moment it was finished....

When Pi Sheng died, his font of type passed into the possession of my nephews, and up to this time it has been kept as a precious possession.

Earthenware types of the nineteenth century survive in China, but it became more common to use wood, enamelware, or metal in later times. The use of wooden type was perfected two-and-a-half centuries after Pi Sheng by Wang Chen, which he did, intending to use it to print his classic, the *Treatise on Agriculture*, in 1313. The wooden type was made in the years 1297 and 1298, and Wang Chen has left an account of the process.

More noteworthy than perfecting wooden type were Wang Chen's storage and handling arrangements. The type was stored in revolving tables 7 feet in diameter, supported by central legs 3 feet high. Ingenious sorting and classifying schemes were used to enable workers to find the necessary characters quickly. In the end, the *Treatise on Agriculture* was printed with bronze characters. But Wang Chen's 60,000 wooden type characters were used to print the local book, *Gazetteer of Ching-Te County*. One hundred copies of it were printed in less than a month in the year 1298.

Movable type continued to be used sporadically throughout Chinese history. It was revived under the Mongols, when a councillor of Kublai Khan decided to use the 'movable type of Shen Kua' to print books of philosophy. A nineteenth-century teacher named Chai Chin-Sheng, born in 1784, spent thirty years making a font of earthenware type, using everyone in his family to help him. By 1844 he had finally made over 100,000 sets in five sizes, which he used to print his own collected poems under the title *First Experimental Edition with Earthenware Type*. He is the earliest and perhaps the only author-printer known in China.

Movable wooden type dating from about the year 1300 survives from Eastern Turkestan in the region known as Turfan, where it was introduced by the Chinese after the Mongol conquest. Movable type was much easier to use for the Uigur scripts of Turfan than for the Chinese language. Printing is thought to have spread to Europe through Turfan, and then through Persia, which the Mongols also conquered. Paper money was printed following the Chinese system in Tabriz in Persia in 1294. It was even called by the Chinese name, *ch'ao*, which then entered the Persian language. Fifty printed pieces of Islamic material printed according to the Chinese method were excavated in Egypt in the nineteenth century; they cannot be precisely dated, but were printed sometime between 900 and 1350. Generally speaking, however, printing was frowned upon by the Muslims.

The Mongol armies pressed ever further westwards, overrunning Russia in 1240, Poland in 1259, and Hungary in 1283. They reached the borders of Germany not long before printing made its appearance in that country – block printing appeared suddenly in Europe early in the fourteenth century. Although no hard evidence exists for its transmission from China, the circumstantial evidence is strong enough to support it. In 1458, Johannes Gutenberg 'invented' movable type. But as Juan Gonzalez de Mendoza said in 1585 in his book, *The History of the Great and Mighty Kingdom of China and Situation Thereof*, speaking of Gutenberg's 'invention' in 'Almaine' (Germany) the century before:

But the Chinos doo afffirme, that the first beginning [of printing] was in their countrie, and the inventiour was a man whome they reverence for a saint: whereby it is evident that manie years after that they had the use thereof, it was brought into Almaine by the way of Ruscia and Moscouia [Russia and Moscow], from whence, as it is certaine, they may come by lande, and that some merchants that came from Arabia Felix, might bring some books, from whence this John Cutembergo [Johannes Gutenberg], whom the histories dooth make author [inventor], had his first foundation [got the idea].

PLAYING-CARDS

NINTH CENTURY AD

Paper was invented in China, and it is therefore not surprising that the Chinese were the first to invent paper playing-cards. By the ninth century at the latest these were in use. The first known book on card games was written by a woman in the ninth century, but it is lost. The scholar Ouyang Hsiu (1007–72) recorded that the use of paper playing-cards arose in connection with the change of book format from paper rolls to paper sheets and pages. The playing-cards were printed by woodcut blocks, and many specimens survive. They were often coloured by hand, and popular designs for the backs were drawn by famous artists of fictional characters from the well-known novel *The Water Margin*. The shape of the cards was generally more elongated than those we use today, being about 2 inches high and only about 1 inch wide. They were

LEFT (84) A Chinese playing-card of about 1400 found near Turfan, photographed before the Second World War at the Museum für Völkerkunde, Berlin, but lost during the war.

LEFT (85) Chinese playing cards were traditionally longer and narrower than European ones. These stencilled cards date from about 1870. The human figures on some of the cards are characters from the famous Ming Dynasty popular novel *The Water Margin* (also called *Outlaws of the Marsh*, or in Pearl Buck's translation, *All Men Are Brothers*), which was set in the early twelfth century and is greatly beloved by all Chinese. (National Museum of Playing Cards, Turnhout, Belgium.)

of fairly thick paper, which made them more durable than those of today, though doubtless they were more difficult to shuffle.

The Chinese enthusiastically proclaimed the advantages of playing-cards over all other pastimes, pointing out that they were 'convenient to carry, could stimulate thinking and could be played by a group of four without annoying conversation, and without the difficulties which accompanied playing chess or meditation'. Furthermore, cards 'could be played in almost any circumstances without restriction of time, place, weather, or qualification of partners'. But the Chinese passion for gambling led to the promulgation during the eighteenth century of laws against gambling by officials, and against the manufacture and sale of more than one thousand paper playing-cards by a single person.

Playing-cards spread to the West from China either through the Arabs or through the travellers such as Marco Polo who circulated during the Mongol Dynasty, when there was such freedom of travel between Europe and Asia. In the seventeenth century, Valère Zani claimed that Venice was the first European city to have playing-cards from China. This may well have been the case, but the earliest appearance of playing-cards in Europe of which we can be certain was in Germany and Spain by the year 1377. By 1379, we know they were being used in Italy and Belgium, and by 1381 in France.

Even Johannes Gutenberg, renowned as the fifteenth-century European inventor of printing by movable type (but see page 127), was involved in playing-card manufacture. He actually developed some of the mechanical means of their production, and when his financial affairs became so desperate that he was forced to close his Mainz workshop, the figures which his artists had prepared to illustrate his famous Bible were used to print the backs of playing-cards. This extraordinary fact certainly highlights the close connection that has always existed between playing-cards and book printing.

PAPER MONEY

NINTH CENTURY AD

The Chinese invented paper money at the end of the eighth or beginning of the ninth century AD. Its original name was 'flying money' because it was so light and could blow out of one's hand. The first paper money was, strictly speaking, a draft rather than real money. A merchant could deposit his cash in the capital, receiving a paper certificate which he could then exchange for cash in the provinces. This private chant enterprise was quickly taken over by the ernment in 812. The technique was then used for forwarding of local taxes and revenues to the tal. Paper 'exchange certificates' were also in use. se were issued by government officials in the tal and were redeemable elsewhere in commodities as salt and tea.

Real paper money, used as a medium of exchange backed by deposited cash, apparently came into g early in the tenth century, in the southern ince of Szechuan, as a private enterprise. Early in eleventh century the government authorized een private businesses or 'banks' to issue notes of hange; but in 1023 the government usurped this ate enterprise and set up its own official agency to bank notes of various denominations which backed by cash deposits. We can thus probably the world's first governmental currency reserve k to 1023.

The money issued by this bank had printed on it a e to the effect that it was good for only three years, and gave the dates. Such a time limit was to be a re feature of Chinese paper money up until the ninete century. By 1107, notes were being printed multiple blocks in no less than six colours.

The issuing of paper money by the govern took on enormous proportions. By 1126, sev million strings (each string being equal to thousand pieces of 'cash') had been officially is Vast amounts of this paper money were not backe any deposits, and a horrifying inflation occu Inflation may be looked upon as a phenom which accompanies paper money, arising from it being backed by anything more substantial. C forms of 'inflation' in history should probabl differently described. Accelerating price incr before the use of paper money, such as occurre scandalously under the Roman Empire, was caused by debasement of the coinage. The Ro Empire is infamous for having produced valuation coinage out of increasingly cheap worthless metals.

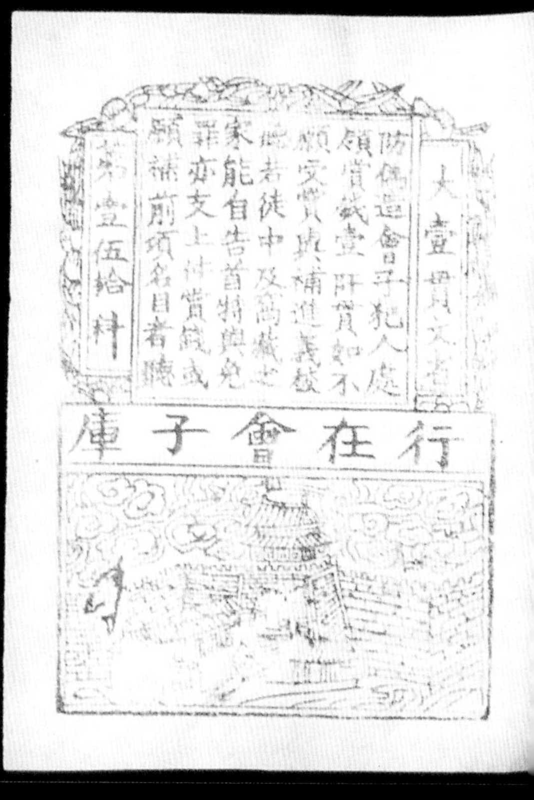

E (86) A copper plate for printing paper money, from the Southern Sung Dynasty capital of Hangchou, and dating from between 112
). Next to it is a print made from the plate as a modern example. Paper money had been invented in China by the late eighth or early

There was also the deleterious process known as 'clipping' whereby not only individuals but sometimes governments themselves issued and circulated coins with pieces cut off them. The real value of such clipped coins was of course diminished, though the fiction was supposed to be maintained that they were worth the same as ever. But debasements and clipping of coinage should surely be differentiated from true inflation, which resulted from the issuance of paper money and was thus 'invented' in China along with the paper money that gave rise to it.

Another problem which soon arose in China was counterfeiting. If precious metal coins are in circulation which are intrinsically worth their true value, the only kind of counterfeiting possible is with false, disguised metals. This has often been done, and was a major impetus to alchemy, and the manufacture of spurious gold and silver. But paper money invited counterfeiting by its very nature, since the essence of it is not its inherent substance but the authority on which it was issued. Paper money is a symbol. To counterfeit is therefore not to fabricate a substance but to impersonate the authority issuing it. Since anyone can print on pieces of paper, the authority must make the processes of manufacture of its paper money so intricate that they cannot be exactly reproduced. Complex manufacturing secrets were thus adopted quite early, and included multiple colourings, immensely complex designs, and a mixture of fibres in the paper. The basic material for the paper of paper money was the bark of mulberry trees, and silk was sometimes incorporated. One could hand in soiled or worn-out notes for new ones, but had to pay the small cost of the printing of the replacement.

A detailed case of a counterfeiter of 1183 survives, which tells us that he printed 2600 false notes during the six-month period before he was caught. He was a master block-cutter, and he cut a block of pearwood in direct imitation of the design on a real note. It was a three-colour process, involving sequential serial numbers in blue and seals in red. It took the counterfeiter only ten days to cut the block. But counterfeiters did not have an easy time of it when they were caught, for their crime carried the death penalty.

When the Mongols came to power in China, they issued a quaint form of paper money called 'silk notes'. The deposits behind this currency were not precious metals but bundles of silk yarn. All older money had to be cashed in and exchanged for silk notes, and the Mongols spread this unified currency all over the Empire and even beyond it. By 1294, Chinese silk notes were being used as money as far afield as Persia. In 1965, two specimens of 'silk notes' were found by archeologists.

When Marco Polo visited China, he was so impressed by paper money that he wrote a whole chapter about it, describing everything about its manufacture and circulation. He described the manner in which it was issued:

> **All these pieces of paper are issued with as much solemnity and authority as if they were of pure gold or silver; and on every piece a variety of officials, whose duty it is, have to write their names, and to put their seals. And when all is duly prepared, the chief officer deputed by the Khan smears the Seal entrusted to him with vermilion, and impresses it on the paper, so that the form of the Seal remains printed upon it in red; the Money is then authentic. Anyone forging it would be punished with death.**

Paper money under the later Ming Dynasty was not so effective. The Ming issued in 1375 a new note called the 'Precious Note of Great Ming'. It was issued in one denomination only throughout the two hundred years in which it was the legal tender. This was naturally very inconvenient for all commercial purposes, although copper coins were permitted to circulate, and these must have provided the small change necessary in everyday life. Through inflation, the Precious Note gradually lost its value and was replaced by silver. In the middle of the seventeenth century, the Ming tried to reinstate paper money after a lapse of about two centuries, but it was badly implemented, resulting in great inflation, and failed. Paper money on a national scale and a regular basis died out until European influence brought it back in modern times.

When the older methods of paper money issuance became known in the West, they had a profound influence on Western banking. The old Hamburg Bank and the Swedish banking system were set up on Chinese lines. Thus, some of the fundamental banking procedures of the Western world came from China directly. The first Western paper money was issued in Sweden in 1661. America followed in 1690, France in 1720, England in 1797, and Germany not until 1806.

'PERMANENT' LAMPS

NINTH CENTURY AD

The Chinese developed the simple oil-and-wick lamp to its furthest possibilities. First of all, they regularly used wicks that do not burn, as described in the book *Memoirs on Neglected Matters*, written about 300 AD:

> **In the second year of King Chao of Yen, the sea-people brought oil in ships, having used very large kettles for extracting it, and presented it to him. Sitting in the Cloud-Piercing Pavilion he enjoyed the brilliant light of the lamps in which the dragon blubber was burnt. The light was so brilliant that it could be seen a hundred li away [somewhat more than 30 miles]; and its smoke was coloured red and purple. The country people, seeing it, said, 'What a prosperous light!', and worshipped it from afar. It was burnt with wicks of asbestos.**

This was in either 598 BC or 308 BC. We cannot be sure of the date, for there were two kings of Yen named Chao. Needham comments on this passage: 'Whatever the date to which this really refers, it must surely imply that some kind of primitive sealing or whaling was going on in Han or pre-Han times, and that the oil or blubber was consumed in the courts of coastal princes with unburning wicks.' Needham has gathered much material on the history of asbestos in ancient China, and mentions that the famous general Liang Chi, who died in 159 AD, had an incombustible gown of asbestos which he used to throw on the fire at parties. The asbestos wick for lamps essentially meant a permanent wick which did not need to be replaced and would burn as long as the oil was replenished.

However, asbestos wicks were by no means the limit of ancient Chinese ingenuity with regard to the economical use of lamps. The Chinese were determined to squeeze fire out of stone and get something for nothing. Having developed the inexhaustible wick, they turned their attention to the nearest thing to inexhaustible oil. Their lamps were simple dishes or cruses full of oil with wicks sticking out of them. The Chinese noticed that the heat of the burning wicks made much of the oil evaporate before it could be productively burnt. In order to counteract this, they devised a way to cool the lamps and thus prevent the evaporation, as we are told by the author Lu Yu, in his book *Notes from the Hall of Learned Old Age*, published about 1190 AD:

> **In the collected works of Sung Wen An Kung there is a poem on 'economic lamps'. One can find these things in Han-chia; they are actually made of two layers. At one side there is a small hole into which you put cold water, changing it every evening. The flame of an ordinary lamp as it burns quickly dries up the oil, but these lamps are different for they save half the oil. When Shao Chi was the Prefect at Han-chia, he sent several of them to scholars and high officials at court. According to Wen An one can also use dew. Han-chia has been producing these for more than three hundred years.**

So, underneath the oil there was a reservoir, into which cold water was poured, resulting in a saving of half the oil used. Since in 1190 the lamps had been manufactured for more than three centuries, their production on a mass scale must have begun no later than the beginning of the tenth century AD and, as Needham says, perhaps early in the ninth century AD. He adds: 'It was an interesting anticipation of the water-jacketing of the chemical condenser in distillation, and of the steam and water circulatory systems of all modern technology.'

The 'economic lamps' were generally made of glazed earthenware, and several good specimens are preserved in the Chungking Museum in China. Since simple wick lamps have remained in use well into the latter half of this century in China, perhaps a reintroduction of the 'economic' water-cooled version would be useful in parts of rural China today.

THE SPINNING-WHEEL

ELEVENTH CENTURY AD

The homely device of the spinning-wheel, which evokes images of European cottage life and Indian rural productivity alike, had its origin in China. The earliest known European reference to a spinning-wheel is an indirect one in the statutes of a guild at Speyer, Germany, in about 1280. Needham believes spinning-wheels and other machines to do with

LEFT (87) A photograph taken by Cecil Beaton during the Second World War, showing an old woman who sought refuge in the Poor People's Hostel at Changsha. Her spinning-wheel is a traditional Chinese form with a double wheel. Between the wheels is strung a cat's cradle of string; the belt-drive (driving-belt) runs round the centre of the cat's cradle rather than along a wheel rim.

textiles were introduced to Europe by Italians who travelled to China during the Mongol Dynasty. He points out that, 'since we find very soon afterwards at such cities as Lucca in Italy silk filatures using machinery closely similar to that of China, the presumption is that one or other of the European merchants who travelled East in those days brought back the designs in his saddle-bags.'

Spinning-wheels derived from Chinese machinery for processing silk fibres. A single continuous strand of silk runs for several hundred yards and has a tensile strength of 65,000 pounds per square inch. This is stronger than any other plant fibre known, and approaches the strength of some engineering materials. The domestication of the silkworm and the development of the silk industry in China had taken place by the fourteenth century BC at the latest. Although it was apparently many centuries after this that they were developed, the silk industry obviously had a need from the beginning for silk-winding machines to deal with these enormously long fibres. Such machines are mentioned in the *Analytical Dictionary of Characters* of 121 AD, and again in the *Enlargement of the Literary Expositor* of 230 AD. These are first depicted in print in the books *Pictures of Tilling and Weaving*, published in 1237.

Quilling-machines of this sort for winding silk onto bobbins also made their way to Europe, and seem to have preceded spinning-wheels slightly. The textile-wheels depicted in windows of Chartres Cathedral and datable to between 1240 and 1245 are quilling-machines, and a clearer illustration of one may be seen in the Ypres *Book of Trades*, dated about 1310.

Quilling-machines go back at least to the first century BC in China. It is not clear how soon spinning-wheels derived from them, and if we wished to be conservative we could say that they did so by the eleventh century. Cotton culture had spread across China, and spinning-wheels were evidently derived from the silk-winders to deal with it. The use of the driving-belt to cause spindles to turn at great speed by connecting them with large wheels was most ingenious. The Chinese invention of the belt-drive, or driving-belt, is discussed on page 59.

Part 5
MEDICINE AND HEALTH

CIRCULATION OF THE BLOOD

SECOND CENTURY BC

Most people believe that the circulation of the blood in the body was discovered by William Harvey, and that it was he who first brought the idea to the attention of the world when he published his discovery in 1628. Harvey was, however, not even the first European to recognize the concept, and the Chinese had made the discovery two thousand years before.

In Europe, Harvey was anticipated by Michael Servetus (1546), Realdo Colombo (1559), Andrea Cesalpino (1571) and Giordano Bruno (1590). These men had read of the circulation of the blood in the writings of an Arab of Damascus, al-Nafīs (died 1288), who himself seems to have obtained the idea from China. The writings of al-Nafīs translated into Latin were lost, and rediscovered by a scholar as recently as 1956, establishing the source for Europe.

In China, indisputable and voluminous textual evidence exists to prove that the circulation of the blood was an established doctrine by the second century BC at the latest. For the idea to have become elaborated by this time, however, into the full and complex doctrine that appears in *The Yellow Emperor's Manual of Corporeal Medicine* (China's equivalent of the Hippocratic writings of Greece), the original notion must have appeared a very long time previously. It is safe to say that the idea occurred in China about two thousand years before it found acceptance in the West.

The ancient Chinese conceived of two separate circulations of fluids in the body. Blood, pumped by the heart, flowed through the arteries, veins and capillaries. *Ch'i*, an ethereal, rarefied form of energy, was pumped by the lungs to circulate through the body in invisible tracts. The concept of this dual circulation of fluids was central to the practice of acupuncture.

The Chinese traditionally identified twenty-eight different types of pulse, which they recognized as emanating from the pumping heart. The entire view of the body and its functioning was that of a dual circulation theory of blood (which was *yin*) and *ch'i* (which was *yang*). The two were interrelated. As a text dating from about the time of Christ says: 'The flow of the blood is maintained by the *ch'i*, and the motion of the *ch'i* depends on the blood; thus coursing in mutual reliance they move around.' *The Yellow Emperor's Manual* says: 'The function of the tract-channel system of the human body is to promote a normal passage [circulation] of the blood and the *ch'i*, so that the vital essentials derived from man's food can nourish the *yin* and *yang* viscera, sustain the muscles, sinews and bones, and lubricate the joints.'

The *Manual* also says: 'What we call the vascular system is like dykes and retaining walls forming a circle of tunnels which control the path that is traversed by blood so that it cannot escape or find anywhere to leak away.' The Chinese, always so methodical at measuring and weighing things, carried out investigations in which they removed the blood vessels from corpses, stretched them to their full

OPPOSITE (88) The internal organs, featuring the heart at the centre from which the blood flows. This is a detail from a large nineteenth-century medical poster printed on rice paper, but the original drawing from which it derives was printed near the end of the Ming Dynasty in 1624, in *The Classics Classified: A System of Medicine* (*Lei Ching*) of Chang Chieh-Pin, to show 'the twelve viscera'. However, this later version is expanded and contains considerably more anatomical (or imagined) detail than the original. (Collection of Robert Temple.)

腦者髓之海諸髓皆屬於腦
上至腦下至尾骶腎主之也

髓海
至陰之在
直通尾骶

臟六腑百骸九竅脈絡盡皆貫通節
無有間斷今畫共大略使觀者便覽

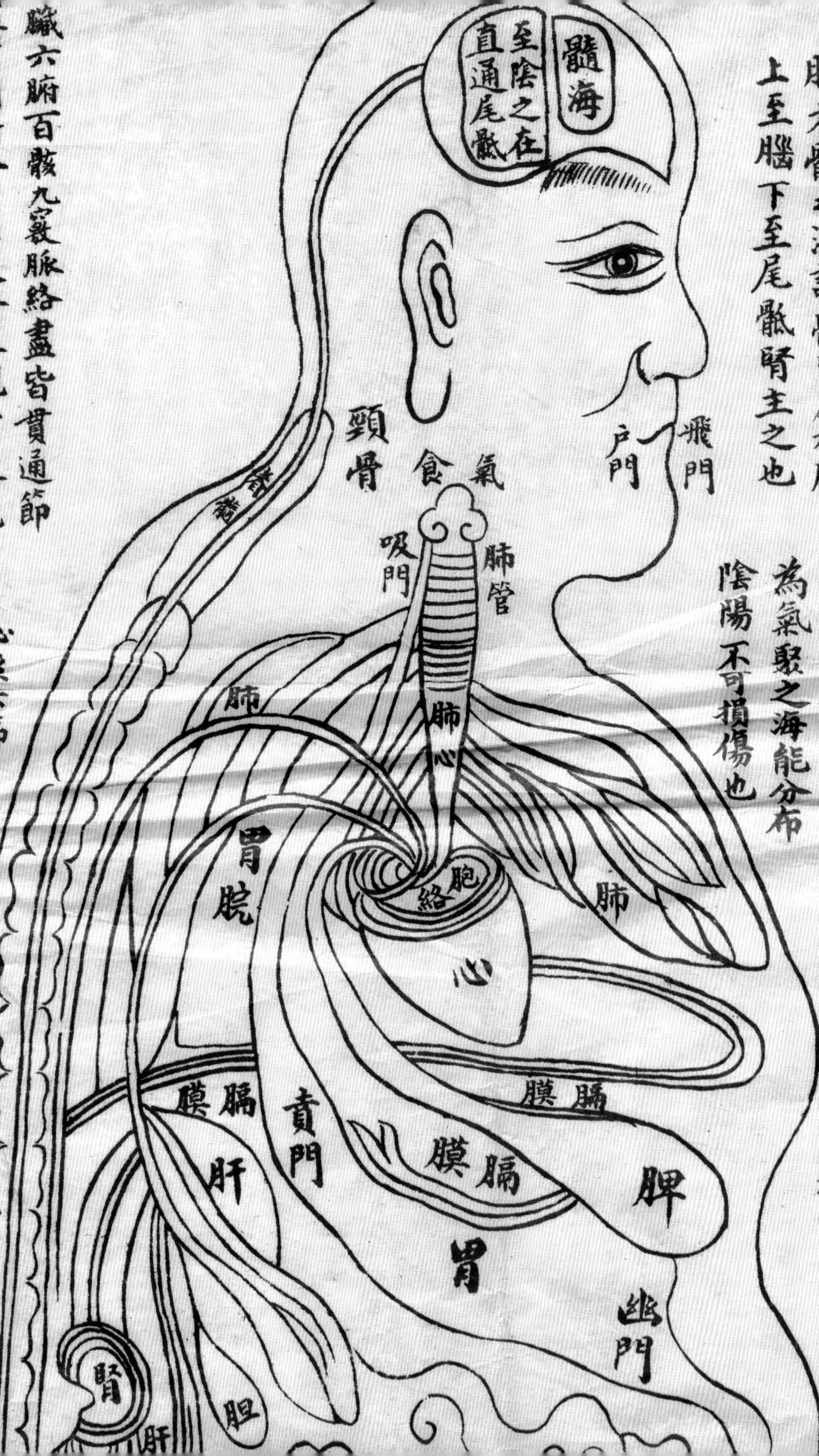

兩乳中間名曰膻中
為氣聚之海能分布
陰陽不可損傷也

膈膜在心肺之下肝腎
之上周回相隔如幕以
遮濁氣使不薰蒸於上

飛門
户門
氣
食
頸骨
脊
吸門
肺管
肺
肺
心
胞絡
胃脘
心
肺
膈膜
膈膜
膈膜
賁門
肝
脾
胃
幽門
闌門在
腎
胆
肝
心系六節
腎脈系七節

ABOVE (89) The frontispiece to the manuscript of Andreas Cleyer's book of 1682 about Chinese pulse lore, *Specimen Medicinae Sinicae*. The Chinese doctor is taking the patient's pulse while his boy assistant stands ready with acupuncture needles, drugs and 'moxa' (powdered mugwort tinder, used for burning at selected acupuncture points). A portable stove stands in the foreground ready for use, possibly to sterilize the acupuncture needles (steam sterilization was traditional in China for several centuries). Cleyer was editor, not author, of this book, which included Latin translations of old Chinese medical texts mentioning the circulation of the blood. (Staatsbibliothek Preussischer Kulturbesitz, Berlin.)

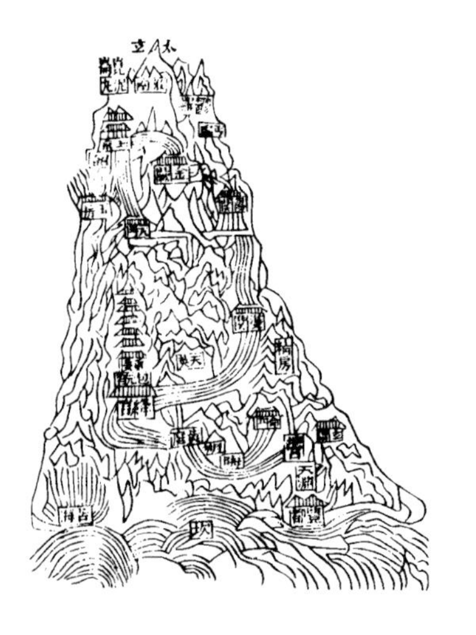

length, and made measurements of the total distance travelled by the blood in one circuit. This was estimated by these measurements to be 162 feet.

Once every 24 hours, the blood circulation and the *ch'i* circulation 'met' again in the wrist, having completed fifty blood circuits, so that the circulations coincided. The Chinese thus computed that the blood flowed 8100 feet per day. During this time, 13,500 breaths were supposed to take place; this meant that the blood flowed 6 inches for every breath. By making all these calculations, the Chinese imagined themselves to be pinning down the phenomenon quite comfortably.

The heart was clearly conceived of as pumping the blood. Indeed, Chinese doctors used in their classrooms an extraordinary system of bellows and bamboo tubes to pump liquid in demonstrations for their pupils, showing how the heart and blood circulation worked.

In the calculations of the flow of the blood in the body, each circulation was estimated as taking 28.8 minutes. We know through medical research that this is too slow by sixtyfold, the true time being only 30 seconds. William Harvey had not come to any conclusion about this, speculating that the time taken might be 'half an hour ... an hour, or even ... a day'.

The Dutch East India physician, Willem ten Rhijne, stated in his book of 1685, *Mantissa Schematica de Acupunctura*, that the circulation of the blood was one of the basic tenets of the whole of Chinese medicine. He wrote: 'The Chinese physicians ... perhaps devoted more effort over many centuries to learning and teaching with very great care the circulation of the blood, than have European physicians, individually or as a group. They base the foundation of their entire medicine upon the rules of this circulation, as if they were oracles of Apollo at Delphi.'

LEFT (90) The body as a mountain: a drawing of the tenth century by Chang Po-Tuan in the version of 1333 as it appeared in Ch'en Chih-Hsü's *Illustrations for the Main Essentials of the Metallous Enchymoma, the True Gold Elixir*. In this 'body-mountain', the rivers represent *ch'i* (life-force currents) and blood flowing around the body. At the top is K'un-Lun mountain peak, representing the head and containing 'the ball of mud' (Taoist nickname for the brain). The large pagoda in the lower central portion of the 'body-mountain' is the Yellow Court, where all rivers of blood and *ch'i* circulation meet.

In the very same year, the renowned scholar Isaac Vossius wrote that the Chinese had known of the circulation of the blood for four thousand years. As Needham says: 'He was of course taking the legendary date of the Yellow Emperor. But some 2000 years would have been right enough.' We thus see that 300 years ago, it was widely realized in Europe that the Chinese had originated the idea of the circulation of the blood. But since that time, Europeans have reverted to a state of ignorance on the subject and forgotten this entirely.

CIRCADIAN RHYTHMS IN THE HUMAN BODY

SECOND CENTURY BC

Even in the mid-1960s it was considered very daring for a scientist to suggest that the body contained 'biological clocks'. Doctors and zoologists put their reputations and careers at risk by holding such opinions. Now, with hundreds, or perhaps thousands, of scientists all round the world routinely studying these 'biological clocks', their existence is universally accepted.

They have come to be known as 'circadian rhythms', from the Latin *circa diem*, meaning 'about a day'. Most of the internal rhythms of the body – many of which concern hormone secretions – are on an approximately 24-hour 'clock'. The pineal gland in the head is suspected of being the actual main body 'clock'. Experiments have been carried out at the University of Texas in which the pineal glands of chickens were kept alive in test tubes and continued to act as 'biological clocks' for four days entirely on their own, without being connected to the rest of the body.

The menstrual and ovulation periods of women are possibly the most obvious of the body's longer internal rhythms. But 2200 years before European and American scientists risked being considered cranks for suggesting the existence of the 24-hour rhythms in human beings, the Chinese had observed and accepted them.

In the classic ancient medical text, *The Yellow Emperor's Manual of Corporeal Medicine*, from the second century BC, appears the following passage:

BELOW (91) A rubbing from the tomb of the Wu family in Shantung Province, showing two trees representing biological rhythms. One tree grows a leaf each day for a week and then loses a leaf a day for the next week. The other grows a leaf each day for 15 days and then loses a leaf a day for the next 15 days, being thus a 'monthly' tree. Sensitivity to body rhythms and 'biological clocks' seems to have been a natural Chinese attitude related to their tendency to view all natural processes as being caused by 'fields' rather than by 'particles'.

ABOVE (92) A pair of tiles from the Han Dynasty (207 BC–220 AD) representing the guardian spirits of two of the divisions of the day and night. *Left: tzu*, guardian of midnight (from 11 p.m. to 1 a.m.). *Right: mao*, guardian of the period 5 to 7 a.m. The Chinese were the first to realize that the body's sensitivity to drugs and proneness to symptoms vary with the hours of the day and night. (National Museums of Scotland, Edinburgh.)

Those who have a disease of the liver are animated and quick-witted in the early morning. Their spirits are heightened in the evening and at midnight they are calm and quiet.... Those who suffer from a sick heart are animated and quick-witted at noon, around midnight their spirits are heightened, and in the early morning they are peaceful and quiet.... Those who suffer from a disease of the spleen are animated and quick-witted around sunset, their spirits are heightened around sunrise and towards evening they become quiet and calm.... Those who suffer from a disease of the lungs are animated and quick-witted during evening, their spirits are heightened at noon and they are calm and peaceful at midnight.... Those who suffer from a disease of the kidneys are quick-witted and active at midnight, and their spirits are heightened during the entire days of the last months of Spring, Summer, Autumn, and Winter, and they become calm and quiet toward sunset.

Modern medical practice has shown that there are indeed quite drastic variations in the severity of diseases and symptoms at different times of the day. For instance, it has been found that between 9 p.m. and midnight, the disabling symptoms of Parkinson's disease frequently disappear altogether. The worst paroxysms of asthma generally occur during the night, when the secretion of the hormones from the cortexes of the adrenal glands drops to its minimum. Sensitivity to histamines is greatest at 11 p.m., when the body levels of certain cortico-steroid hormones are at their lowest. The affliction of cholecystitis (inflammation of the gall bladder) is always worst in the early hours of the morning. Fevers and body temperatures increase toward evening, and our sensitivity to pain varies considerably at different times of day and night. These are all phenomena resulting from the body's internal circadian rhythms.

Acupuncture, which is thought to go back to 1500 BC in China, was practised with circadian rhythms in mind. Hence the name of one medieval acupuncture manual, *Noon and Midnight Manual.* Another work on acupuncture was entitled *Mnemonic Rhyme to Aid in the Selection of Acu-points According to the Diurnal Cycle, the Day of the Month and the Season of the Year.* The diurnal cycle refers to the daily rhythms. This work is said to have been written about 419 AD, though it may date from 930 AD.

There were many other cycles besides the circadian rhythms noted by the Chinese doctors. Some were sensible, others nonsensical. Today, through modern medicine, we know that duodenal ulcers have cycles of 139 days, that Hodgkin's disease has a cycle of 21 days, and so on. The Chinese practitioners seem to have noticed some of these things; but most of their cyclic lore was superstitious numerology. What was soundest of all was the brilliant early insight into the existence of those circadian rhythms which even today we cannot yet fully explain or understand.

THE SCIENCE OF ENDOCRINOLOGY

SECOND CENTURY BC

The Chinese anticipated modern biochemistry to such an extent that by the second century BC they were isolating sex and pituitary hormones from human urine and using them for medicinal purposes. The crystals which they obtained were traditionally called 'autumn mineral', being likened to the hoar-frost of the autumn. The Prince of Huai-Nan seems to have coined this term some time before 125 BC 'to express its white colour and its solidity'. This prince was one of the foremost proto-scientists of early China, who had a school of Taoist adepts and philosophers where experiments on many physical processes were carried out.

Opposition to experimentation and study by the Taoists was voiced in 25 BC by Ku Yung, a high official whose speech against all magicians and alchemists is preserved in the official history of the time. He makes the following attack:

> **The Taoists say that by fusing cinnabar they can transform it and make yellow gold, and that from dark and muddy [that is, concentrated] urine they can made a hard white ice-like [crystalline] substance.**

LEFT (93) A painting from *Essentials of the Pharmacopoeia Ranked According to Nature and Efficacy: Imperially Commissioned*, of the year 1505, edited by Liu Wen-T'ai for private deposit in the Imperial Library and not intended for publication. We see here the process of sublimation of calomel (mercurous chloride). The figure on the left gently brushes the crystals from the sublimation lid with a feather. This same process was used for collection of pure crystals of human sex and pituitary hormones from human urine by the second century BC. From 150 gallons of human urine, 2 to 3 ounces of hormone crystals could be produced by this method. (Biblioteca Nazionale Centrale Vittorio Emanuele II, Rome.)

The great poet Pai Chü-I (772–846 AD) wrote a poem entitled 'Thinking of Old Friendships' in which he mentioned that his friend Yuan Chen, another poet (779–831), had 'prepared the "autumn mineral" drug / yet while still young encountered sudden death.'

Several centuries later, explicit recipes for the preparation of 'autumn mineral' begin to appear in print. To our knowledge the first such recipe to be published was in the book *Valuable Tried and Tested Prescriptions*, written by Chang Sheng-Tao and published in 1025. Between 1025 and 1833, at least ten different methods of obtaining sex and pituitary hormones from urine were published in thirty-nine different books. Production of these hormones took place on an enormous scale, using hundreds of gallons of human urine for each batch, and manufacturing countless thousands of doses of the drugs for medicinal use.

It was only in 1927 that S. Ascheim and B. Zondek published the discovery in Europe that the urine of pregnant women was rich in steroid sex hormones. Subsequently, the discovery was made that urine contains androgens and oestrogens (male and female sex hormones), and pituitary hormones known as gonadotrophins, which stimulate the sex glands (gonads), in profusion. Fertility drugs given to women today to make them produce more eggs are extracted from the urine of menopausal nuns in Italy. The derivation of sex hormones from human urine is today a standard practice, and modern medicines could not do without human urine as their source.

It is therefore remarkable that the Chinese anticipated this practice by 2200 years. The oldest published recipe for these hormones, dating from 1025, says:

Collect ten *tan* [over 150 gallons] of male urine and set up a large evaporating pan in an empty room. Fix on top of it a deep earthenware still, luting the edges together with paper-pulp and lime so that when it has dried no steam can escape. Fill the evaporating basin 70 to 80 per cent full with urine, and heat strongly from below, setting a man to watch it. If it froths over, add small amounts of cold urine. It must not be allowed to overflow. The dry residue is *jen chung poi*. Put some of this, finely powdered, into a good earthenware jar and proceed according to the method of sealing and subliming by placing the whole in a stove and heating with charcoal. About two or three ounces [of sublimate] will be obtained. Grind this to a powder, and mix with date-flesh to make pills the size of a mung bean. For each dose take five to seven pills with warm wine or soup before breakfast.

The process commences with simple evaporation, of the kind practised on a huge scale in the industry for producing salt by the evaporation of brine (see the account of deep drilling for natural gas on page 51). This left the dried solids of the urine, which had to be further processed to rid them of the urea, salts, and so on. The Chinese then turned to the process of sublimation, which was familiar from alchemy. The simplest form of sublimation is just to invert a pot over a glowing fire. Substances are thrown onto the red hot charcoal or hot ashes and sublimed upwards where they form a coating of material on the underside of the lid. An improvement of this process is to set a removable lid into the pot, which can be taken out and the powder coating brushed off with feathers. But in the urine process just described, the sublimation is a more sophisticated type. It is noteworthy that from 150 gallons of urine only two or three ounces of the precious sublimate of hormone crystals were yielded.

The Chinese stumbled upon a discovery which Europeans made only in this century, namely that steroid hormones are stable at temperatures below their melting-points, and can be successfully sublimed at temperatures varying between 130° and 210°C. There is no decomposition of steroids up to 260°, and many of them will sublime successfully at temperatures up to 300°C. Not all the Chinese methods involved sublimation, but of those that did, the process would have taken place at temperatures carefully controlled between 120° and 300°C – the very temperatures modern scientists now know to be appropriate. By this means, the hormone substances were separated from a large amount of extraneous matter and rendered relatively pure. They would still have contained amounts of some other substances, including cyanuric acid; but this has no known effects and would simply have been neutral in the preparation. Small amounts of other matter, such as indol, skatol, mercaptans, volatile fatty acids and non-steroid phenols would have mattered little, as they are all non-toxic. The hormone crystals obtained by the Chinese would not have been pure according to our modern standards, but they were without doubt highly effective, concentrated hormone substances with pronounced biological effects on patients.

LEFT (94) Often mistaken for lamps, elegant gilt bronze objects such as this, dating from the second century BC, are actually single-tube subliming vessels used for chemical and alchemical purposes, such as the preparation of sex and pituitary hormones. This particular sublimer was more probably used in preparing alchemical elixirs involving mercury, however, since it is so ornamental. The manner in which the subliming tube is disguised as the capacious right-hand sleeve of a young woman is ingenious, to say the least. This type of sublimer is called a rainbow vessel, or *teng*. It formed part of the dowry of Princess Tou Wan, and was found in her husband's tomb in 1968: he died in 113 BC. (Hopei Provincial Museum, Tientsin.)

The sublimation techniques of the ancient Chinese were crucial in many, though not all, of the processes of hormone extraction from urine. In China sublimation equipment was mainly used for the extraction of mercury from the mineral cinnabar, large deposits of which occur in various parts of the country. Mercury was much used for various purposes, most importantly as the essential ingredient of many elixirs of immortality. (Since mercury is a deadly poison, this resulted in many premature deaths of those who thought they were extending their lives.) Another common use of sublimation was in the extraction of camphor from chopped-up pieces of camphor wood. Small camphor sublimatories were apparently quite common in China, and featured side-tubes which arched in such a way that the sublimatories were often called 'rainbow vessels'. Archeologists often confuse them with lamps. A beautiful example may be seen in Plate 94 (above), of a single side-tube disguised as the arm and sleeve of a young girl. This gilt bronze object was excavated in a prince's tomb and can be dated precisely to 113 BC. Objects similar to this were used for the preparation of the sex hormones from urine.

Another astonishing technique used in extracting the hormones was the use of chemicals to precipitate the hormones out of the urine. Gypsum (calcium sulphate) was used in this way, a technique which probably was derived from the bean curd industry, as gypsum is used in its production. But the most surprising and impressive substance used to precipitate the hormones out of solution was 'the juice of soap beans'. This was an extract from a saponin-containing plant, the beans of *Gleditschia sinensis*. The saponins, or natural soap, and also the proteins of the soap-beans, had remarkable success in precipitating the hormones out of the urine into sediment. Adolf Windaus only discovered in 1909 that such substances can be precipitated by natural soaps (in his case, one called digitonin). Once again, the Chinese anticipated a modern scientific discovery by many centuries. (The earliest published method using natural soaps in this way dates from 1110, though the technique may be centuries older than that.)

There were other methods of extracting the hormones from urine which involved neither sublimation nor forced evaporation by fires. These used natural evaporation by the heat of the sun, after the addition of distilled water (sometimes called 'autumn dew water').

A great deal of attention was paid to whether the urine came from males or females, and it was mixed in varying quantities according to which kind of hormone one wanted, so that the effect was inevitably to produce 'autumn mineral' predominantly of androgens (male hormones) in some cases and predominantly of oestrogens (female hormones) in others.

In addition to the sex hormones which were isolated, there were pituitary hormones. The pituitary gland is a tiny gland in the brain which is the master gland for the entire body. It secretes a wide variety of hormones which activate or de-activate the body's other glands. Although it is only about the size of a pea, the pituitary is the most important gland we have. Its anterior lobe secretes hormones called gonadotrophins, which stimulate the gonads (sex glands). These occur in the urine, and some of the Chinese processes would have isolated gonadotrophins as well as steroids produced by the sex glands themselves. The gonadotrophins themselves actually stimulate the production of steroids by the gonads, so that by giving patients 'autumn mineral', the ancient Chinese doctors were giving double treatment – the steroids themselves and also stimulants for the patient to produce more of his own steroids.

The Chinese were predisposed by their manner of thought and a long tradition to view the body as producing powerful biologically active materials. There was an entire branch of Chinese alchemy which was devoted exclusively to this. On the one hand, there was the *wai tan* school of alchemy which attempted to produce elixirs of immortality, from minerals such as cinnabar, by straightforward chemical means in the laboratory. But there was on the other hand the more esoteric school of alchemy, the very existence of which was masked by veiled references and unknown to all but the initiated – the *nei tan* school. It pursued the achievement of immortality by transformations of bodily substances through various physiological techniques. (A simple example was the practice of refusing to spit, since saliva was considered to be a precious bodily fluid which must not be wasted.)

Prominent in *nei tan* alchemy were various peculiar sexual practices. Asceticism was never popular in China, and Taoist monks were so far from being celibate that they pursued extravagant careers of sexual intercourse as part of their devotions. Taoist nuns were anything but virgins. The *yang* and *yin* ideas in China led to a prominence of sexual activity in Chinese culture which contemporary Westerners would have found incomprehensible, having been brought up in religious traditions where sex was generally viewed as sinful.

It is important to realize the essential and central role of sex in Chinese life and culture in order to view in proper perspective the Chinese achievement in isolating and using the sex hormones and gonad-stimulating pituitary hormones as medicines. Far from shunning sexual matters, the Chinese thought about sex and practised sexual activity far more than Westerners. It is doubtful whether, in the West, it would have been considered proper or even admissible to pursue the isolation of such substances from urine, even had the idea occurred to anyone. The mass-production of sex hormones in the West would at most points in history simply have been unacceptable.

How did the Chinese use these hormones? They were used to treat a wide variety of ailments relating to the sex organs. Among those conditions treated were hypogonadism, impotence, sex reversals (where males spontaneously turned into females or vice versa – a phenomenon well known in ancient China), hermaphroditism, spermatorrhea, dysmenorrhea, leucorrhea, sexual debility, and even apparently stimulating the growth of the beard (since the Chinese knew that men grew beards as a result of having testicles and ceased to do so when castrated). As for the matter of swallowing oestrogens, it is known that they tend to be rendered inactive by the liver, but the Chinese seem to have indulged in such large doses that it is thought that oral consumption of them would have been effective. Perhaps the pituitary hormones were more important than the oestrogens, since they would have stimulated the patient's own oestrogens. There are still a number of unanswered questions about the ancient Chinese hormones. For instance, no one has yet tried to reproduce the processes of extraction described in the Chinese texts, to see just what is actually produced. It is clearly a field for experimentation and investigation in the future. But there is no doubt whatsoever that the Chinese actually founded the science of endocrinology, and are entitled to the credit for that great achievement.

DEFICIENCY DISEASES

THIRD CENTURY AD

By the end of the nineteenth century in the West, medical men had come to realize that many diseases, such as beriberi, scurvy and rickets, were deficiency diseases. That is, they were caused by the lack of certain items in the diet. In this century, we have identified the missing dietary constituents as vitamins.

But the Chinese were aware of deficiency diseases centuries before the West. This awareness goes back into indefinite antiquity, and seems to have been a natural attitude based on the Chinese view of balance in nature, and consequently of balance in diet. As early as the fourth century BC there were Imperial Dietitians. But textual evidence of overt awareness of deficiency diseases commences in the book, circa 200 AD, of the famous doctor Chang Chi (often called 'the Galen of China'), entitled *Systematic Treasury of Medicine*. Chang gives vivid accounts of deficiency diseases and suggests dietary treatments. Although he did not have a knowledge of vitamins, the food he recommends would have been rich in the vitamins necessary to cure the patients. He presumably knew which foods to recommend as a result of much trial and error over the years.

The well-known literary figure Han Yü (762–824) observed in one of his essays that the disease beriberi (which is caused by deficiency of Vitamin B1) was more prevalent south of the Yangtze river than north of it. A twentieth-century study of the incidence of beriberi in China found that this was quite true. The observation of Han Yü twelve hundred years earlier was thus verified in modern times. The normal human daily requirement of Vitamin B1 is 300 to 350 international units. North of the Yangtze the study found an average intake of 450 to 690, but south of the Yangtze it was only 250 to 322. This was because in the southern area people eat rice, which when polished and washed has no husk and hence no Vitamin B1 whereas in the northern area people eat wheat and millet, and thus consume plenty of the vitamin.

The most prominent author on the subject of dietary deficiencies and diseases was Hu Ssu-Hui, who was Imperial Dietitian between 1314 and 1330. He wrote a book called *The Principles of Correct Diet* which is the Chinese classic in this field, and is collected from previous works, many of which have since been lost. In this book, Hu describes the two types of the deficiency disease beriberi (recognized today as the 'wet' and 'dry' types), and the remedies which he proposes are essentially diets rich in Vitamin B1, as well as many other vitamins:

> **For the cure of 'dry' beriberi:**
>
> **Cook one big carp with half a pound of small red beans, two-tenths of an ounce of *ch'eng*-fruit skin, two-tenths of an ounce of small dried peppers, and two-tenths of an ounce of dried grass seed. Let the patient eat it.**
>
> **For the cure of 'wet' beriberi:**
>
> **a) Make a soup of rice and horsetooth vegetable, and let the patient drink it on an empty stomach early in the morning.**
>
> **b) Cook 16 ounces of pork with one handful of onion, three dried grass seeds, pepper, fermented beans and rice (half a pound) and let the patient eat it in the morning.**

LEFT (95) Deficiency diseases were recognized in China by at least the third century AD. Here we see the frontispeice to the classic discussion of the subject, *Principles of Correct Diet* by Hu Ssu-Hui, published in 1330. The picture shows nutrition specialists giving a consultation to a patient. The caption reads: 'Many diseases can be cured by diet alone.'

In Plate 90 (page 138) we see the frontispiece of Hu's book, published about 1330. Two dietitians are seen in conference about a patient's diet. The motto in the upper right hand corner states: 'Many diseases can be cured by diet alone.'

DIABETES

SEVENTH CENTURY AD

Diabetes was originally called *hsiao k'o* in China, which means 'dissolutive thirst'. This was very appropriate, because diabetics have an unnatural thirst and pass vast amounts of urine. In the *Yellow Emperor's Manual* of the second century BC, diabetes is described at length. And the book pertinently says: 'A patient suffering from this disease must have been in the habit of eating many sweet delicacies and fatty foods.' Even from this period, the Chinese showed astonishing diagnostic acumen about diabetes.

We do not know when the Chinese first noticed that diabetics had excess sugar in their urine, but this was mentioned in the seventh century AD by the physician Chen Ch'üan, who died in 642. His book, *Old and New Tried and Tested Prescriptions*, is apparently lost, but key passages are quoted in a later book by Wang T'ao, dating from 752 and entitled *Important Medical Formulae and Prescriptions Now Revealed by the Governor of a Distant Province.* Wang quotes Chen as follows:

> **The *Old and New Tried and Tested Prescriptions* says that there are three forms of diabetic affection. In the first of these the patient suffers intense thirst, drinks copiously, and excretes large amounts of urine which contains no fat but flakes looking like rolled wheat bran, and is sweet to the taste. This is diabetes (*hsiao k'o ping*). In the second form the patient eats a great deal but has little thirst.... In the third form there is thirst but the patient cannot drink much; the lower extremities are oedematous [swollen with excessive tissue fluids] but there is wasting of the feet, impotence and frequent urination.**

The first form above is the common *diabetes mellitus*, and the second is the form of it where the patient eats vast quantities of food. As for the last form, this may simply refer to diabetes in obese patients, for obesity is a complicating factor for diabetics. The reference to the feet may be an observation of the problem diabetics have with poor circulation: if the diabetes gets too serious or if the sufferer wears shoes which are too tight and indulges in the use of hot-water-bottles on his feet, he can develop complications so serious in terms of boils, inflammations and so on that these can even lead to gangrene.

Also in the seventh century, the physician and bureaucrat Li Hsüan wrote an entire monograph on diabetes, and attempted to explain the reason for the sweetness of the urine in diabetic patients. He wrote:

> **This disease is due to weakness of the renal and urinogenital system. In such cases the urine is always sweet. Many physicians do not recognize this symptom ... the cereal foods of the farmers are the precursors of sweetness ... the methods of making cakes and sweetmeats ... mean that they all very soon turn to sweetness.... It is the nature of the saline quality to be excreted. But since the renal and urinogenital system at the reins is weak it cannot distil the nutrient essentials, so that all is excreted as urine. Therefore the sweetness in the urine comes forth, and the latter does not acquire its normal colour.**

Another seventh-century physician, Sun Ssu-Mo, wrote about the year 655 in his book *A Thousand Golden Remedies* that with diabetes, 'three things must be renounced, wine, sex, and eating salted, starchy cereal products; if this regimen can be observed, cure may follow without drugs.'

Thus, by the seventh century AD, the Chinese had published their observations on the sweetness of urine of diabetics, tried to come up with an explanation for it, and proposed a dietary regimen for control of diabetes which was not far from the modern method, of avoiding alcohol and starchy foods.

By 1189, in *Medical Discourses*, the physician Chang Kao also noted the importance of skin care in diabetics and the danger of the slightest skin lesions: 'Whether or not such patients are cured, one must be on the watch for the development of large boils and carbuncles; should these develop near the joints the prognosis is very bad. I myself witnessed my friend Shao Jen-Tao suffering from this disease for several years, and he died of the ulcers.'

It should be mentioned that many of the cases of diabetes which occurred in Chinese history, especially

LEFT (96) Jujube dates (*Zizyphus vulgaris*). Their crushed pulp made the body of the pills in which patients swallowed thyroid hormone to treat goiter in the seventh century.

of prominent people, were evidently caused by metallic poisoning. This was yet another of the dangers of taking the notorious elixirs of immortality, which tended to be full of lead, mercury, and even arsenic.

The sweetness of the urine of diabetics was also known to the Indians, though it is difficult to date the Indian texts, unlike the Chinese. The sweetness of the urine of diabetics was only discovered in Europe about 1660 by Thomas Willis, and published in 1679. In 1776, Matthew Dobson identified this sweetness with sugar, and only in 1815 was the sugar specified as glucose. In terms of their anticipation of the West, the Chinese were ahead of Europeans by over a thousand years in identifying and attempting to control diabetes, though they never connected the disease with the pancreas or had any knowledge of insulin (first isolated in 1921 in the West). It was quite an achievement for the seventh century for the Chinese to arrive at this insight: 'All those who pass urine that tastes sweet but has no fatty flakes floating on it are suffering from diabetes.'

USE OF THYROID HORMONE

SEVENTH CENTURY AD

Goiter is an enlargement of the thyroid gland, seen as a swelling in the neck. By the seventh century AD at the latest, the Chinese were using thyroid hormone to treat goiter. This practice may have started earlier, but in the seventh century Chinese physicians described the technique in writing. The first, Chen Ch'üan, died in 642 AD. His book, *Old and New Tried and Tested Prescriptions*, which some attribute to his younger brother, Chen Li-yan, gives three uses for the hormone obtained from thyroid glands of gelded rams. In one prescription, he recommends that the physician wash one hundred thyroid glands in warm water, remove the fat, dry them and chop them up. He should then mix them with jujube dates and make them into pills to be swallowed by the patient.

In another prescription, he suggests that a single thyroid gland be removed from a sheep, the fat taken off, and the gland put raw into the mouth. It should be sucked by the patient until all the juice possible has been extracted and swallowed. Then the gland itself should be eaten.

Since the goiter may on occasion be a tumour, rather than an enlargement of the thyroid gland caused by gland malfunction, it is necessary to distinguish between the two conditions. This the Chinese did at the time. A near contemporary of Chen Ch'üan, Ts'ui Chih-T'i, who flourished about 650 AD, wrote a book in which he clearly distinguished between solid neck swellings which could not be cured (tumours) and movable ones which could be cured (real goiter).

Wang Hsi used the thyroid glands of various animals, including pigs and sheep, for the treatment of goiter. His method involved air-drying the glands to

ABOVE (97) Sargassum seaweed. It was used to treat goiter in the first century BC, according to the ancient book *Classical Pharmacopoeia of the Heavenly Husbandman*. Seaweed, with its high iodine content, is very effective for goiter. Knowledge of this reached Europe in the twelfth century from China.

reduce them to powder, which was to be taken every night in cold wine. He mentions drying fifty pigs' thyroid glands at once for such a preparation. A contemporary of his with the same surname, Wang Ying, wrote a book in which he gave a prescription for extracting the active principles of seven pigs' thyroid glands in wine, evaporating down the result, and then taking it mixed with dew. But would thyroid hormone from animals be of use to humans? It seems so. Insulin from animals treats human diabetics today. *The Great Pharmacopoeia* of 1596 states that it does not matter from which animals the thyroid glands are taken, as thyroid glands have the same function in all animals. The glands of pigs, sheep, water buffalo and sika deer were all in use by that time. And despite their animal source, they appear to have been effective with people.

The transmission of the knowledge of seaweed as a treatment for goiter reached the West from China along with the knowledge of the magnetic compass, the stern-post ship's rudder, and paper-making. It is thought that this entire body of knowledge was transmitted by sea during the twelfth century AD. This use of seaweed is first mentioned in a Western source by Roger of Palermo, in his *Practica Chirurgiae* of 1180, who suggested the use of ashes of seaweed as a medicine for goiter.

Western understanding of goiter was about two-and-a-half millennia behind China's. About the beginning of the nineteenth century, Westerners began to realize that there were environmental causes of goiter to do with the nature of the water and the soil. Awareness of this, however, is found in a Chinese book dating from 239 BC, *Master Lu's Spring and Autumn Annals*, in which it is stated that 'in places where there is too much light water there is much baldness and goiter'.

It is only in 1860 that Chatin proved clearly that goiter was related to a lack of iodine in the soil and water. Iodine was discovered in the thyroid gland in 1896 by Baumann. Murray and others began to administer thyroid extract to patients in the year 1890.

IMMUNOLOGY

TENTH CENTURY AD

Traces of smallpox have been found on an Egyptian mummy of the twentieth dynasty, so the disease was with us for a long time. The disease was called smallpox in Europe to differentiate it from 'grand pox', or syphilis of the skin. The official name of the disease is *Variola*.

It was under the desperate pressure to escape from the disease that humanity was introduced to that most fundamental of medical treatments, which has saved far more millions of lives than were ever lost to smallpox: vaccination. And this breakthrough occurred in China, from where it spread to Europe and led to the modern science of immunology.

The origins of inoculation against smallpox in China are somewhat mysterious. We know that the technique originated in the southern province of Szechuan. In the south-west of that province there is a famous mountain called O-Mei Shan which is known for its connections with both Buddhism and the native Chinese religion of Taoism. The Taoist alchemists who lived as hermits in the caves of that mountain possessed the secret of smallpox inoculation in the tenth century AD. How long before that they had it we shall never know.

The technique first came to public attention when the eldest son of the Prime Minister Wang Tan (957–1017) died of smallpox. Wang desperately wished to prevent this happening to any other members of his family, so he summoned physicians, wise men and magicians from all over the Empire to try to find some remedy. One Taoist hermit came from O-Mei Shan, described variously as a 'holy physician', a 'numinous old woman' (in which case a nun), and a 'ouija board immortal' (ouija boards, or planchettes, were widely used in China, where whole books were written through 'spirit dictation'). This monk or nun brought the technique of inoculation and introduced it to the capital.

One account describes the hermit as a 'three-white adept of the school of the ancient immortals', and we may safely assume he or she was one of the Taoist alchemists specializing in 'internal alchemy', by which the elixir of immortality was meant to be concocted not in a laboratory but inside the adept's own body. The elaborate techniques used for this led to the discovery of the sex and pituitary hormones in human urine (see page 141). Inoculation is thus another esoteric product of the quest for immortality. And it certainly did give the gift of life to many.

Inoculation has certain dangers, as well as certain advantages, which set it apart from the later technique of vaccination. When one is inoculated, one has the live virus inserted into one's body. When the process is successful, one is certainly immune for life. But the process can simply be one of direct exposure to the disease, so that one ends up with smallpox. With vaccination, the immunity conferred is only temporary, so that vaccinations have to be given every few years as 'boosters'. This is because vaccination uses dead viruses or some other kind of denatured virus (perhaps a related one) which cannot actually give one the disease.

At first sight it looks as if inoculation against smallpox must have been madness. Were not people just being given smallpox every time? The answer is no. And here we find the subtlety of the Chinese inoculators to be truly astounding. They practised a variety of methods for the attenuation of the deadly virus, so that the chances of getting the disease were minimized and the chances of immunity were maximized.

First of all, there was a strong prohibition against taking the smallpox material from people who actually had the disease. It was recognized that this would simply transmit it. They conceived of the inoculation as a 'transplant' of poxy material imagined as being like beansprouts which were just germinating. 'To inoculate' in Chinese was called *chung tou* or *chung miao*, meaning 'to implant the germs', or 'implant the sprouts'.

The method used was to put the poxy material on a plug of cotton, which was then inserted into the nose. The pox was thus absorbed through the mucous membrane of the nose and by breathing. (The technique of scratching the skin and putting the pox on the scratch seems to have developed long afterwards, possibly in Central Asia as the technique spread westwards.)

Ideally, inoculators chose poxy material not from smallpox patients but from persons who had been inoculated themselves and had developed a few scabs. They also knew the difference between the two types of smallpox, *Variola major* and *Variola minor*, and they chose poxy material from the latter, which was a less virulent form. Indeed, the favourite source of poxy material was from the scabs of someone who had been inoculated with material from somebody who had been inoculated with material from somebody who had been inoculated.... In other words, a several-generations attenuation of the virus through multiple inoculations.

But there were other artificial methods used to attenuate the virus even further, so that it would be safer still. Here is one account from a work on *Transplanting the Smallpox* by Chang Yen in the year 1741:

> Method of storing the material. Wrap the scabs carefully in paper and put them into a small container bottle. Cork it tightly so that the activity is not dissipated. The container must not be exposed to sunlight or warmed beside a fire. It is best to carry it for some time on the person so that the scabs dry naturally and slowly. The container should be marked clearly with the date on which the contents were taken from the patient.
>
> In winter the material has *yang* potency within it, so it remains active even after being kept from thirty to forty days. But in summer the *yang* potency will be lost in approximately twenty days. The best material is that which had not been left too long, for when the *yang* potency is abundant it will give a 'take' with nine persons out of ten; but as it gets older it gradually loses its activity, giving perhaps a 'take' with only five out of ten people – and finally it becomes completely inactive, and will not work at all. In situations where new scabs are rare and the requirement is great, it is possible to mix new scabs with the more aged ones, but in this case more of the powder should be blown into the nostril when the inoculation is done.

Needham comments on this and similar passages:

> Thus the general system was to keep the inoculum sample for a month or more at body temperature (37°C) or rather less. This would certainly have had the effect of heat-inactivating some 80 per cent of the living virus particles, but since their dead protein would have been present, a strong stimulus to interferon production as well as antibody formation would have been given when inoculation was done.

In other words, 80 per cent of the smallpox viruses with which the Chinese were inoculated would have been dead ones which could not have given anyone

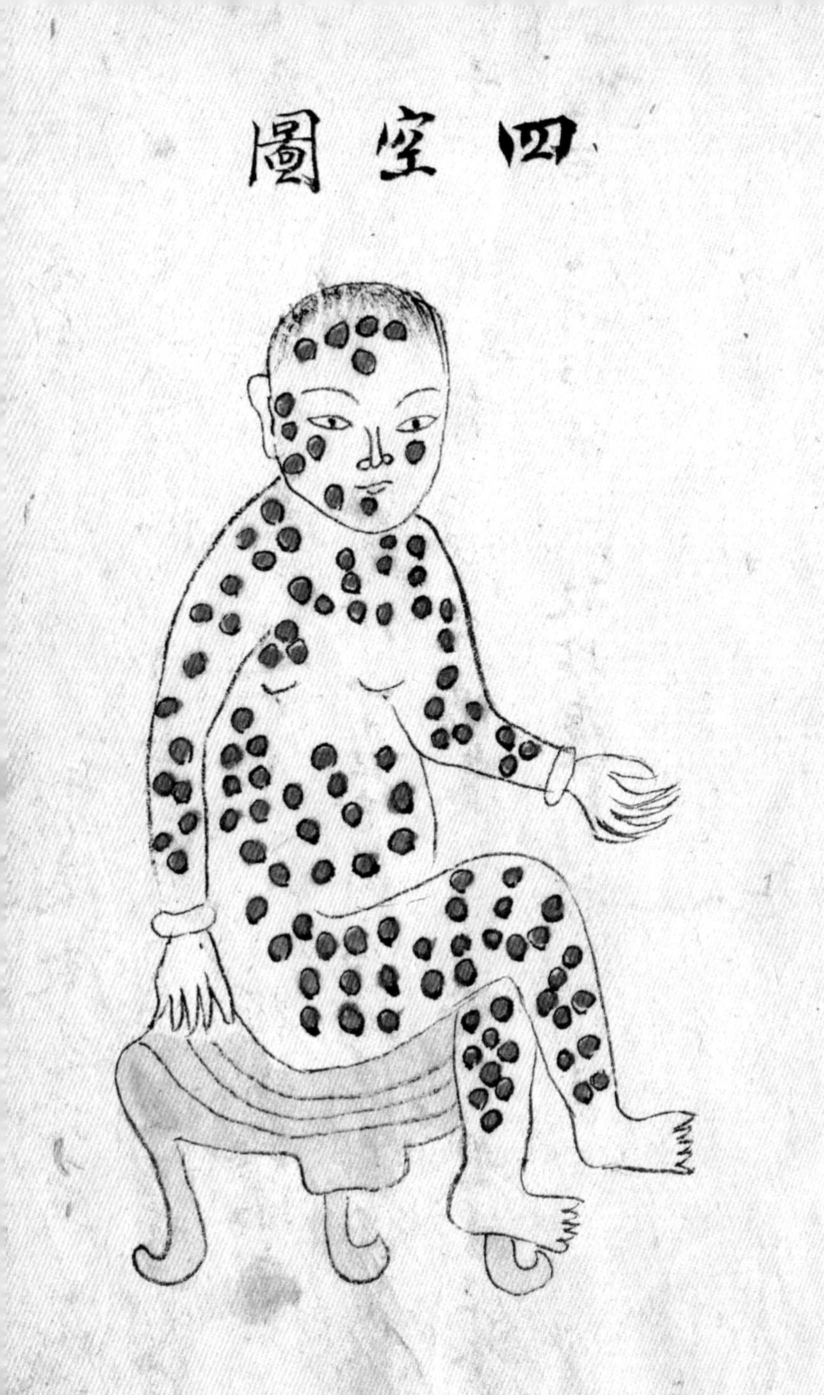

LEFT (98) This watercolour of a child with smallpox is from a late seventeenth- or eighteenth-century manuscript entitled *Methods of Treating Smallpox*. This apparently unpublished work by an unknown author appears to have been used as a source for the book published later (in 1743), entitled *Golden Mirror of Medicine*. The three Chinese characters read, 'Drawing of the Four Emptinesses/Voids/Holes', and must refer to some technical medical terminology of that time, as the meaning is otherwise unclear, unless it refers to the poxes covering all four limbs. (Collection of Robert Temple.)

smallpox. Instead, they would (as with vaccination) have stimulated the body to produce antibodies against smallpox, as well as the substance interferon which assists the immune system in general. So only about 20 per cent of the poxy material was 'live', and that was in as attenuated a form as it was possible to obtain, and of the *Variola minor* variety. Thus, traditional Chinese smallpox inoculation was about as safe as it could be, and every conceivable dodge was used to minimize the risk of its actually giving anyone smallpox.

Inoculation against smallpox in China did not become widely known and practised until the period 1567–72, according to the author Yü T'ien-Ch'ih. Vivid descriptions of the practice are recorded by Yü Ch'ang in his book *Miscellaneous Ideas in Medicine*, of 1643.

During the seventeenth century, the practice spread to the Turkish regions, and it was there that it came to the attention of Europeans. The wife of the British Ambassador to Constantinople, Lady Mary Wortley Montagu (1689–1762), allowed her family to be 'variolated' in 1718. Four years before this, E. Timoni had published an account of the practice in the *Philosophical Transactions of the Royal Society* in London, and two years after that, J. Pilarini published a further account in the same source. So the process was being much discussed in London and Lady Mary must have been encouraged by that to take her bold step. By 1721, variolation (called then 'engrafting') began to be widely practised in Europe as protection against smallpox. We owe to this transmission from China the later developments of vaccinations and the science of immunology itself.

Part 6 MATHEMATICS

THE DECIMAL SYSTEM

FOURTEENTH CENTURY BC

The decimal system, now fundamental to modern science, originated in China. Its use can be traced back to the fourteenth century BC, the archaic period known as the Shang Dynasty, though it evidently was used long before that.

Evidence of the use of the decimal system has also been tentatively reported from the ancient Harappan, or Indus Valley, civilization, which was situated on the border of modern India and Pakistan and existed earlier than China's Shang Dynasty. However, the lack of written records from that most ancient Indian culture may mean that we can never be certain of this, and whether it may thus have preceded the earliest use of decimals in China. We will also never know whether the system may have spread from one to the other or whether it evolved independently in both. Too little is also known about the shadowy Chinese dynasty prior to the Shang, and when it comes to that era in the whole of Asia, we are really lost in the mists of time. The only certainty which we have, therefore, is the extreme antiquity of the decimal system in China, and that it goes back to the earliest recorded period.

An example of how the ancient Chinese used the decimal system may be seen from an inscription dating from the thirteenth century BC, in which '547 days' is written 'Five hundreds plus four decades plus seven of days'.

From these early times, then, Chinese mathematics had the great advantage of using decimal place value in the expression of numbers and the carrying out of computations. One reason for this may be that the Chinese wrote in characters rather than using an alphabet. With an alphabet, which is inevitably more than just nine letters, there is the temptation, when using the letters to represent numbers, not to stop after 'nine', but to go on. But when that is done, there can be no decimal system, for it means giving 'ten' its own symbol rather than using the symbol for 'one' and moving it into a new column. 'Eleven' will also have its own symbol, so that it cannot be expressed as 'one ten and one unit', and so on. The ancient Greeks used their first letter, *alpha*, to represent 'one'; but they did not stop with 'nine', which was the letter *theta*; they continued, and for 'ten' used the letter *iota*.

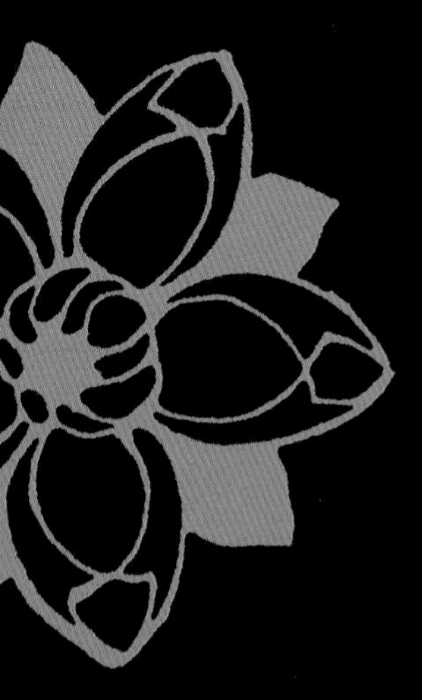

In computation, the Chinese used counting rods on counting boards. To 'write' ten involved placing a single rod in the second box from the right, and leaving the first empty, to signify zero. To change the ten to eleven, a single rod was added in the first box. To 'write' 111, single rods were placed in the first, second and third boxes. Apparently from the earliest times, the decimal place system for numbers was literally a *place* system; the Chinese *placed* counting rods into actual boxes.

The fact that the decimal system existed from the very beginnings of mathematics in China gave the Chinese a substantial advantage, laying a foundation for most of the advances they later made. It was an advantage lacking in the West. The first evidence of the proper use of decimals in Europe is found in a Spanish manuscript of 976 AD, approximately 2300 years later than the earliest Chinese evidence.

OPPOSITE (99) A modern Chinese bank, where the traditional methods of computation by abacus are still used. The Chinese used decimal mathematics for at least 2300 years before the system was adopted in the West, and invented the negative numbers so dreaded on today's bank statements.

ABOVE (100) This large, sturdy nineteenth century abacus, strengthened with brass fittings and made of strong teak, would have been used in a bank, large business, or official office, enduring heavy daily use as the 'office calculator'. It is can be used very rapidly (often faster than one's fingers could enter the same numbers into a modern electronic calculator), and the result can instantly be seen at a glance from the positions of the counters. (Collection of Robert Temple.)

A PLACE FOR ZERO

FOURTH CENTURY BC

Most of us take zero for granted. But if it were not used every day by millions of people all round the world, modern technology would collapse and be unable to function. Zero is essential to the efficient carrying out of mathematical computations. Its invention, therefore, was of the utmost importance in history. But when we speak of its invention, we may speak of two different things. First, a blank space was left for zero. This was obviously the far more important step forward. Later, an actual symbol for zero was written in the blank space. But this was largely a formality. For, once the space was there, it hardly mattered whether or not it had something written in it.

ABOVE (101) Frontispiece to Ch'eng Ta-Wei's *Systematic Treatise on Arithmetic*, published in 1593. The picture shows 'discussions on difficult problems between Master and Pupil', with the Master seated in front of a traditional Chinese counting-board, as described in the account of the decimal system. A natural consequence of the counting-board method was the leaving of a blank space for zero. We have evidence of this as a working concept by the fourth century BC, though it must have existed long before. The actual symbol '0' for zero is also thought to have originated in China, as a representation of the blank space.

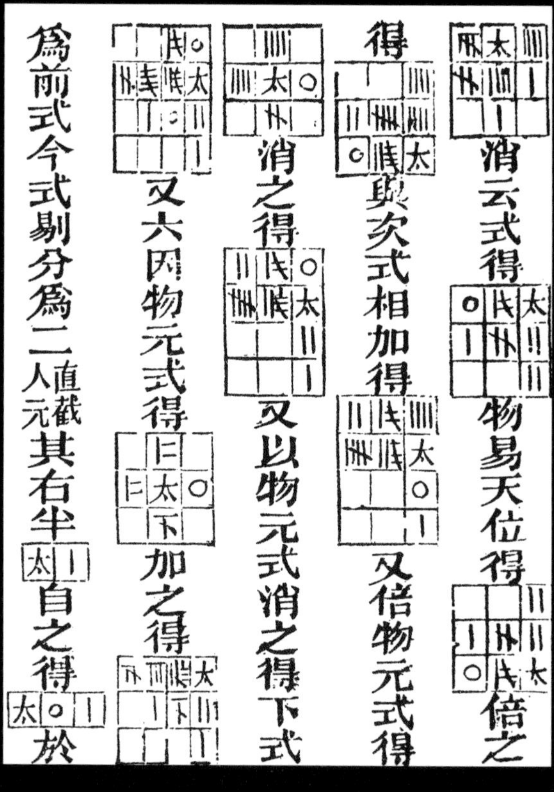

ABOVE (102) Both zeroes and negative terms occur in this page from Chu Shih-Chieh's book on algebra, *Precious Mirror of the Four Elements*, published in 1303. Each box, consisting of a group of squares containing signs, represents a 'matrix' form of writing an algebraic expression. The frequent occurrence of the sign '0' for zero may be clearly seen. (In these cases it means that terms corresponding for those squares do not occur in the equation.) The diagonal lines slashed through some of the numbers in the squares indicate that they are negative terms. (The number 'one' is one vertical line, the number 'two' is two vertical lines, etc.)

We know for certain that it was the Chinese who invented the use of a blank space for zero. As for the actual symbol '0', we suspect that it too first began in China, but the evidence is less clear for that. The blank space on the Chinese counting board, representing zero, dates at least as far back as the fourth century BC. Its use was perfectly simple. The number 405, for instance, would be 'written': 'four blank five', or 'four hundreds, no tens, five units'.

The traditional story for the origin of the symbol '0' for zero, as told in the West, is that it was invented in India in the ninth century AD. It may be seen in an inscription at Gwalior dated 870 AD. However, in actuality, the zero can be traced back earlier than this. It may be seen in inscriptions in Cambodia and Sumatra, both dated 683 AD, and on Bangka Island, off Sumatra, dated 686 AD. Experts believe that these inscriptions, which antedate the use of the zero in India, indicated that the zero came to India from China by way of Indo-China.

For many centuries the Chinese seem to have felt no need for an actual symbol for zero. On the counting board, the blank space was entirely adequate. It was the Chinese tradition not to include calculations in mathematical treatises, but merely to give problems and their solutions with little or no indication of the methods used; thus it was at first only in the actual recording of numbers that the need arose, and then, simply, a blank was left. When this was eventually seen to be insufficient, it is probable that the zero symbol evolved as a circle drawn round the blank space.

Absolute priority for the zero symbol is not claimed for China, since its first appearance in print is not until 1247, though there are firm grounds for believing it was used at least a century earlier. No one knows, and perhaps no one will ever know, where the actual symbol was first used, or when.

NEGATIVE NUMBERS

SECOND CENTURY BC

Westerners had a great mental struggle to accept negative numbers. These are numbers with a minus sign in front of them. To the ordinary man in the street who is not concerned with mathematics, negative numbers may still be insignificant or impossible. 'How can minus ten exist?' someone might say. 'Or if it does, it has no relation to the real world.'

But this is not true. Minus ten dollars in your bank account is real enough: you will have to pay in eleven dollars before you have a dollar in credit. But it is better to be minus ten dollars than minus a hundred. From this example we can begin to appreciate that as soon as anything is expressed in mathematical language, negative numbers become essential for a proper description. And yet before the Renaissance, the West did not have them.

What was lacking in the West, however, was common in China. Negative numbers were recognized and used there by the second century BC. They occurred frequently on the counting board, and were represented by black rods, positive numbers being represented by red ones. Alternatively, square-sectioned rods represented negative numbers while triangular-sectioned ones represented positive numbers. If no special rods were available, a number on the counting board would be represented in a slanting position to set it apart from the positive numbers, as was done by the mathematician Liu Hui in the third century AD.

The law of signs, for plus and minus, was partially stated in a Chinese mathematical classic of the first century AD, but occurs full-blown on numerous occasions in medieval Chinese writings, stated overtly for instance in the book *Introduction to Mathematical Studies* of 1299. The Chinese found no difficulty in the idea of negative numbers. In this, they were seventeen hundred years ahead of the West. It was not until 630 AD that the mathematician Brahmagupta began to use negative numbers in India. In Europe, negative numbers first appear in a book by the Greek mathematician Diophantus (about 275 AD), only to be dismissed as 'absurd' when occurring as the solution of an equation. The first acceptance of negative numbers in Europe was in the mid-sixteenth century when the great Renaissance genius, Jerome Cardan (Girolamo Cardano), published his book on algebra entitled *The Great Art*, in 1545. In the book Cardan recognized the negative numbers he obtained in solutions of various equations, and he stated simply but clearly the laws of negative numbers. For instance, in speaking of square roots, he said 'there are two solutions, one plus and one minus, which are equal to each other'. He called a negative number *debitum*. But the positive numbers occurring as solutions to equations, Cardan called 'true solutions', whereas the negative ones he called 'fictitious solutions'. Therefore, though Cardan fully accepted negative numbers in mathematical operations and was prepared to treat them as normal for the sake of computation and calculation, he did not take the modern view that negative numbers can represent non-fictitious phenomena. Cardan was polite to them, but he did not take them seriously.

EXTRACTION OF HIGHER ROOTS AND SOLUTIONS OF HIGHER NUMERICAL EQUATIONS

FIRST CENTURY BC

David S. Smith, the great historian of mathematics, wrote in reference to the solution of higher numerical equations: 'Indeed, this is China's particular contribution to mathematics.' This is certainly true. China seems to have been the birthplace of such mathematical processes, which did not arise in Europe until the fourteenth or fifteenth century, and even then remained largely undeveloped for some time.

In the first century BC the Chinese mathematical classic *Nine Chapters on the Mathematical Art* was compiled. One of the algebraic problems given in this work is to find the cube root of the number 1,860,867. (The answer is 123.) The method used is similar to 'Horner's method', developed by W.G. Horner in 1819 in Europe.

Horner's process is a simple and elegant method of numerical calculation. It is so fundamental that it is generally used as the standard method of extraction of a square root. It is a technique of estimating a root of an equation by approximation again and again, each time more accurately than in the preceding step. Horner did this by increasing decimals, whereas earlier (in 1767) the Comte Lagrange had done it by continued fractions. Lagrange's method had been more cumbersome because the resulting fraction had to be converted to a decimal. This affords an interesting parallel to what happened in China: a 'Lagrange method' using fractions was developed there by the first century BC (nineteen hundred years ahead of Lagrange), and it was improved to a 'Horner method' in the third century AD by Liu Hui (sixteen hundred years before Horner; see page 142). Their methods were essentially specific applications of those generalized in Sturm's Theorem, which was formulated by J.C.F. Sturm in 1835, and based on their work.

Numerical equations of higher degrees than the third (that is, involving powers higher than cubes) make their appearance in print in China in the year 1245, in the *Mathematical Treatise in Nine Sections*, by Ch'in Chiu-Shao. In this work are included equations involving higher powers, such as:

$$-x^4 + 763{,}200x^2 - 40{,}642{,}560{,}000 = 0$$

It should be said that the Chinese did not write such equations in symbols, but spelled them out. They were, nevertheless, proper equations. In two books published in 1248 and 1259, *Tse Yuan Hai Ching* and *I Ku Yan Tuan,* the mathematician Li Yeh had equations like these:

$$ax^6 + bx^5 + cx^4 + dx^3 + ex^2 + fx + e = 0$$

and:

$$-ax^6 - bx^5 - cx^4 - dx^3 - ex^2 - fx - e = 0$$

As Needham says: 'Early Greek and Indian mathematics seem to have contributed little or nothing to the solution of higher numerical equations.' In the mid-sixteenth century, Nicolo Tartaglia and Jerome Cardan were able to solve the cubic equations. But both of them considered that equations of higher degrees were not relevant to the real world, Cardan believing that algebra was complete with the cubic equations, which related to the third dimension (than which he believed there could be none higher). Cardan's pupil Lodovico Ferrari did, however, work on equations of the fourth degree. But, by and large, Europeans were far less willing to consider higher equations in the sixteenth century than the Chinese were in the thirteenth. The first actual solutions of cubic equations in China were by Wang Hsiao-T'ung in the seventh century AD, and the first in Europe, in the thirteenth century, were by Leonardo Fibonacci, who is thought to have been influenced by Chinese sources.

DECIMAL FRACTIONS

FIRST CENTURY BC

It is hardly surprising that, with the decimal place system so well established in China, decimal fractions first occurred there. The earliest definite traces of them appear in connection with measurements. Decimal systems of measurement implied that when something exceeding a unit of whatever size was encountered, the fraction left over could be expressed in terms of the units smaller by a power of ten. This did not require a great mental leap, and was in fact a more or less routine matter. The system probably

RIGHT (103) Europe's least appreciated scientific genius, Simon Stevin (1548–1620). He began his career as an accountant, later becoming an engineer and mathematician and serving as Quartermaster General to the Dutch armies. Possibly influenced by Chinese sources, Stevin made innovations in the area of higher equations: he insisted on the reality of negative roots and that equations could be complete even with missing terms, and suggested that signs could be attached to numbers. He also gave rules for the solution of equations of any degree. He published the first tables of compound interest of sufficient extent to be useful to bankers and merchants, and introduced double-entry book-keeping to Holland from Italy. But the greatest triumph of his life was his introduction and defence of decimal fractions in 1585, which led to their adoption in Europe. (Artist unknown, Bibliothek der Rijksuniversiteit, Leiden.)

existed several centuries BC. A surviving inscription by Liu Hsin on a standard measure of volume can be precisely dated to the year 5 AD and speaks of a length correct to 9.5 of certain units.

The first occurrence of decimal fractions in a surviving work of mathematical literature is to be found in the writings of Liu Hui in the mid-third century AD. The fractions occur in two contexts: with measurements and in solutions to equations. In his commentary on the classic *Nine Chapters on the Mathematical Art* (first century BC), Liu Hui expresses a diameter of 1.355 feet. The *Nine Chapters* itself had spoken of extracting square roots and getting results which were not integral, that is, which left fractions. But fractions were not good enough for Liu Hui, who was, as Needham says, concerned about what he called these 'little nameless numbers', and stated that the answers should be expressed as a series of decimal places. Needham adds: 'These decimal roots were undoubtedly computed with counting rods and the results expressed in decimal fractions.'

Although developed quite clearly by the third century AD, decimal fractions were not universally adopted in China. The major obstacle to this was that the Chinese were so advanced and skilled at manipulating ordinary fractions that many of them simply did not feel the need for decimal fractions.

But decimal fractions do continually crop up in the literature. The official history of the Sui Dynasty in 635 AD expresses the value of *pi* as the decimal fraction 3.1415927, written in words. The first person to drop the descriptive words and merely write the number as in modern decimal notation was apparently Han Yen at the end of the eighth century.

Full-blown decimal fractions applied to all operations generally, and constituting a genuine system and approach, appeared in the thirteenth century. Two of the mathematicians who were prominent in this development were Yang Hui and Ch'in Chiu-Shao. From the Chinese, the idea of the decimal fraction spread to the Arab al-Kashi who was director of the astronomical observatory at Samarkand, and who died in 1436. And, according to the historian of mathematics D.E. Smith, 'The first man who comprehended the significance of all this [in Europe] ... seems to have been Christoff Rudolff, whose *Exempel-Buechlin* appeared at Augsburg in 1530.' Smith adds that 'the first to show by a special treatise that he understood the significance of the decimal fraction was Stevin, who published a work upon the subject in Flemish, followed in the same year (1585) by a French translation.' Simon Stevin is known to have introduced other Chinese notions into Europe (see pages 217 and 234), so this is not surprising. But what is surprising is how late both the Arabs and the Europeans were at appreciating decimal fractions. Europe lagged behind China by over sixteen hundred years in this.

USING ALGEBRA IN GEOMETRY

THIRD CENTURY AD

Algebra and geometry developed independently. Today, we could not possibly do without their intimate partnership. To deprive ourselves of the use of them *together* would render modern technology impossible at once. But the connection between algebra and geometry was not always obvious. Far from it, in fact.

We now routinely use equations (algebra) to describe shapes (geometry). Everything from buildings to aeroplanes is constructed not just from blueprint drawings but from sets of equations describing the contours, surfaces, and structures. But the first people to do this sort of thing, expressing geometrical shapes by equations, were the Chinese.

A Chinese book of the third century AD called the *Sea Island Mathematical Manual* gives a series of geometrical propositions in algebraic form and describes geometrical figures by algebraic equations. Throughout Chinese history after that, if one wanted to consider geometry, algebra was regularly employed.

These techniques spread westward to the Arabs when the famous mathematician al-Khwarizmi was sent by the Caliph to be ambassador to Khazaria during the years 842 to 847. (Khazaria lay on the main trade routes between China and the West.) The first European who adopted the methods appears to have been Leonardo Fibonacci, who in his *Practica Geometriae* of 1220 used algebra in solving geometrical problems relating to the area of a triangle.

Since the Chinese for so many centuries used algebra to study geometry, why did they not go on to invent analytic geometry, which is the great expression of those principles, in which every geometrical object and every geometrical operation can be referred to the realm of numbers? It was developed in Europe by the mathematicians Pierre Fermat and René Descartes in the seventeenth century. The reason seems to have been that the Chinese, strangely enough, never made the necessary study of conic sections, which gives such basic forms as ellipses, parabolas, and hyperbolas. This was one of the Chinese blind spots. And furthermore, Needham says of the Europeans who did develop analytic geometry that they were 'reasoning from equations to geometrical figures; what the Chinese had always done was to transform geometrical figures into equations'.

It is curious that the Chinese should have been a thousand years ahead with the basic idea but that they should fail to push it home. It has often been said that the key to modern science was the applying of mathematics to every aspect of the physical world – what is called 'the mathematization of nature'. Nowhere did the Chinese come so close as in using algebra to study geometry, and perhaps nowhere was their failure to follow through more crucial in dooming them never to achieve 'modern science'.

折抵地爲弦以句及股弦并求股故先令句自乘見矩
冪令如高而一凡爲高一丈爲股弦并之以除此冪得
差所得以減竹高而半其餘即折者之高也此率與係
索之類更相返覆也亦可如上術令高自乘爲股弦并
冪去本自乘爲矩冪減之餘爲實倍高爲法則得折之
高數也

去根三尺

去根如勾折處
如股折梢如弦
通長如股弦和

股弦和與勾求股法曰勾自乘爲實變股弦較乘股弦
和如股弦和而一正除得股弦較以減股弦和餘二股

LEFT (104) The Problem of the Broken Bamboo, from Yang Hui's book *Detailed Analysis of the Mathematical Rules in the 'Nine Chapters' and Their Reclassification*, published in 1261. Yang Hui was one of China's leading mathematicians, and in his works quadratic equations with negative coefficients appear for the first time. Here the broken bamboo, which forms a natural right-angled triangle, is discussed as an example of the properties of such triangles, their expression in algebra, and the use of such expressions for measuring heights and distances.

THIRD CENTURY AD

The irrational number *pi* can be computed to an infinite number of decimal places. It expresses the ratio of the circumference of a circle to its diameter, a relationship which cannot be framed in terms of whole numbers. (*Pi* is needed to compute the area of a circle or volume of a sphere.) The value of *pi* was computed by Archimedes to three decimal places, and by Ptolemy to four decimal places. But after that, for 1450 years, no greater accuracy was achieved in the Western world. The Chinese, however, made great strides forward in computing *pi*.

One way in which the ancient mathematicians tried to approach an accurate value for *pi* was to inscribe polygons with more and more sides to them inside circles, so that the areas of the polygons (which could be computed) would more and more closely approach the area of the circle. Thus, they could try to find a value for *pi*, since the circle's area was found by using the formula containing it. (They could measure the diameter, and squeeze a polygon whose area they knew into the circle; the only unknown number would be *pi*, which could then be calculated.) Archimedes used a 96-sided polygon, and decided that *pi* had a value between 3.142 and 3.140.

The Chinese tried to sneak up on *pi* in the same fashion, but they were better at it. Liu Hui in the third century AD started by inscribing a polygon of 192 sides in a circle, and then went on to inscribe one of 3072 sides which 'squeezed' even closer. He was thus able to calculate a value of *pi* of 3.14159. At this point, the Chinese overtook the Greeks.

But the real leap forward came in the fifth century AD, when truly advanced values for *pi* appeared in China. The mathematicians Tsu Ch'ung-Chih and Tsu Keng-Chih (father and son), by means of calculations which have been lost, obtained an 'accurate' value of *pi* to ten decimal places, as 3.1415929203. The circle used for the inscribing of the polygons is known to have been 10 feet across. This value for *pi* was recorded in historical records of the period, but the actual books of those mathematicians have vanished over the centuries, and the greatest loss of these is perhaps that of their book *Chuei Shuei*. Nine hundred years later, the mathematician Chao Yu-Ch'in (about 1300 AD) set himself to verify this value of *pi*. He inscribed polygons in a circle with the enormous number of 16,384 sides. By this means he confirmed the value given by the Tsu family.

The Tsu family had a lead in the computation of *pi* of about 1200 years. Even by 1600 AD in Europe the celebrated calculation of the value of *pi* by Adriaen Anthoniszoon and his son only gave 3.1415929, an approximate value extending to seven places, which still fell three short of the value found by the Tsu family. It is not clear when a computation of *pi* in the West actually equalled that of the Tsu family. Abraham Sharp, in about 1717, found the value to 72 places, and a computation to 136 places was made by Georg Vega (1756–1802). Sometime between 1600 and 1700, Europeans appear to have equalled the precision of the Chinese of the fifth century.

Therefore, although the Greeks were the first recorded mathematicians to compute *pi* to four decimal places (in equivalent fractions, since they did not use decimal fractions), the first advanced computations of *pi* beyond that were made in China, and were not equalled in the West for about twelve hundred years.

ABOVE (105) The first advanced knowledge of the value of *pi* originated in China, but was forgotten there in the fourteenth century. When the Jesuits went to China in the seventeenth century, the Chinese were impressed by the European knowledge of *pi*. Here we see a diagram explaining Liu Hui's exhaustion method in 264 AD for finding the value of *pi*. By inscribing 3072 sides of a polygon in a circle, Liu Hui was able to overtake the Greeks and compute the value to a fifth decimal place at 3.14159. By the fifth century, the value was computed to ten decimal places. In Europe, *pi* was only approximately calculated to seven places by the year 1600, a full twelve hundred years later.

'PASCAL'S' TRIANGLE

ELEVENTH CENTURY AD

Blaise Pascal (1623–62) gave his name to a triangular array of numbers such as the one below:

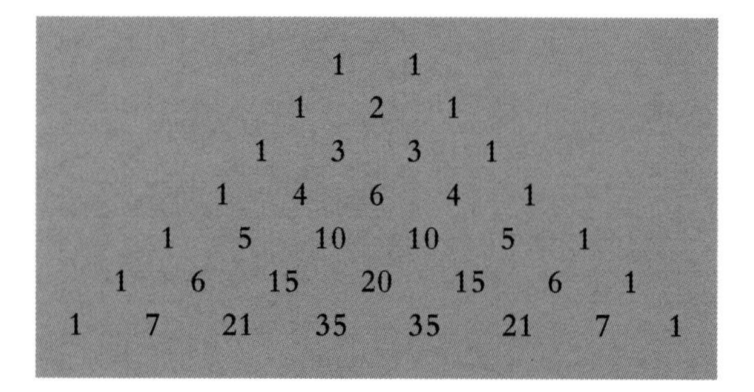

1 1
1 2 1
1 3 3 1
1 4 6 4 1
1 5 10 10 5 1
1 6 15 20 15 6 1
1 7 21 35 35 21 7 1

If you study this array of numbers, you will see that every number in it is equal to the sum of the two numbers above it on either side, except for the 1's of course. Thus, 15 is the sum of the 10 above it to the left and the 5 above it to the right. And 35 is the sum of the 15 above it to the left and the 20 above it to the right.

But 'Pascal's' Triangle is not just an intriguing oddity for people who like to play around with numbers. It actually gives the numerical coefficients (the numbers which go beside algebraic letters) of the series of solutions to the raising to successive higher powers of a binomial. A binomial consists of two numbers added together, represented as (a + b). In the triangle, each successive line across gives the numbers which go with the solutions. Thus, if (a + b) is raised to the power of one, it stays exactly as it is, and the first line gives the coefficients of the result: 1 and 1, for a plus b. But if (a + b) is raised to the second power, meaning that it is squared, (a + b) times (a + b) gives us the answer $a^2 + 2ab + b^2$, and it will be immediately obvious that the numbers beside the letters of this answer are given by the second line across in the triangle, namely, 1, 2, 1. And so on as (a + b) is cubed, then raised to the fourth power, and fifth power, on indefinitely.

Now, it may seem that this too is merely an oddity. But not so. As one raises a binomial to higher and higher powers, one soon can lose one's way and wonder what the numerical coefficients are actually going to be in the solution. But a glance at an extended 'Pascal' Triangle can give the answer instantly, thus providing one with solutions to the problems without having to multiply them out. The Triangle is a wonderful time-saver, and one of the fundamental steps in getting mathematics really on its feet. But although it bears the name 'Pascal's' Triangle, it was by no means invented by Pascal. He merely put it in a newer form in the year 1654. In fact, this Triangle was invented in China. It may be seen depicted in a Chinese book of 1303 AD by Chu Shih-Chieh, entitled *Precious Mirror of the Four Elements*. Even here, it is called 'The Old Method'.

And old indeed it was, for it was known in China by 1100. The mathematician Chia Hsien expounded it at that time as 'the tabulation system for unlocking binomial coefficients'; but its first appearance is thought to have been in a book of that date, now lost, entitled *Piling-up Powers and Unlocking Coefficients*, by Liu Ju-Hsieh.

The mathematician and poet Omar Khayyam discussed the 'Pascal' Triangle somewhat indirectly about 1100. We do not know whether he got it from China or invented the elements of the system independently. But the first appearance of the Triangle in print in Europe was on the title page of the book on arithmetic of Petrus Apianus in 1527. Several succeeding mathematicians, such as Michael Stifel, considered it. And the Italian Nicolo Tartaglia, who was something of a scoundrel, claimed it as his own invention. But as far as we know, the inventor was indeed Liu Ju-Hsieh, 427 years before the appearance of the 'Pascal' Triangle in Europe.

ABOVE (106) Blaise Pascal (1623–1662), French mathematician and scientist after whom 'Pascal's Triangle' is named, because he wrote about it in 1654. It was known in China more than five and a half centuries before this.

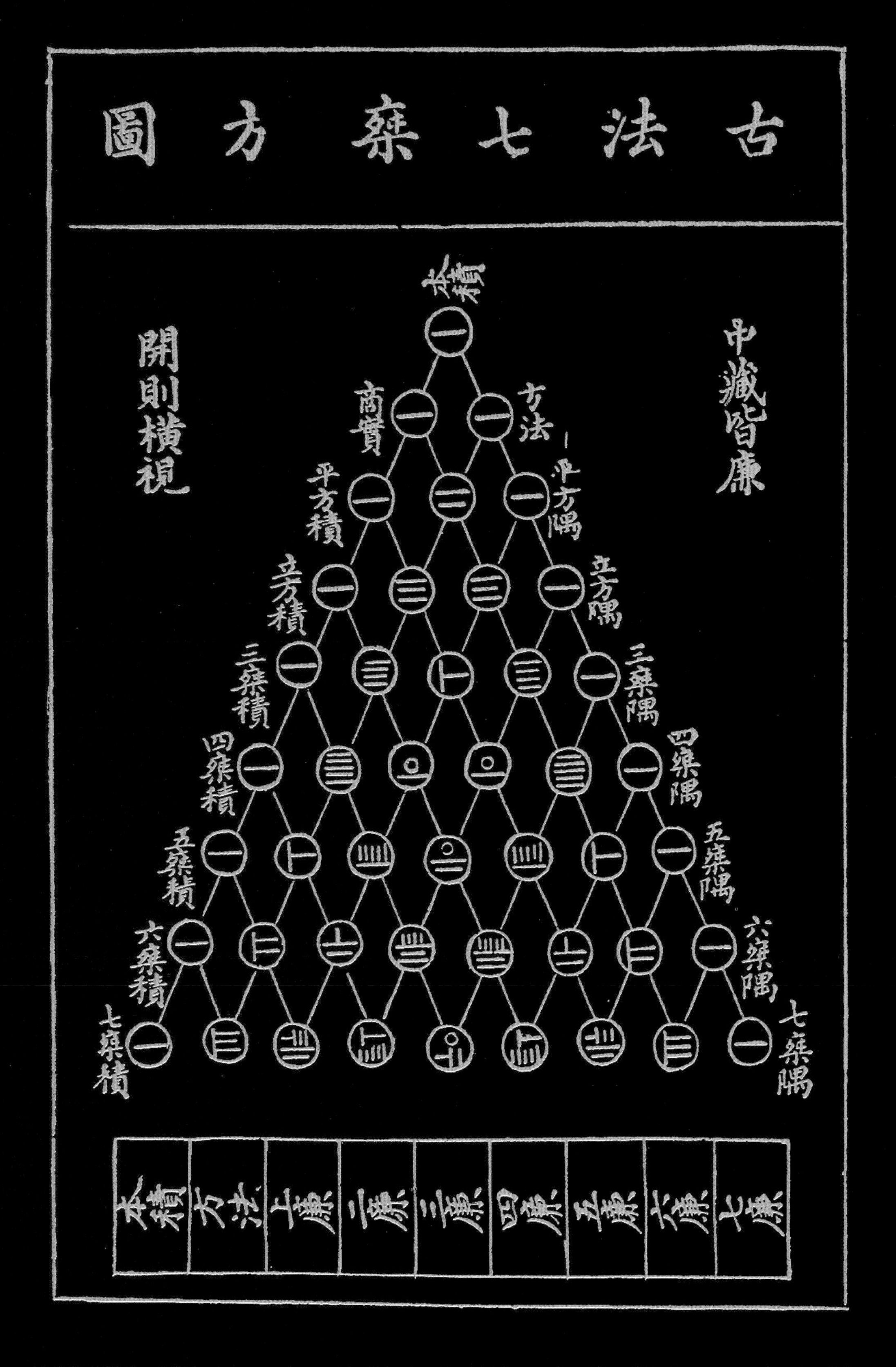

ABOVE (107) 'Pascal's' Triangle was not invented by Blaise Pascal in 1654: it came from China. This diagram comes from Chu Shih-Chieh's *Precious Mirror of the Four Elements*, published in 1303. The caption refers to the triangle as the 'Old Method'; it had been expounded by the year 1100 by the mathematician Chia Hsien, who called it 'the tabulation system for unlocking binomial coefficients'.

Part 7
MAGNETISM

THE FIRST COMPASSES

FOURTH CENTURY BC

Needham has been able to establish that Europe acquired the compass from the Chinese. The first mention of the magnetic compass in European writings occurred in the year 1190, when, in his *De Naturis Rerum*, Alexander Neckam wrote:

> **The sailors, moreover, as they sail over the sea, when in cloudy weather they can no longer profit by the light of the sun, or when the world is wrapped up in the darkness of the shades of night, and they are ignorant to what point of the compass their ship's course is directed, they touch the magnet with a needle. This then whirls round in a circle until, when its motion ceases, its point looks direct to the north.**

The compass does not seem to have reached Europe via the Arabs. There is no mention of the compass in Arabic writings until approximately 1232, when sailors are described as finding their way by means of a fish-shaped piece of iron rubbed with a magnet. The fish-shape was typically Chinese. The Europeans and the Arabs thus seem to have adopted the magnetic compass for sailing at roughly the same time, through nautical contact with China; but it may be that the Europeans had the compass some decades before the Arabs.

Contemporary texts support this theory. In his famous book *Dream Pool Essays*, of about 1086, the medieval Chinese scientist Shen Kua clearly wrote:

> **Magicians rub the point of a needle with the lodestone; then it is able to point to the south.... It may be balanced on the fingernail, or on the rim of a cup, where it can be made to turn more easily, but these supports being hard and smooth, it is liable to fall off. It is best to suspend it by a single cocoon fibre of new silk attached to the centre of the needle by a piece of wax the size of a mustard-seed – then, hanging in a windless place, it will always point to the south. Among such needles there are some which, after being rubbed, point to the north. I have needles of both kinds by me.**

This was written a full century before the first mention of the magnetic compass in Europe. Needham comments:

> **The two needles mentioned by Shen Kua may of course have been magnetized at different poles of the lodestone ... Shen Kua's experimental conditions indicate a considerable amount of careful investigation. The**

OPPOSITE (108) A Chinese mariner's compass of the nineteenth century. It has a wooden lid which fits snugly over the top to protect it, so that it can be carried in the sea captain's pocket. This was the most valuable possession of the captain of a Chinese junk, as his life depended on never losing it. The characters for north and south are here both painted red. Traditionally, south was the primary bearing, and I have a somewhat damaged sixteenth century mariner's compass where only south is painted red. By the time this compass was made, however, Western influence had become evident, and the numbers are not shown in their Chinese characters but rather as Western numerals. (Collection of Robert Temple.)

LEFT (109) A working model of the oldest instrument in the world, which is known to be a compass. The spoon or ladle is of magnetic lodestone, and the plate is of bronze. The circular centre represents Heaven, and the square plate represents Earth. The handle of the spoon points south. The spoon is a symbolic representation of the Great Bear. The plate bears Chinese characters which denote the eight main directions of north, north-east, east, etc., and symbols from the *I Ching* oracle books which were correlated with directions. Separately marked are the finer gradations of twenty-four compass points, and along the outermost edge are the twenty-eight lunar mansions. This type of compass has been scientifically tested and found to work tolerably well. It was used not for navigation, but for quasi-magical purposes.

use of a single thread only for the suspension would avoid twisting effects. That the thread should be of silk meant that it would be a continuous fibre, unlike a thread of hempen yarn (cotton was almost certainly not known in China in his time), in which shorter fibres would be spun together under tension. That the silk thread should be new would imply an evenly distributed elasticity.

We also have an explicit account of the use of the magnetic compass for navigation, dating from before the first European mention. It is from a book by Chu Yü, son of a former high port official and then governor of Canton. The book is quaintly called *P'ingchow Table Talk*, and dates from 1117. Chu writes:

According to the government regulations concerning sea-going ships, the larger ones can carry several hundred men, and the smaller ones may have more than a hundred men on board.... The ship's pilots are acquainted with the configuration of the coasts; at night they steer by the stars, and in the day-time by the Sun. In dark weather they look at the south-pointing needle. They also use a line a hundred feet long with a hook at the end, which they let down to take samples of mud from the sea-bottom; by its appearance and smell they can determine their whereabouts.

The author Meng Yuan-Lao wrote in his *Dreams of the Glories of the Eastern Capital* about 1126: 'During dark or rainy days, and when the nights are overclouded, sailors rely on the compass. The Mate is in charge of this.' And in the book *Illustrated Record of an Embassy to Korea in the Hsüan-Ho Reign-Period*, from 1124, the author Hsü Ching wrote: 'During the night it is often not possible to stop because of wind or current drift, so the pilot has to steer by the stars and the Great Bear. If the night is overcast then he uses the south-pointing floating needle to determine south and north.' These references are also both prior to any Western mention of the compass.

The compass had in fact existed for centuries before this in China, but Needham has established that it was only late in its history that it came to be used for navigation at sea, the most probable period being between 850 AD and 1050. Before we discuss the history of the compass prior to its use in navigation, we should note a point of technological detail. Needham mentions that the French scholar L. de Saussure, who was himself a keen sailor:

RIGHT (110) A simple mariner's compass – a bowl of water on which floats a magnetized steel needle, pointed north and south. The main marking, for South, is usually painted in red, but is not in this instance. It is the fourth from the top, counting clockwise; this compass is not oriented properly for the photograph.

RIGHT (111) A detail from a Han Dynasty stone relief of imperial magicians and conjurors dated to the year 114 AD. The seated figure is looking down at a ladle compass of the sort shown in Plate 109. (Museum Rietberg, Zurich)

> ...pointed out that the use of the compass in navigation depended to some extent upon metallurgical procedures for the production of steel. Soft iron does not retain its magnetism long, or show it strongly; for extended voyages magnetized needles of good steel would have been desirable. De Saussure considered that the Chinese narratives of deep-water sailing, which we have from the thirteenth century AD, such as the embassy to Cambodia, would not have been possible without steel needles.... It may well be ... that good steel needles were available to the Chinese several centuries before Europe had them.... Failing them, lodestones had to be carried on board every ship for remagnetization, as Bromehead has described, quoting from a book on navigation of 1597. Thus for 600 years the lodestone was an economic mineral. [Magnets are remagnetized by being stroked appropriately by naturally magnetic lodestones.]

The Chinese were far advanced in their steel manufacturing capabilities (see pages 53 and 76).

The earliest and simplest form of compass was obviously a naturally magnetic piece of lodestone used to indicate direction, and this long preceded the more advanced idea of using needles. How far back did such lodestone compasses go in China, and if they were not immediately used for navigation, to what uses were they put?

We can trace the Chinese use of the lodestone compass back as far as the fourth century BC with certainty. The textual mention of it occurs in the *Book of the Devil Valley Master*, which dates from that time. The author is unknown but is thought possibly to have been the philosopher Su Ch'in. The passage says: 'When the people of Cheng go out to collect jade, they carry a south-pointer with them so as not to lose their way.' Another text from the third century BC occurs in the *Book of Master Han Fei*, written by the philosopher of that name. There, we read:

> Subjects encroach upon the ruler and infringe his prerogatives like creeping dunes and piled-up slopes. This makes the prince forget his position and confuse west and east until he really does not know where he stands. So the ancient kings set up a south-pointer, in order to distinguish between the directions of dawn and sunset.

These are the two earliest known references in world literature to a compass, with the possible exception of some ancient Egyptian texts, the interpretation of which could be disputed and cannot be discussed here. In neither case is the 'south-pointer' (and it should be remarked that all Chinese compasses pointed southwards rather than northwards if possible, by preference) mentioned as something new or novel. On the contrary, it is credited to 'ancient kings' even in the third century BC.

In Plate 109 (opposite) may be seen a model of an early Chinese compass, where the pointer is in the shape of a spoon or ladle, which represents the polar constellation of the Great Bear (Big Dipper). Another early form of compass had the pointer shaped as a fish. But with the advent of steel needles, these ladles and fish were superseded, at least as far as navigation was concerned. It should be noticed, however, that far more important than the use of the compass for navigation to most Chinese was its use for geomancy. This was a magical technique for aligning houses and cities harmoniously with the breaths and currents of the Earth's forces, which were partly detected with the aid of the compass. The ladle-shaped pointer in Plate 109 is one of these geomantic compasses, and it is doubtful that it could ever have served on board a ship.

We should just mention in passing that what may have been a compass dating from 1000 BC was excavated in Olmec ruins in 1967 at San Lorenzo in southern Veracruz, Mexico. If it was a compass, it is six hundred years older than the earliest Chinese evidence. Needham disputes the interpretation of the object as a compass, and we cannot enter into the controversy here.

DIAL AND POINTER DEVICES

THIRD CENTURY AD

The earliest Chinese compasses did not have needles. The 'pointers' were shaped as spoons, fish, or even sometimes as turtles. The introduction of a needle was a refinement which made possible a much greater precision of readings on the dials surrounding the pointer. It was at this stage of development that we can say that the Chinese pioneered the world's first dial and pointer devices, which are absolutely fundamental to modern science.

Needham says of this development: 'Probably by the seventh or eighth century AD the needle was replacing the lodestone, and pieces of iron of other shapes, on account of the much greater precision with which its readings could be taken.'

Needles were also used as pointers on computing machines. The use of a needle in a calculating device can be traced back to at least 570 AD in China; it seems to have been a form of abacus based on compass readings. A description of it survives in a book entitled *Memoir on Some Traditions of Mathematical Art*, and its accompanying Commentary by Chen Luan (flourished 570 AD). Chen Luan writes: 'In this kind of computing, the digits are indicated by the pointing of the sharp end of a needle. The first digit occupies the *Li* position, that is, pointing full south; second, or two, is *K'un*, south-west; the third, or three, is at *Tui*, full west…'. And so on. Chen Luan adds that a digit to be multiplied is indicated by the point of the needle, whereas a digit to be divided is indicated by the needle's (differently shaped) tail.

Needham says of this:

> **This technique, which would seem to have been a simple sort of abacus-like device, arising out of the old diviner's board, is elsewhere attributed to, or associated with, the name of Chao Ta, a famous diviner of**

the Three Kingdoms Period (221–65 AD). But the remarkable thing is that a needle is said to be used as a pointer, and the series starts from full south. It seems hard to believe that this can have had no connection with the magnetic compass, and it must be at least of 570 AD if not earlier.

Therefore, dial and pointer devices were in use in China by the sixth century AD at least, and quite possibly by the third century. Needham rightly points out that these Chinese devices were 'the most ancient of all pointer-readings, and... the first step on the road to all dials and self-registering meters.'

The Chinese dial and pointer devices became amazingly complex. The most elaborate ones were without question the geomancers' compasses. Some of these had dials with as many as forty concentric circles containing different sets of numbers measuring different phenomena, to be read off as required. One such may be seen in Plate 112 shown here.

LEFT (112) The dial and pointer devices upon which we depend in the modern world had originated in China by the third century AD. They were geomantic compasses, used for consultation on such questions as to where a house should be built, or a city laid out. Of course, much was superstition, but at the basis of the practice was the phenomenon of the north–south alignment of the magnetized magnetic needle. The fantastic array of readings which were possible to a geomancer's compass may be seen here, though this is by no means the most complicated. Some are known with forty concentric circles of readings. The outermost circle here marks the twenty-eight lunar mansions. The next circle is marked in the 'New Degrees' of 360° adopted for the circle under Jesuit influence, indicating that this compass cannot be earlier than the seventeenth century. A full description of the readings is obviously impossible in the space available. (Science Museum, London.)

BELOW (113) A Chinese geomancer had to take special care of his geomantic compass, and I believe this may be the only old compass-carrying pouch to survive into our times. This compass is 6½ inches in diameter and dates from the eighteenth century. It fits snugly into its pouch, decorated with an embroidered *yin–yang* double-fish design and trigrams from the divination book, the *I Ching* (*Book of Change*), for luck. The pouch is fastened by three cloth buttons, so that the compass cannot slip out when travelling. (Collection of Robert Temple.)

BELOW (114) A detail of an illustration showing the selection of the site for a new city. The geomancer studies his geomantic compass, which rests on a folding table. From *Imperial Illustrated Edition of the Historical Classic.*

MAGNETIC DECLINATION OF THE EARTH'S MAGNETIC FIELD

NINTH CENTURY AD

The Earth's magnetic field is oriented slightly askew from what might be expected. The magnetic North Pole is about 1200 miles from the geographical true north of the planet. The difference between a finger pointing at the true geographical North Pole and a compass needle pointing at the magnetic North Pole is known as the angle of declination (or, in the USA, variation); it is not constant, and continually shifts. By the eighth or ninth century AD at the latest, the Chinese had discovered this magnetic declination.

As Needham says, in doing so the Chinese were 'antedating European knowledge of the declination by some six centuries. The Chinese were theorizing about the declination before Europe even knew of the polarity.... The magnetic compass and the polarity of the Earth's field are not mentioned in any Western writing until 1190 AD, and the Chinese had had the compass for a good fifteen hundred years before that. In our account of the compass (page 162) we quote the medieval Chinese scientist Shen Kua, from his *Dream Pool Essays* of 1086. On magnetic declination he wrote: 'Magicians rub the point of a needle with the lodestone; then it is able to point to the south. But it always inclines slightly to the east, and does not point directly at the south.'

Here is another passage describing the declination, from K'ou Tsung-Shih's *Meaning of the Pharmacopoeia Elucidated*, of 1116 AD: 'When one rubs a pointed iron needle upon the lodestone proper, it acquires the property of pointing to the south, yet it inclines always towards the east, and does not point due south.'

The oldest precise and thoroughly explicit description of the declination as shown by a needle dates from about 1030–50 AD, and appears in a poem on the magnet written by Wang Chi, founder of the Fukien school of geomancers. (Fukien, as a coastal province and home of sailors, gave birth to a school of geomancy which put more emphasis on the compass than elsewhere.)

The geomancers' compasses of China (discussed under 'Dial and pointer devices' above) preserve centuries-old knowledge of two different sets of compass points showing the magnetic declination at two different times in the past. Two concentric circles of compass points were traditionally embodied in the many circles which such devices bore, showing the magnetic declination at two separate times in history.

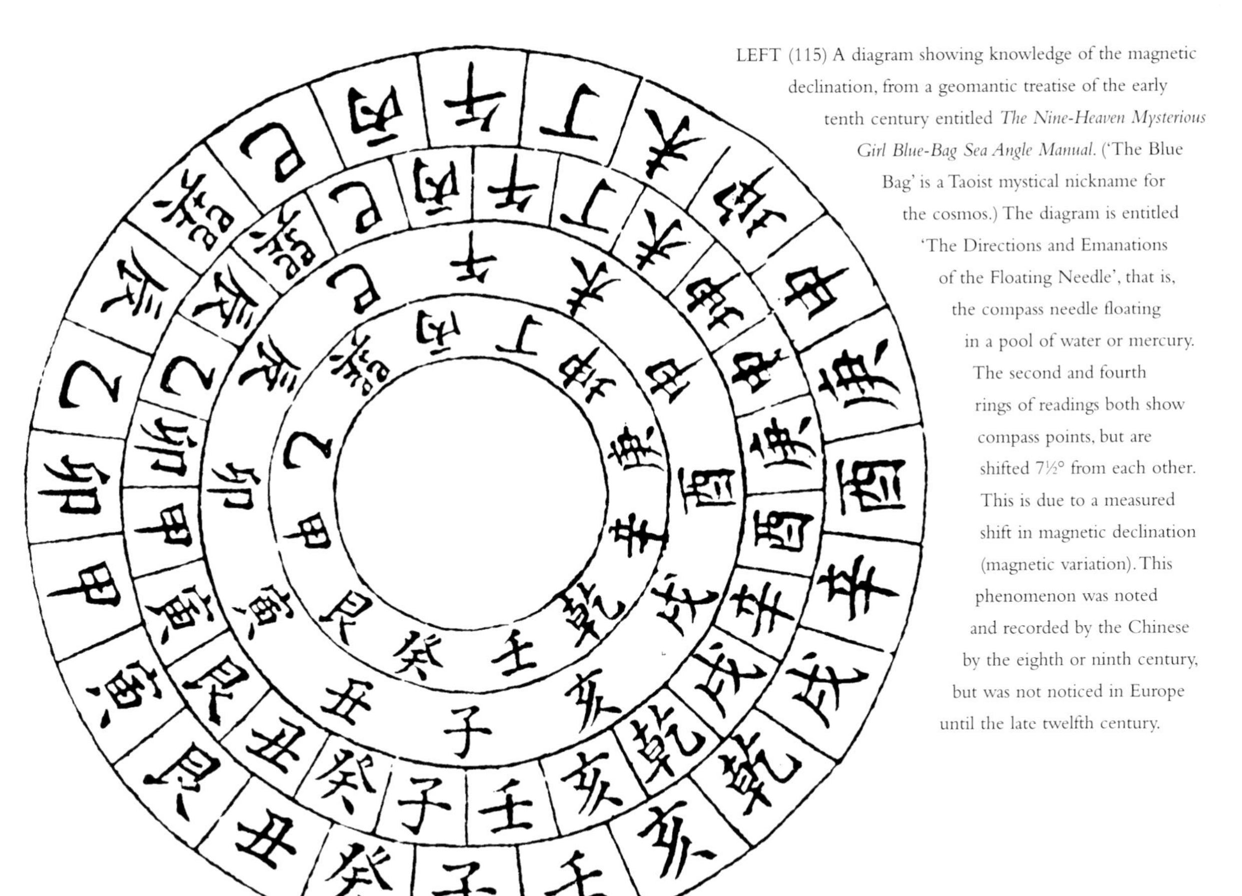

LEFT (115) A diagram showing knowledge of the magnetic declination, from a geomantic treatise of the early tenth century entitled *The Nine-Heaven Mysterious Girl Blue-Bag Sea Angle Manual*. ('The Blue Bag' is a Taoist mystical nickname for the cosmos.) The diagram is entitled 'The Directions and Emanations of the Floating Needle', that is, the compass needle floating in a pool of water or mercury. The second and fourth rings of readings both show compass points, but are shifted 7½° from each other. This is due to a measured shift in magnetic declination (magnetic variation). This phenomenon was noted and recorded by the Chinese by the eighth or ninth century, but was not noticed in Europe until the late twelfth century.

Europeans do not seem to have known about the magnetic declination until the early fifteenth century. By 1450, German sundials were being made with a set of markings rather like those on the Chinese geomancers' compasses, indicating empirical knowledge of the declination. These date from before the voyage of Columbus to America in 1492, and show to be false the traditional story that Columbus discovered the declination when he crossed the Atlantic.

MAGNETIC REMANENCE AND INDUCTION

ELEVENTH CENTURY AD

Just as boiled water turns to steam and undergoes a change of state, so magnets can become demagnetized by heating. There is a temperature known as the Curie point (after the famous nineteenth-century scientist) at which a magnet loses its magnetism.

But just as steam can recondense into water when it cools, so iron, after being heated to a very high temperature, can take on magnetic properties as it cools down through the Curie point. It does this under the influence of a magnetic field in which it may happen to lie. This is described as magnetic remanence acquired through the magnetic induction of the cooling iron. Since the Earth has a faint magnetic field, usually called the geomagnetic field, a piece of cooling iron can take on a slight magnetism from the Earth's field alone, as long as the piece is oriented north-south as it cools. The Chinese were aware of this. As Needham says:

> **Some time before the eleventh century AD it was discovered in China that magnetization could be carried out not only by rubbing the pieces of iron on the lodestone, but by cooling them (quenching) from red heat, through the Curie point, while held in a north–south direction (the Earth's magnetic field).**

Wang Ch'i wrote in his *Universal Encyclopedia* of 1609 about magnets and the destruction of magnetization by heating:

> **The *ch'i* [emanation or subtle force] and influence of the lodestone is truly as if it were living. It has a head and a tail. Its head points towards the north and its tail towards the south. The force of the head is superior to that of the tail. If it is broken into pieces, all of them also have heads and tails.... If it is heated in the fire, it 'dies' and can no longer point to the south.**

But the text which explicitly describes the induction of remanent magnetism in iron is found in the *Compendium of Important Military Techniques*, by Tseng Kung-Liang, published in 1044. He refers to a fish-shaped magnet – which also comes into the account of the compass on page 162. The south-pointing carriage in the following account is not a magnet, but a mechanical navigation device such as has already been described (page 70). Tseng writes:

> **When troops encountered gloomy weather or dark nights, and the directions of space could not be distinguished, they let an old horse go on before to lead them, or else they made use of the south-pointing carriage, or the south-pointing fish to identify the directions. Now the carriage method has not been handed down, but in the fish method a thin leaf of iron is cut into the shape of a fish two inches long and half an inch broad, having a pointed head and tail. This is then heated in a charcoal fire, and when it has become thoroughly red-hot, it is taken out by the head with iron tongs and placed so that its tail points due north. In this position it is quenched with water in a basin, so that its tail is submerged for several tenths of an inch. It is then kept in a tightly closed box. To use it, a small bowl filled with water is set up in a windless place, and the fish is laid as flat as possible upon the water-surface so that it floats, whereupon its head will point south.**

The tightly closed box is thought to have contained a lodestone base. Needham points out that compass needles are placed in boxes with lodestone floors even today, and that a Chinese professor he knew had seen this process, called 'nourishing the needle', practised about 1950 in Anhui Province. It is sometimes necessary to 're-charge' a compass needle by contact with a lodestone. And a needle with magnetism induced by the thermo-remanence clearly described above could be 'strengthened', or 'nourished', by the additional contact with the lodestone.

The Curie point is not an invariable temperature; it varies according to the material. It is doubtful that the Chinese could actually have measured it, therefore. They merely knew of it empirically; they had observed the results obtained by passing through it, and therefore had some conception of it through knowing the necessity for 'red heat'. Obtaining a magnet through thermo-remanence had the great advantage that a compass could be made without a lodestone. It is easy to see why this was of particular interest to those concerned with military techniques, for soldiers on campaign continually have to make shift by expedients, through lack of essential supplies. And losing one's lodestone must have been common in the heat of battle or, what is more, in retreat.

The recognition that the iron being magnetized had to be oriented to the Earth's magnetic field indicates some awareness of the existence of that field. As Needham laconically remarks: 'Remanent magnetism is a surprise to meet with in the early eleventh century. Thermo-remanence was of course known to William Gilbert, who illustrated the forging of a steel rod in the direction of the Earth's field.' Gilbert published his book in 1600, and in it declared the Earth to be a magnet. How much of his ideas Gilbert drew from China we do not know, but he was sufficiently aware of Chinese origins to state in *De Magnete* that Marco Polo had brought the magnetic compass to Europe (though, in fact, it had come earlier). And in recognizing the Chinese origin of the compass for Europe, Gilbert was better informed than many a twentieth-century scholar has been.

It must be stressed that by merely using the compass, and realizing that it pointed roughly north and south, the Chinese could have interpreted the whole process as a mere cosmic affinity. But magnetic induction by the Earth's field proved that that field was an *active force*. This was one of the greatest of scientific insights achieved by the Chinese.

OPPOSITE (116) Artisans drawing steel wire and making steel needles to be magnetized and used for compasses. From *Exploitation of the Works of Nature*, published in 1637. The needles would be magnetized by thermo-remanence – heating until red-hot, aligning north–south, and quenching in a basin of water. The magnetism thus induced was discovered by the Chinese before the eleventh century; it was unknown in Europe until William Gilbert published his book *De Magnete* in 1600.

抽線琢鍼圖

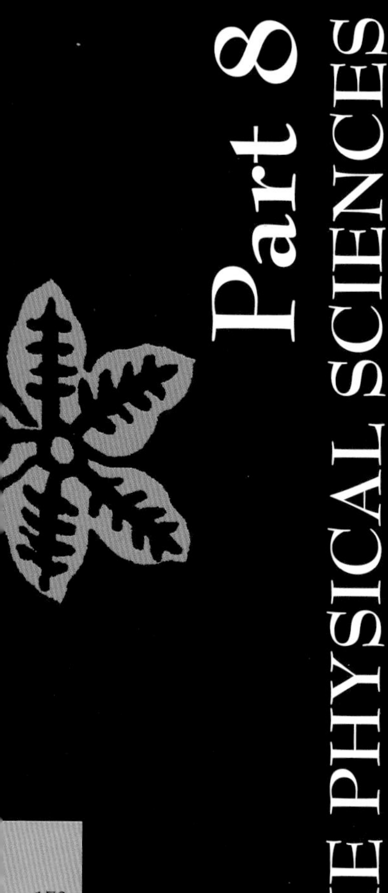

Part 8
THE PHYSICAL SCIENCES

GEO-BOTANICAL PROSPECTING

FIFTH CENTURY BC

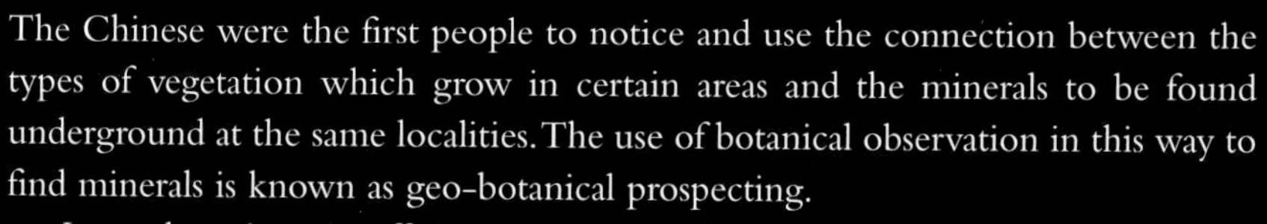

The Chinese were the first people to notice and use the connection between the types of vegetation which grow in certain areas and the minerals to be found underground at the same localities. The use of botanical observation in this way to find minerals is known as geo-botanical prospecting.

In modern times insufficient attention has been paid to this practice, and many of the ancient Chinese findings have not been investigated. There are, however, some widely recognized examples of plants which grow in soil too rich in certain minerals to be tolerated by other plants. There is, for instance, a wild pansy (*Viola calaminaria*) which is zinc-loving, 1 per cent of its ash constituting zinc. Even more zinc-loving is the pennycress (*Thlaspi*), 16 per cent of whose ash can constitute zinc. The tragacanth (source of gum tragacanth, once widely used in pharmacy) is quite insensitive to selenium in the soil, which poisons other plants. A particular type of alyssum (*Alyssum bertoloni*) is highly tolerant to nickel, and is a classic indicator of that metal's presence. A particular type of grass (*Panicum crusgalli*) indicates lead in the soil. There are several plants which indicate copper, and so on.

The oldest traces of this knowledge in China would seem to go back several centuries BC; but it is first found, substantiated by texts, in the third century BC. A curious ancient book called *The Classic of Mountains and Rivers*, made up of material from between 600 BC and 100 BC, mentions that the plant *hui-t'ang* grows near gold ore. This plant cannot be identified with certainty; it could be a type of orchid, basil, hawthorn, or wild pear or plum, all of which have names somewhat similar to this archaic one.

The origins of geo-botanical prospecting in China go back to the preoccupation with the nature of different types of soil and their suitability for crops. By at least the fifth century BC, geo-botany was in full swing, as testified by the treatise 'The Tribute of Yü' of that date. This text describes the natures of the soils in the different regions of China in terms which Needham has recently been able to demonstrate are technical to a degree not previously appreciated. It is therefore safe to assume that, although the primary interest was agricultural, geo-botany was beginning by that time to be used for prospecting as well. However, for the whole of Chinese history we are short of texts actually giving accounts of prospecting by these methods. Possibly the reason was secrecy. But it is also probable that such accounts as do survive still await discovery and are to be found in the thousands of old regional histories and gazetteers which have not yet been studied with any thoroughness by either Western or Chinese scholars.

The Book of Master Wen, compiled about 380 AD but containing material of the third century BC, says that in areas where jade is found, tree branches tend to droop. It is clear that the Chinese noticed not merely the occurrences of certain plants, but their physiological condition, with relation to mineral deposits. In the first half of the sixth century AD, there were at least three manuals devoted entirely to systematic accounts of geo-botanical mineral prospecting, and listing the varieties of plants and their associated

OPPOSITE (117) Purslane (*Portulaca oleracea*) is still sometimes eaten in salads in the West. High in vitamin C and the richest leafy-plant source of Omega-3 fatty acids, it also naturally contains potentially dangerous levels of mercury, a fact discovered by Su Sung in the eleventh century. In 1597, John Gerard in his famous *Herbal* said of purslane that it 'is much used in sallades, with oyle, salt and vinegre: it cooleth an hot stomacke, and provoketh appetite; but the nourishment which commeth thereof is little, bad, cold, grosse, and moist.... The leaves eaten raw, take away the paine of the teeth, and fasten them...'.

minerals. One of these, *Illustrated Mirror of the Earth*, says: 'If the stalk of [a certain] plant is yellow and elegant, copper will be found below.' It also says: 'If the leaves of [a certain] plant are green, and the stalks red, much lead will be found below.'

The early scientist and poet Chang Hua noted about 290 AD that 'where the smartweed grows abundantly, there must be plenty of haematite [ferric oxide] below.' And about 800 AD Tuan Ch'eng-Shih in his *Miscellany of the Yu-Yang Mountain Cave*, wrote: 'When in the mountains there is a ciboule onion, then below silver will be found. When in the mountains there is the *hsiai* plant [a kind of shallot], then below gold will be found. When in the mountains there is the ginger plant, then below copper and tin will be found.'

None of the above signs has ever been tested in modern times, and there is scope for research in this field. Definite awareness that mineral trace elements actually occurred in and could be extracted from certain plants is seen in the year 1421 in a book called *Precious Secrets of the Realm of the Keng and Hsin* (symbols of metals and minerals). There we are told quite specifically that gold occurs in the rape turnip, silver in a type of weeping willow, lead and tin in mugwort, chestnut, barley, and wheat, and copper in the Indian sorrel (*Oxalis corniculata*).

A later book quotes a lost work of 1062 AD called *The Illustrated Pharmacopoeia*, as saying that a particular species of purslane (still sometimes used today in the West as a herb and vegetable) contains enough mercury for the metal to be extracted from it by careful pounding, drying, and natural organic decay. The author of this lost work, Su Sung, claimed to have acquired 8 to 10 ounces of mercury in this way from 10 catties' weight of dried purslane. Mercury is a deadly poison, so purslane would appear to be dangerous to eat. The Chinese plant containing the mercury was 'green purslane' (*Portulaca oleracea*), the very same salad herb eaten in the West.

European awareness of geo-botanical matters was much delayed. Needham could not find an instance before G.F. Henckel in 1760, in his book *Flora Saturnisans*. However, about 1600 there was a famous English example. About that time, Sir Thomas Challoner and his first cousin, another Thomas Challoner, discovered the first alum mines in England on the former's property of Belman Bank, Guisborough, Yorkshire. A manuscript in York

RIGHT (118) Round-leaved pennycress (*Thlaspi rotundifolium*). Pennycress is zinc-loving, and as much as 16 per cent of its ash can constitute pure zinc. Soil rich in zinc is indicated by this plant, one of the few which can tolerate such amounts. Knowledge of plants as indicators of mineral deposits goes back many centuries BC in China, but is first documented in *The Classic of Mountains and Rivers* about the third century BC.

Minster informs us: 'Sir Tho. Challoner Observed the Leaves of the Oak Trees where the Mines are, to be of a deeper green than elsewhere and the Bough more spreading the Boles Dwarfish but strong having little sap, and not deep rooted ... on which he conceited there was included some valuable minerals especially Alum.' This proved to be true. Sir Thomas had visited the alum mines of Italy and noticed similar vegetation there. This appears to be the first example of geo-botanical prospecting in Europe, and is about 2100 years later than the beginnings of such methods in China.

THE FIRST LAW OF MOTION

FOURTH CENTURY BC

Isaac Newton formulated his First Law of Motion in the eighteenth century. It stated that 'every body continues in its state of rest, or of uniform motion in a right line, unless it is compelled to change that state by forces impressed upon it.

Needham's researches have now established that this law was stated in China in the fourth or third century BC. We read in the *Mo Ching*: 'The cessation of motion is due to the opposing force.... If there is no opposing force... the motion will never stop. This is as true as that an ox is not a horse.'

The book *Mo Ching* is the collection of writings of a school of philosophers called Mohists, after their founder and sage Mo Ti (more commonly known as Mo Tzu, which means 'Master Mo'). The Mohists disappeared completely from Chinese history after only a moderate time, and most of their writings remained unread and almost forgotten until recently. Their brilliant scientific insights were also largely lost, and made very little lasting impact on later Chinese history. The Mohists were also the only ancient Chinese to consider the subject of dynamics in the theoretical sense, though practical dynamics was continuously applied in the great strides made by Chinese technology and invention.

Sadly, although Newton's First Law of Motion was anticipated two thousand years earlier by the Mohists, nothing seems to have come of it. It was only in 1962 that Needham, in the first proper study of Mohist scientific doctrines, published the fact that this great step had ever been taken – but it was a step in soft sand, quickly washed away by the advancing tide of history.

THE HEXAGONAL STRUCTURE OF SNOWFLAKES

SECOND CENTURY BC

The knowledge that snowflakes are hexagonal in shape may seem a commonplace to us today. But when the Scandinavian bishop Olaus Magnus wrote about snowflakes in 1555, he believed – despite the abundance of snow in his homeland – that they could be shaped like crescents, arrows, nails or bells, or even like the human hand. Perhaps he did see such bizarre clumps of snow, but the fact that the flakes themselves are always hexagonal eluded him. It was only in 1591 that the Elizabethan genius Thomas Hariot seems to have recognized the hexagonal nature of snowflakes, for the first time in European history. But he only jotted his findings down in his private manuscripts, and did not publish them.

The first European publication announcing the hexagonal nature of snowflakes was in 1611 (English translation first published in 1966). This was a fifteen-page account by the astronomer Johannes Kepler of his observations of snowflakes during the winter of 1610–11, entitled *A New Year's Gift, or On the Six-Cornered Snowflake.*

How impressive it is, then, to discover that the Chinese knew of the hexagonal nature of snowflakes in or before the second century BC. About the year 135 BC, Han Ying wrote a book entitled *Moral Discourses Illustrating the Han Text of the 'Book of Songs'*. In this work he incidentally referred to what appears at that time to have been common knowledge, as follows: 'Flowers of plants and trees are generally five-pointed, but those of snow, which are called *ying*, are always six-pointed.'

Thus we see that understanding of the hexagonal nature of snowflakes in China preceded such knowledge in Europe by at least 1726 years, and probably far more. Knowledge of hexagonal patterns in nature seems to have been fundamental to the earliest Chinese proto-science several centuries before the time of Han Ying. The knowledge may extend back to the earliest levels of advanced civilization in China, but Han Ying's mention is the earliest textual evidence in an explicit form connecting hexagonal patterns with snow.

Chinese literature is full of mentions of hexagonal snowflakes. For instance, the sixth-century poet Hsiao T'ung wrote in one of his poems:

The ruddy clouds float in the four quarters of the caerulean sky / And the white snowflakes show forth their six-petalled flowers.

The Chinese were traditionally keen on number mysticism. The number six was correlated with the element water. The greatest Chinese medieval philosopher, Chu Hsi, wrote in the twelfth century: 'Six generated from Earth is the perfected number of Water, so as snow is water condensed into crystal flowers, these are always six-pointed.' Even this late reference is still four-and-a-half centuries before Kepler's book on the snowflake.

The hexagonal nature of snowflakes was very much part of a Chinese cosmic scheme of nature. It was seen as an example of the most extreme form of the *yin* force of the Universe, and as a manifestation of the number associated with the element of water. It was also clearly believed that in forming snow, water condenses into the crystals. T'ang Chin, in his book *Records of My Daydreams*, spoke of water's 'congealing' into snow: 'That flowers of plants and trees are always five-pointed and snow crystals six-pointed is a saying of the old scholars, for, since six is the true number of Water, when water congeals into flowers they must be six-pointed.'

We even have textual evidence of systematic examination of snowflakes in the sixteenth century. For, about the year 1600 AD, Hsien Tsai-hang wrote in his book *Five Assorted Offering Trays*: '...every year at the end of winter and the beginning of spring I used to collect snow crystals myself and carefully examined them; all were six-pointed...'. He was probably examining his snowflakes under the magnifying power of lenses, which were in use in his day.

Although the Chinese were thus fully aware of the hexagonal structure of snowflakes, they seem never to have investigated it as a mathematical problem in the manner of Kepler. As Needham says, 'the Chinese, having found the hexagonal symmetry, were content to accept it as a fact of nature.'

OPPOSITE (119) A snowflake, photographed through a microscope. The Chinese recognized that snowflakes had six sides, or points, by the second century BC, nearly two thousand years before this was realized in Europe.

THE SEISMOGRAPH

SECOND CENTURY AD

China has always been plagued by earthquakes. The high mountains and precipitous gorges of its crumpled, twisted surface are frequently disturbed by tremors which wreak immense havoc among the people who live there. More than 800,000 people are said to have been killed in three provinces by the great earthquake of 2 February 1556. Full records exist in the Chinese histories of all the major earthquakes over the centuries. These events were often the trigger for food riots, or attempts at rebellion. The imperial government had every reason to want to know as soon as possible when there had been an earthquake in a distant province. First of all, it would mean that grain shipments would be interrupted, which was relevant since taxes were paid in grain. But it would also mean that both food aid and extra military forces would be needed in the afflicted area without delay. Early warning was essential.

A solution was provided by the brilliant scientist, mathematician and inventor Chang Heng, who was Astronomer-Royal during the later Han Dynasty. He also wrote a number of books of which one, *Spiritual Constitution of the Universe*, survives only in fragments; but these are sufficient to show that he envisaged the Earth as a spherical ball suspended in infinite space, with nine continents. Chang Heng was the first in China to introduce the crisscrossing grid of latitudinal and longitudinal lines in geography – 'throwing a net over the Earth', as it was called.

But what amazed the court and all officialdom was Chang Heng's spectacular earthquake detector, or seismograph. 'Nothing like this had ever been heard of since the earliest records began,' was the comment of the official historian for the year 132 AD, speaking of this invention.

At first the court officials could not really believe that Chang Heng's invention could work, even though they had all inspected it and seen it demonstrated. The signal for an earthquake was the falling of a bronze ball into the open mouth of a bronze toad, from the mouth of a bronze dragon above. The official historian describes how the sceptical were converted as follows:

On one occasion one of the dragons let fall a ball from its mouth though no perceptible shock could be felt. All the scholars at the capital were astonished at this strange effect

occurring without any evidence of an earthquake to cause it. But several days later a messenger arrived bringing news of an earthquake in Lung-Hsi [Kansu, about 400 miles away to the north-west]. Upon this everyone admitted the mysterious power of the instrument. Thenceforward it became the duty of the officials of the Bureau of Astronomy and Calendar to record the directions from which earthquakes came.

What did this extraordinary machine look like, and how did it work? It consisted of a 'fine cast bronze' vessel, rather like a wine-jar, 6 feet across, with a domed lid. The outer surface of the vessel was decorated with motifs of mountains, tortoises, birds, animals and antique writing. All around the vessel was a series of eight dragons' heads, equally spaced, holding bronze balls in their mouths. The balls would drop out if the dragons' mouths opened, or if pushed.

Round the base of the vessel sat eight corresponding bronze toads, looking upwards, with their mouths wide open. They were positioned directly beneath the dragon mouths, ready to catch the falling balls. Obviously, a bronze ball dropping into a bronze toad would make a great deal of noise; people would be alerted by the resounding clang. This principle in itself is an interesting innovation, being a kind of predecessor

ABOVE (120) A modern reconstruction of Chang Heng's seismograph of 132 AD. A bronze ball falling from the mouth of a dragon into the waiting open mouth of a bronze toad made a loud noise and signaled the occurrence of an earthquake. By looking to see which ball had been released, one could determine in which direction the epicentre of the earthquake lay. (Science Museum, London.)

of the alarm clock. It may well be that Chang Heng invented the idea of the dropped ball independently, but the principle had been used earlier in the West. Needham says: 'The principle of recording by means of dropping balls was one in which Chang Heng had been anticipated by Heron of Alexandria (fl. 62 AD), who used them in some of his hodometers.' Needham was unaware, however, of an even earlier tradition in Greece, according to which the falling ball alarm device appears to be an invention of Aristotle. Towards the end of his life Aristotle suffered from either a severe gastric ulcer or stomach cancer. In order to alleviate the terrible pain he would lie in bed with a skin filled with hot oil placed on his stomach – a kind of ancient hot-water bottle. If he drifted off to sleep, though, he would be badly burned. So he devised a system whereby he held a bronze ball in his hand over a bronze bowl, so that if he dozed off he would be awakened immediately by the sound of the ball dropping into the bowl.

This may be where Heron got the idea, and it is possibly also the origin of Chang Heng's falling-ball arrangement. For this is just the sort of popular tale about a famous sage which would have circulated round the ancient world, eventually becoming part of the lore brought to China from the Near East during the three centuries before Chang Heng's time. (In the stable period of the Han Dynasty there was considerable trade and travel between East and West.)

But how precisely were the balls released, and how could the machine be restricted to dropping only a single ball? The ancient sources are rather vague. Little is said about the workings of the dragon-mechanism: 'The toothed machinery and ingenious constructions were all hidden inside the vessel, and the cover fitted down closely all round without any crevice.'

We are also told: 'Inside there was a central column capable of lateral displacement along tracks in eight directions, and so arranged that it would operate a closing and opening mechanism.' This 'central column' was quite clearly a pendulum of some kind. It would move when there was an earth tremor and release a ball. As soon as the ball was dropped, an immobilization of the mechanism took place, preventing secondary earth tremors releasing all the balls and thus ruining the point of the device.

The account adds: 'Now although the mechanism of one dragon was released, the seven other heads did not move, and by following the azimuthal direction of the dragon which had been set in motion, one knew the direction from which the earthquake shock had come. When this was verified by the facts there was found an almost miraculous agreement.'

LEFT (121) A cut-away view of the modern model of Chang Heng's seismograph of 132 AD in the Science Museum, London, showing one version of the interior apparatus. In this version, designed by the Chinese scholar Wang Chen-to in 1936, the long cylindrical object in the middle is a pendulum, which is imagined as tilting slightly with an earth tremor. However, an inverted weighted bob is more likely to have been used by Chang Heng than a pendulum, and later reconstructions are closer to the surviving descriptions of how the machine actually worked.

In later centuries, similar instruments were constructed in China using the principles of Chang Heng's machine. The mathematician Hsintu Fang wrote about the machine, giving diagrams, three hundred years later, and is presumed to have made one. The same is true of Lin Hsiao-Kung, some time between 581 and 604 AD. But by the time of the Mongol rule in the thirteenth century, the principles of the seismograph had been lost. Thus did the Chinese often forget their own achievements.

Several attempts have been made in this century to reconstruct the machine's mechanism. Japanese scientists led by Imamura Akitsune built what is probably the most accurate version at the Seismological Observatory of Tokyo University in the 1930s. It worked as an earthquake detector, but they:

> **...found that the ball was usually released not by the initial longitudinal wave, but by secondary transverse waves; though if the first shock was very strong, it would do.... In some circumstances one had to take the direction at right angles to the dropped ball as that of the earthquake's epicentre. Imamura points out that by estimating the direction of the preliminary tremors, and consequently the focal distance, one could give a rough estimate, not only of the direction, but also of the distance, of the earthquake.**

Imamura's reconstruction may be seen in the accompanying diagrams, which have been redrawn with slight amendments by the author. An inverted pendulum with a weighted bob near the top is sensitive to earth tremors, and tilts over when it perceives one. A long sharp pin at the top moves along one of the eight channels between two plates, pushing a 'slider'. This 'slider' is in contact with the ball in the dragon's mouth at the end of the channel, and pushes the ball out of the mouth so that it drops (or alternatively, it eases the dragon's mouth open slightly, thus releasing the

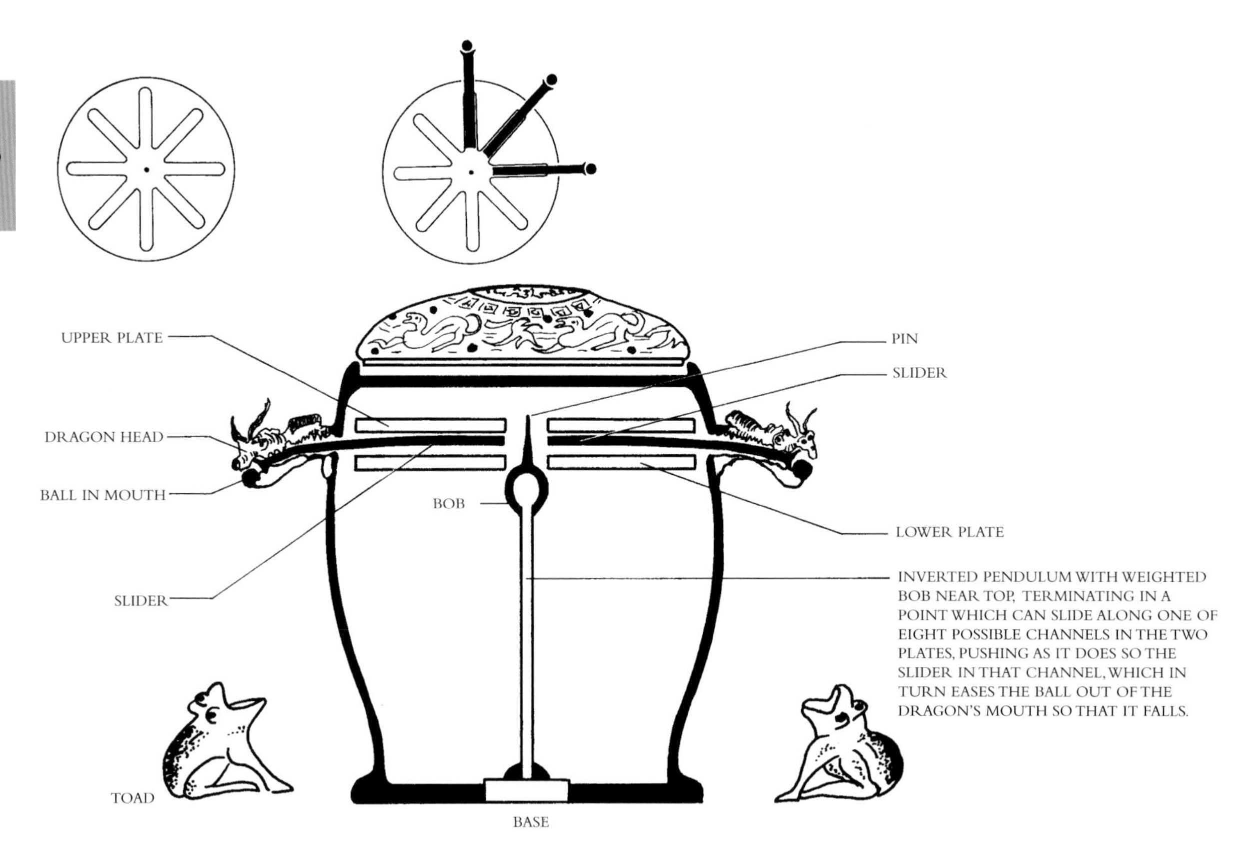
UPPER PLATE
DRAGON HEAD
BALL IN MOUTH
SLIDER
TOAD
BOB
BASE
PIN
SLIDER
LOWER PLATE
INVERTED PENDULUM WITH WEIGHTED BOB NEAR TOP, TERMINATING IN A POINT WHICH CAN SLIDE ALONG ONE OF EIGHT POSSIBLE CHANNELS IN THE TWO PLATES, PUSHING AS IT DOES SO THE SLIDER IN THAT CHANNEL, WHICH IN TURN EASES THE BALL OUT OF THE DRAGON'S MOUTH SO THAT IT FALLS.

ABOVE (122) The author's own reconstruction of the most likely mechanism of Chang Heng's seismograph, an improved version of Imamura Akitsune's attempt in 1939. Imamura adopted the principle of the inverted weighted bob, shown here, and the pin which entered one of eight grooves and pushed sliders, which released the balls. The top view shows the two plates from above, with three of the sliders and balls represented (the sliders would be between the plates, and hence are only shown superimposed on the lower one). The side view shows the full apparatus.

ABOVE (123) A modern Chinese seismograph in operation. The senior seismologist at Taiwan's Central Weather Bureau points out on a chart an earthquake measuring 6.2 on the Richter Scale, which struck Taipei on June 14, 2001. No one was injured.

ball). The immobilization is simply obtained because the pin is in a channel and would have to slide back first, before it could enter another channel. This version seems to offer the most likely explanation of the original mechanism.

The instruments we use today only began their development in 1848. The first modern seismograph of any kind was designed by De la Hautefeuille in 1703, making Chang Heng exactly 1571 years in advance of the West with his ingenious machine.

SPONTANEOUS COMBUSTION

SECOND CENTURY AD

The Chinese were apparently the first people to become aware of the phenomenon of spontaneous combustion. The earliest surviving account of it in a Chinese text is found in a book which may be called either *Record of the Investigation of Things,* or *Record of Strange Things*, depending on how the title is translated. It was written before 290 AD by the statesman, philosopher and poet Chang Hua.

He collected tales of bizarre and supernatural phenomena, and studied them in a more or less scientific fashion. His biographer says of him: 'All sorts of rare books and objects whose very existence was thought to be doubtful could have been found in Hua's household. So huge was his collection, and so wide his knowledge of the world, that his erudition was regarded as peerless in his time.'

Many of the phenomena he studied involved 'strange radiances', 'strange phenomena', 'mysterious purple vapour', 'luminous objects', 'five-coloured lights', and resonance puzzles. It is therefore hardly surprising that, with such interests, Chang Hua should have preserved the earliest account of spontaneous combustion. Here is the note he made of it, in describing an episode which took place during his own lifetime:

> If ten thousand piculs of oil are accumulated in store, the oil will ignite itself spontaneously. The calamitous fire which occurred in the arsenal in the time of the Emperor Wu [of the Chin Dynasty] in the T'ai-Shih reign-period [265–74 AD] was caused by the stored oil.

It took a great deal of perspicacity to recognize and acknowledge the phenomenon of spontaneous combustion, since the more usual course, often repeated, was simply to blame arsonists. In a thirteenth-century Chinese work entitled *Parallel Cases Solved by Eminent Judges*, we have an account of how spontaneous combustion was only with difficulty accepted as the explanation for a disastrous fire, which had occurred in about 1050, and the lives of those who had been thought responsible were thereby saved:

> When the Director of the Sacrifices Department of the Ministry of Rites, Ch'iang Chih, was serving as Intendant of the Imperial Guards at the palace of K'aifeng, oiled curtains had been left piled up in the open air, and one night they caught fire. According to the law the men responsible for looking after them all incurred the penalty of death. But at the preliminary hearing of the case, Ch'iang Chih conceived doubts about the cause of the fire, so he summoned the workmen who had made the curtains and questioned them. These artisans said that during the manufacture of the curtains a certain chemical was added to the oil, and that if they were left for a long time piled up, then on getting damp they might start burning. When Ch'iang Chih reported this to the Emperor Jen Tsung [reigned 1023–63] the Emperor was suddenly struck with an idea and said, 'The fire which recently occurred in the mausoleum of

the Emperor Chen Tsung [died 1022] started in oil garments. So that was the cause!' The guards were let off with a lighter punishment.

The author of this refers to the previous account by Chang Hua, and adds: 'Chang Hua thought that the fire which occurred formerly during the Western Chin Dynasty in the arsenal originated from the oil which was stored there, but in fact it must have been from the same cause as mentioned here [the spontaneous ignition of oiled cloth].'

The oiled cloths in question were probably material for tents. Needham points out that the only chemical which would seem to suit the description and which could have had the effect described is quicklime, which was possibly used as a whitening agent. It is also interesting, he notes, that oil and quicklime were once thought to be the recipe for 'Greek fire', the burning liquid which the Byzantines used in warfare (and which is now known to have been petroleum).

The first Western recognition of spontaneous combustion was apparently by J.P.F. Duhamel, in a scientific paper published (in French) in 1757. He discussed how a stack of canvas sails treated with ochre in oil and left to dry in a July sun was found to be burning at its centre within only a few hours. We now know that oiled cloths must never be stacked up and stored because they will ignite spontaneously; army and navy storekeepers, for instance, are issued with instructions never to allow this to happen. The processes of auto-oxidation, desaturation, aldehyde formation and so on can give out sufficient heat to create the ignition temperature for the surrounding oil. The cellulose fibres of the textiles provide ample wick, and the small amounts of air between the layers provide oxygen for the burning. The result is a gigantic conflagration.

'MODERN' GEOLOGY

SECOND CENTURY AD

James Hutton (1726–97) was the father of modern geology. In 1785 he published his *Theory of the Earth*, in which the principles of modern geology were for the first time enunciated in the West. Because his writings were difficult, his work received little attention until a more lively summary of it was published in 1802 by his friend John Playfair, entitled *Illustrations of the Huttonian Theory of the Earth*.

It may seem strange to us now to realize that, before Hutton, Europeans did not understand what we today

LEFT (124) James Hutton, founder in the West of 'Huttonian geology', which unknown to him had been explained by the Chinese 1600 years earlier. Hutton's views were fiercely opposed by his contemporaries, and he is one of the heroes of Western science for advocating them in the face of vicious and even hysterical opposition, especially by religious fundamentalists. His views are now the basis for the whole of modern geology, and without them we would understand nothing whatever of how the earth's surface was formed.

ABOVE (125) A painting by Li Kung-Lin, c. 1100. The background is an exposed cliff of twisted rock strata, which in geology is called an 'anticlinal arch'. This one is just north of the Yangtze River between Hankow and Nanking, at Lung-Mien Shan near T'ung-ch'eng in Anhui Province. When James Hutton claimed in 1785 that the Earth's surface had been drastically buckled and twisted by volcanic forces, producing such configurations as this anticlinal arch, his theory was not believed. But the Chinese had understood its principles since the second century AD, long anticipating him in founding 'modern geology'.

think obvious. But it was Hutton who first put forward the idea that sedimentary rocks were laid down under ancient seas, then thrust up, buckled and twisted as mountains, by the expansive effect of subterranean heat. He suggested that intrusions of molten lava were sent shooting through the cracks in the distorted layers, giving us the igneous veins of rock. All the upraised land, he said, was then subject to erosion, and was worn away, making its way to the sea, where it was deposited on the sea bed and one day would in turn be uplifted to form mountains. This, in 1802, was considered revolutionary.

All of these ideas were common currency in China centuries before James Hutton. All of the essentials of the Huttonian system of geology were enunciated in 1086 by Shen Kua in his *Dream Pool Essays*. For instance, he speaks of erosion of mountains and says that there are certain lofty peaks which

> ...are precipitous, abrupt, sharp, and strange.... Considering the reasons for these shapes, I think that (for centuries) the mountain torrents have rushed down, carrying away all sand and earth, thus leaving the hard rocks standing alone.

He describes sedimentary deposition as follows:

> When I went to Hopei on official duties I saw that in the northern cliffs of the T'ai-Hang Shan mountain range, there were belts [strata] containing whelk-like animals, oyster-shells and stones like the shells of birds' eggs [fossil echinoids]. So this place, though now a thousand *li* west of the sea, must once have been a shore. Thus what we call the 'continent' must have been made of mud and sediment which was once below the water. The Yü Mountain, where Yao killed

> Kun, was according to ancient tradition, by the side of the Eastern Sea, but it is now far inland.
>
> Now the Yellow River, the Chang Shui, the Hu T'o, the Cho Shui, and the Sang Ch'ien are all muddy, silt-bearing rivers. In the west of Shensi and Shansi the waters run through the gorges as deep as a hundred feet. Naturally mud and silt will be carried eastwards by these streams year after year, and in this way the substance of the whole continent must have been laid down. These principles must certainly be true.

Shen Kua did not originate all these ideas, but he expounded them well and combined them with his own personal observations in the field. The concept of the seabed's having been lifted up over the course of ages to become mountains went back centuries before Shen Kua's time in China. It seems to have developed under the influence of the Buddhists, who came from India. Since the Buddhists believed that the world was often destroyed and recreated, the Chinese took this notion and combined it with their observations of sea shells on mountain tops, to envisage cataclysmic geological changes. The key Chinese concept was of the *sang t'ien*, or 'mulberry grove'. This was the quaint phrase used to describe patches of dry land in the mountains which had once been under the sea. We find this phrase used by the alchemist Ko Hung about 320 AD in his book *Lives of the Divine Hsien*. (A *hsien* is a man who has become an immortal.) Ko Hung wrote of the legendary immortal Ma Ku, and had her say:

> Since I was last invited here I have seen the Eastern Sea has turned into mulberry groves. This change has occurred three times. The last time I arrived at Mount P'eng-Lai [for an assembly of *hsien* immortals] I noticed that the sea was only half as deep as it had been at the

RIGHT (126) An aerial photograph taken in the province of Shensi, showing the drastic soil erosion in that area of loess soil. The Chinese were the first to appreciate the important part played by erosion in shaping the Earth's surface. As Shen Kua wrote in 1086: 'I think that for centuries the mountain torrents have rushed down, carrying away all sand and earth, thus leaving the hard rocks standing alone.'

> previous meeting. It looks as if the sea will again be turned to mountains and dry land.
>
> Fang P'ing laughed and said, 'The sages all maintained that where the sea is now the dust will one day be flying.'

If in 320 AD an author could say that the sages 'all maintained' this, then the tradition must be far more ancient. There is a reference to the idea in the official history of the Chin Dynasty, where Tu Yü (222–84 AD) is described:

> Tu Yü often used to say that the high hills will become valleys and the deep valleys will become hills. So when he made monumental steles recording his successes he made them in duplicate. One was buried at the bottom of a mountain, and the other was placed on top. He considered that in subsequent centuries they would be likely to exchange their positions.

These concepts were a continuous part of Chinese literary, historical and philosophical tradition. The famous Neo-Confucian philosopher Chu Hsi (1130–1200 AD) wrote:

> The waves roar and rock the world boundlessly, the frontiers of sea and land are always changing and moving, mountains suddenly arise and rivers are sunk and drowned. Human things become utterly extinguished and ancient traces entirely disappear; this is called the 'Great Waste-Land of the Generations'. I have seen on high mountains conchs and oyster shells, often embedded in the rocks. These rocks in ancient times were earth or mud, and the conchs and oysters lived in water. Subsequently everything that was at the bottom came to be at the top, and what was originally soft became solid and hard. One should meditate deeply on such matters, for these facts can be verified.

PHOSPHORESCENT PAINT

TENTH CENTURY AD

The Chinese have been fascinated by natural luminescence since the earliest times. Fireflies appear in one of the oldest surviving texts, the *Book of Odes*, which dates from at least the early centuries of the first millennium BC. 'Flashing go the night-travellers', says an old folk-song in this work. A traditional story is preserved in an early official history of the impoverished but diligent student Ch'e Yin (who died about 397 AD). His family was too poor to buy oil for him to study at night, so he collected a bagful of fireflies each evening and read by their light.

The connection between the bio-luminescence of fireflies and the glowing of certain kinds of rotting vegetation was certainly made by the Chinese. By the seventh century BC, a book entitled the *Monthly Ordinances of the Chou Dynasty* provided the somewhat garbled observation that 'decaying grass becomes fireflies'. Luminescence was recognized as being common to the two, and perhaps at this period the Chinese actually thought that fireflies were engendered in this way. But the luminescence of decaying vegetation was clearly recognized in its own right, as we see from a book of the second century BC which states that 'old pieces of *Sophora* wood shine like fire', and the official history for 466 AD which tells us: 'In the time of the Emperor Ming ... a Taoist, Sheng-Tao, from a mountain temple ... reported that a pillar in an apartment next to the Hall of Salvation was spontaneously shining brightly in the dark. Thus wood had lost its natural properties. Some people said that when wood goes rotten it shines of itself.'

A clear understanding of these phenomena was somewhat impeded and complicated by numerous accounts of marsh-fires. These may be caused by burning methane gas, though even today we are not certain what causes 'marsh lights'. The statesman and poet Chang Hua described in his *Records of Strange Things* (290 AD) both will-o'-the-wisps and static electricity:

> These lights stick to the ground and to shrubs and trees like dew. As a rule they are invisible, but wayfarers come into contact with them sometimes; then they cling to their bodies and become luminous. When wiped away with the hand, they divide into innumerable other lights, giving out a soft crackling noise, as of peas being roasted. If the person stands still a good while, they disappear, but he may then suddenly become bewildered as if he had lost his reason, and not recover before the next day.

LEFT (127) A glow-worm (*Lampyris noctiluca*), whose glowing tail is visible at night over long distances. Glow-worms were described in 1596 by Li Shih-Chen, who correctly differentiated them from fireflies and other apparently luminous creatures such as mayflies or midges infected with luminous bacteria. Luminescence and phosphorescence fascinated the Chinese for centuries, and were called '*yin* fire'. As the poet Ts'ao T'ang wrote in the ninth century: '*Yin*-fires are cool'. This is a fact well known to a culture which collected naturally phosphorescent whale eyes ('night-shining pearls') and glowing luminescent minerals, and was using phosphorescent paint by the tenth century.

> Nowadays it happens that when people are combing their hair, or when dressing or undressing, such lights follow the comb, or appear at the buttons when they are done up or undone, accompanied likewise by a crackling sound.

Chang Hua gives good evidence of the meticulous empirical observational powers of the proto-scientists of early China. This tradition led, by the eleventh century AD at the latest, to the exploitation of natural luminescence in the making of phosphorescent paintings. We are given this account of the acquisition of one of these rare objects in an obscure book called *Rustic Notes from Hsiang-Shan*, written at that time:

> The Provincial Legate Hsu Chi was fond of collecting curios. He paid 50,000 mace [coins] to a barbarian merchant for a stuffed bird's head, very brightly coloured, which he used as a pillow. He also got hold of an extraordinary painting which he presented to Li Hour Chu [the third and last Emperor of the Southern T'ang Dynasty; this can be dated to 977 AD]. This ruler upon the extinction of the Southern T'ang passed it on as tribute to the second Sung Emperor, T'ai Tsung. T'ai Tsung hung it up in the back garden of the palace in order to show it to the court. On the painting there was an ox which during the day appeared to be eating grass outside a pen, but at night seemed to be lying down inside it. None of the officials could offer any explanation for this phenomenon. Only the monk Tsan-Ning, however, said that he understood it.... [He] said that it would be found in a book called the *Hai Wai I Chi* written by Chang Ch'ien [a famous envoy of the second century BC]. Afterwards Tu Hao examined the collections in the imperial library, and found the reference in a manuscript dating from the Liu Ch'ao period [third to sixth centuries AD].

The first production of a phosphorescent substance in the West appears to have been in 1768 by John Canton, who used oyster shells to prepare an impure calcium sulphide by calcining the carbonate with sulphur. It became known as Canton's phosphorus, but appears to owe nothing to China.

Part 9

TRANSPORTATION AND EXPLORATION

THE KITE

FIFTH/FOURTH CENTURY BC

The kite was not known in Europe before the sixteenth century. It is first mentioned (as a 'flying sail') in 1589 by the scientist Giambattista della Porta in *Natural Magic*, which was a popular book of marvels and tricks. Kites existed in China, however, as early as the fifth or fourth centuries BC. The earliest record of the construction of a kite is semi-legendary, but is probably based on fact. A man named Lu Pan of the fifth century BC was the historical basis for a tutelary god of artisans, named in later ages Kungshu P'an. Kungshu P'an is reported to have made kites, and it is thought that this bears relation to something that the real historical figure, Lu Pan, actually did.

Kites were in any case certainly constructed in the fourth century BC by the philosopher Mo Ti. Kungshu P'an and Mo Ti (who died 380 BC) were actually contemporaries. The stories of their kite-making were famous in Chinese tradition. Kungshu P'an made kites shaped like birds which could fly for up to three days; he also made his kites do somersaults. Mo Ti is said to have spent three years building a special kite, and we may safely assume that his followers, the Mohists, were much concerned with kite design.

The Mohists were preoccupied with military technology, and consequently many of the earliest stories of kites in China concern military uses. Indeed, Kungshu P'an is said to have flown one of his kites over the city of Sung during a siege, though how he utilized it is unclear. From the book *Things Uniquely Strange*, which dates from the seventh or eighth century AD, we have this interesting account of the military uses of kites:

> **In the T'ai-Ch'ing reign-period [547–9 AD] of Liang Wu Ti, Hou Ching rebelled, and besieged T'ai-ch'eng [Nanking], isolating it from loyal forces far and near. Chien Wen [later emperor for one year in 550] and the crown prince Ta-Ch'i decided to use many kites flying in the sky to communicate knowledge of the emergency to the army leaders at a distance. The officers of Hou Ching told him that there was magic afoot, or that messages were being sent, and ordered archers to shoot at the kites. At first they all seemed to fall but then they changed into birds which flew away and disappeared.**

This either refers to kites shaped as birds, which were very common, or implies that the besieged forces switched to carrier pigeons as their means of sending messages. An early example of an airborne 'leaflet drop' – perhaps the first in history – occurred in the year 1232 with the aid of kites. The Mongols were laying siege to the Chin Tartars at K'aifeng, and the official history describes it thus:

> **The besieged sent up paper kites with writing on them, and when these came over the northern [Mongol] lines, the strings were cut so that they fell among the Chin prisoners there. The messages incited them to revolt and escape.**

OPPOSITE (128) A nineteenth-century Chinese painting showing a celebration where fireworks are being set off in a small (apparently brick) kiln. Above the festivities, two lanterns are suspended and ornamental kites are being flown. This may be intended to represent the annual Moon Festival, as a full moon is prominent in the twilit scene. (Bridgeman Art Library.)

The Chinese were always most ingenious in their use of kites. They developed a technique of using them for fishing. The hook and bait would be attached to a kite and flown from a boat on a wide river or a lake, and would come to rest at some distance sufficiently far from the boat's shadow to deceive even the canniest of fishes.

The variety of kites in China is extraordinary. Apart from the traditional bird shape there are kites in the forms of centipedes, frogs, butterflies, dragons and hundreds of other creatures, both real and imaginary. Some have rolling eyes or moving paws and tails.

Musical kites seem to have existed in China since the seventh or eighth century. They make whistling, moaning or harp-like sounds. A kite fixed with a single bamboo strip is a 'wind psalter', or a 'wind zither'. A kite with seven strings fixed across a gourd-shaped framework is called a 'hawk lute'. Such Aeolian harps and whistles are not peculiar to kites. Whistles are also often put into the tails of pigeons.

The use of kites in China was intimately connected with a concept of what Needham translates into English as the 'hard wind'. This idea of the Taoist philosophical school partly derived from the early shaman traditions, where mystical journeys through the air were 'experienced' either in trance or with the aid of drugs. In the early *Lieh Tzu* book, parts of which date from the fifth century BC, we read that: 'Lieh Tzu could ride upon the wind. Cool and skilfully sailing, he would go on for fifteen days before returning.'

Kite flying seems to have been a kind of meditative exercise for the early Taoists, and had a similar significance to that which archery had to the later Zen Buddhists. One may imagine an early Taoist meditating upon the Way of the Universe as he flew his kite. His reveries are combined with an intimate contact with the actual aerial phenomenon of the wind, and the need to keep altering the tension of the string and compensating for wind changes. The sage might regard this as an analogy for 'following the Way', which was also often compared to 'following the grain of the Universe', with the grain of wood as the analogy.

The subtle variations in air currents with which the sage would have developed familiarity would have been associated with his reveries and the 'imaginary flights' of his mind. We can thus understand how Lieh Tzu is described as depending upon the wind when he 'rides upon it' in his fancy. But there is the further possibility that the Taoists are referring at this early date to man-lifting kites.

In one of the two greatest classics of Taoism, the book called *Chuang Tzu* ('The Book of Master Chuang'), which dates from about 290 BC, we read of more soaring into the air, and are given some opinions about the nature of air, and how it bears things up:

> **In the Northern Ocean there is a fish, by the name of *k'un*, which is many thousand *li* in size. This fish metamorphoses into a bird by the name of *p'eng*, whose back is many thousand *li* in breadth. When the bird rouses itself and flies, its wings obscure the sky like clouds ... it flaps along the water for three thousand *li*. Then it ascends on a whirlwind up to a height of ninety thousand *li*, for a flight of six months' duration.... Without sufficient density, the wind would not be able to support the large wings. Therefore when the *p'eng* had ascended to ninety thousand *li*, the wind is all beneath it. Then ... no obstacle ahead of it, it mounts upon the wind and starts for the south.**

This is the very bizarre myth with which the *Chuang Tzu* opens. The concept of air density is literally expressed as 'thickness of condensation'. Burton Watson, in his translation of the *Chuang Tzu*, renders it 'piled up deep enough'. The early Taoists were much preoccupied with the actual nature of air and wind, and they formulated a concept of the 'hard wind' to express the aerial capacity to lift a kite. It was in fact the pressure exerted by a moving airstream on the undersurface of an airfoil.

The Chinese did develop cambered-wing kites, with surfaces like wings, curved upwards at the top and either concave or flat underneath. But we cannot fix any date for this. We do not know whether or not the early Western designers of aircraft, such as Sir George Cayley in the nineteenth century, knew of Chinese cambered kites. But the chief influence of the kite on aviation was the invention in 1893 of the box kite by the Australian Lawrence Hargrave. Box kites were copied by bi-plane builders later. And until 1910, books on aviation generally began with introductory chapters on kites because of these origins. Early aviators even tended to use the slang term 'kites' to refer to their aircraft at the beginning of this century. Such was the heritage of the kite, which the Chinese had for two thousand years before the Europeans.

MANNED FLIGHT WITH KITES

FOURTH CENTURY BC

The earliest specific description which gave physical details of manned flight with kites occurs in historical accounts of the short-lived and obscure Northern Ch'i Dynasty. This dynasty existed for only 27 years, between 550 and 577 AD. The incidents described here occurred at the end of a century of disintegration preceding the third unification of China and the start of what is generally thought of as its Golden Age; at this time Europe was still in the throes of the Dark Ages.

Our historical account of man-flying kites comes from a rather gruesome episode in history when the first emperor of the Northern Ch'i, Emperor Kao Yang, who reigned from 550 to 559, carried out a systematic extermination of the entire Yuan family (whose original surname was T'opa, which because it was not Chinese they changed to Yuan to facilitate social assimilation), who had controlled the previous Wei Dynasty. In the last year of his reign alone, no less than 721 members of the family were massacred on his orders. Many of these murders were carried out with a sinister ingenuity.

Having embraced Buddhism, the Emperor Kao Yang went to receive his ordination at the Tower of the Golden Phoenix, north-west of the capital city of Yeh near modern Lin-chang, north of the Yellow River. In the ancient Chinese exercise of piety called 'the liberation of living creatures', fish and birds were released after they had been caught. The Buddhists took on this practice, believing that it conferred merit on a pious man. In a typical example of the perverse and insane behavior of one of China's most evil emperors, Kao Yang celebrated his Buddhist ordination with a ceremony which he also called

ABOVE (129) *Left:* American aeronautical pioneer Colonel Samuel Franklin Cody's man-lifting kite, in England, 1905, carrying Sapper Moreton, officer of the British Royal Engineers. On this occasion, Moreton rose 2,600 feet and remained there for one hour. Cody (1867–1913) was the first person actually to fly in his own kite. *Right:* Lieutenant Bassel in a man-lifting multi-stage kite in France, 1909. (From E. Charles Vivian, *A History of Aeronautics*, London, 1921.)

'the liberation of living creatures'; but his version of it was somewhat different.

The creatures he decided to 'liberate' were his enemies, the T'opa and Yuan families, and his method was to throw them from the top of the 100-foot tower. As the official history tells us: 'He caused many prisoners condemned to death to be brought forward, had them harnessed with great bamboo mats as wings, and ordered them to fly to the ground from the top of the tower. This was called a "liberation of living creatures".' We are also told that 'All the prisoners died, but the Emperor contemplated the spectacle with enjoyment and much laughter.'

The Emperor's aerodynamic fancies then underwent a further development, and he became determined to carry his amusing experiments to more exciting lengths. By 559, the last year of his reign, he was regularly using condemned prisoners to jump from the Tower of the Golden Phoenix as test pilots for man-flying kites. Needham comments: 'The circumstances show that what was going on was not quite simply a cruel emperor's sport with prisoners, for the cables of the kites must have required man-handling on the ground with considerable skill, and with the intention of keeping the kites flying as long and as far as possible.' One prince of the Wei, a prominent member of the Yuan family, was able to make a very successful flight for what seems to have been about two miles. The incident is described in a history of the period called the *Comprehensive Mirror of History for Aid in Government*, which was compiled from official documents of the time:

> **Kao Yang made Yuan Huang-T'ou and other prisoners take off from the Tower of the Golden Phoenix attached to paper kites in the form of owls. Yuan Huang-T'ou was the only one who succeeded in flying as far as the Purple Way, and there he came to earth. But then he was handed over to the President of the Censorate, Pi I-Yün, who had him starved to death.**

The famous Taoist author Ko Hung (283–343 AD) is called by Needham 'the greatest alchemist of his age, and the greatest Chinese alchemical writer of any age'. Ko Hung was an all-round scientist who discussed natural phenomena of all kinds, and he greatly concerned himself with astronomy and geology. He had a love of the exotic and the strange, and repeatedly emphasized that conventional ideas about nature were inadequate. When younger he had achieved prominence and fame as a soldier, suppressing the rebellions of 303 AD. Many years later, he was given the title of Marquess for these early services to

ABOVE (130) Modern hang-gliding has made the activities of the early Taoists seem less strange, since we have plenty of experience today of how it is possible to 'ride the hard wind' with a structure resembling a kite. The Chinese, however, were doing this in the fourth century BC.

the state. But he spurned promotion within the bureaucracy in order to pursue his scientific interests. He was friendly with the best scholars of his day.

Needham wrote of Ko Hung: 'His words about the series of different kinds of animals would be incomprehensible if we did not know well the perennial Chinese tradition of making kites in the shapes of animals. I have no doubt that what he was referring to were man-lifting kites.' Here is the passage, about manned flight in the fourth century, in which Ko Hung is referred to as the Master:

> **Someone asked the Master about the principles of mounting to dangerous heights and travelling into the vast inane. The Master said ... 'Some have made flying cars with wood from the inner part of the jujube tree, using ox leather straps fastened to returning blades so as to set the machine in motion. Others have had the idea of making five snakes, six dragons, and three oxen to meet the "hard wind" and ride on it, not stopping until they have risen to a height of forty *li* [about 65,000 feet]. That region is called the Purest of Empty Space. There the *ch'i* [emanation of the sky, or perhaps wind] is extremely hard, so much so that it can overcome the strength of human beings. As the Teacher [we do not know who is meant by Ko Hung here, but perhaps he refers to the Taoist sage known as Chuang Tzu] says: "The kite [bird] flies higher and higher spirally, and then only needs to stretch its two wings, beating the air**

no more, in order to go forward by itself. This is because it starts gliding on the 'hard wind'. Take dragons, for example; when they first rise they go up using the clouds as steps, and after they have attained a height of forty *li* then they rush forward effortlessly, gliding." This account comes from the adepts and is handed down to ordinary people, but they are not likely to understand it.'

As Needham says: 'For the beginning of the fourth century AD this is truly an astonishing passage.... There can be no doubt that the first plan which Ko Hung proposes for flight is the helicopter top; returning (or revolving) blades can hardly mean anything else, especially in close association with a belt or strap.

On the rest of the passage Needham goes on to say:

> I have no doubt that what he was referring to were man-lifting kites, and though as yet we have no evidence that Ko Hung or any of his contemporaries constructed such large instruments, there would have been really nothing to prevent it. For people expert in kite-flying the possibility was obvious.... Lastly, what is to be said of Ko Hung's 'hard wind'? From the examples he gives of the gliding and soaring of birds, it is obviously nothing else than the property of 'air-lift', the beating or rising of the inclined aerofoil subjected to the forces of an airstream, whether natural or artificial.... Ko Hung applies the concept ... very clearly to gliding flight, as indeed had Chuang Chou [Chuang Tzu] before him, when he wrote [in the fourth century BC] about the wings of the giant *p'eng* bird being airborne upon the density of the wind beneath. Ko Hung ends by attributing to him the idea that flying things rise up 'using the clouds as steps', which may be more than a poetic metaphor, hinting as it does at the existence of those ascending air-currents which modern glider pilots have learnt so well to utilize. Something of these could probably have been observed in the behaviour of smoke, and particularly of the mists and clouds on the lofty mountain heights which the Taoists delighted to frequent.

The Taoists had been taking refuge in obscure mountain sanctuaries for centuries. There they passed on their traditions orally from master to pupil, leaving much of the more esoteric lore unwritten. Kite flying, and flying with the aid of kites, was not only suited to the high and windy locations where these early adepts of aviation lived, but was most probably a pursuit viewed as immensely important, and carried out in great secrecy. A sage would be no more inclined to impart the secrets of how he was imitating the birds and learning how to ascend bodily to heaven, than he would be likely to explain to the man in the street how to make an elixir of immortality, or how to make 'gold' in a secret cave laboratory. By virtue of these pursuits the Taoists are now thought to be the true proto-scientists of early China.

By the thirteenth century, man-lifting kites were widely used throughout China. A dramatic account of the practice was given by Marco Polo in a version of his book called the 'Z' Manuscript:

> And so we will tell you how when any ship must go on a voyage, they prove whether her business will go well or ill. The men of the ship will have a hurdle, that is a grating, of withies [willow stems], and at each corner and side of this framework will be tied a cord, so that there be eight cords, and they will all be tied at the other end of a long rope. Next they will find some fool or drunkard and they will bind him on the hurdle, since no one in his right mind or with his wits about him would expose himself to that peril. And this is done when a strong wind prevails. Then the framework being set up opposite the wind, the wind lifts it and carries it up into the sky, while the men hold on by the long rope. And if while it is in the air the hurdle leans towards the way of the wind, they pull the rope to them a little so that it is set again upright, after which they let out some more rope and it rises higher. And if again it tips, once more they pull in the rope until the frame is upright and climbing, and then they yield rope again, so that in this manner it would rise so high that it could not be seen, if only the rope were long enough. The augury they interpret thus; if the hurdle going straight up makes for the sky, they say that the ship for which the test has been made will have a

quick and prosperous voyage.... But if the hurdle has not been able to go up, no merchant will be willing to enter the ship....

This must surely be one of the strangest forms of divination or fortune-telling in history!

In modern times, men have flown in kites. Since Pocock's attempts in 1895, many more have been made in Europe. B.F.S. Baden-Powell was the first European to achieve full success with man-flying kites in 1894. Therefore, if we take Yuan's successful flight in 559 AD as the first in the world from the point of view of historical record, we have a gap of 1335 years between the Chinese precedent and Europe's first flight.

Nowadays, the widespread use of hang-gliders by ordinary members of the public makes early Chinese manned flight in kites more credible than it would have been even a few decades ago. Skilled hang-glider pilots understand the nature of the air currents referred to by Ko Hung, and know how to use an ascending current of hot air to rise to over 2000 feet, simply by spiraling up with it. This is, in other words, 'using the clouds as steps' and 'flying higher and higher spirally', as described by Ko Hung, even if his figure of 65,000 feet must be a gross exaggeration. It would appear that the early Taoists did achieve dramatic success in manned kite-flying long before Ko Hung – rising to at least 2,000 feet – and probably over two thousand years ago. But even if we take the generation before Ko Hung as a conservative estimate for the date of the first attempt, we must still acknowledge the existence of manned flight with kites as having been no later than 250 AD – or well over seventeen hundred years ago.

ABOVE (131) A bronze incense-burner of the fourth or third century BC representing the mysterious mountain paradise of P'eng Lai, which the Chinese believed existed on the far side of the Pacific. It was customary to represent P'eng Lai in this way, with three-dimensional peaks and contours, often inlaid with precious or semi-precious stones. Such censers were a powerful influence in the development of relief maps during the third century BC. The earliest relief map in Europe dates from 1510. (Freer Gallery of Art, Smithsonian Institution, Washington D.C.)

THE FIRST RELIEF MAPS

THIRD CENTURY BC

The Chinese were the first to use relief maps, where the contours of the terrain were represented in models. Relief maps in China go back at least to the third century BC. In the great historical classic, the *Shih Chi (Historical Records)* of Ssuma Ch'ien (90 BC), there is preserved an account of probably the most famous relief map ever made. It was built in 210 BC for the tomb of the unifier of China, the megalomaniac emperor Ch'in Shih Huang Ti, and showed his conquests and empire. The history says:

In the tomb-chamber the hundred water-courses, the Yangtze River and the Yellow River, together with the great sea, were all imitated by means of flowing mercury, and there were machines which made it flow and circulate. Above [on the ceiling] the celestial bodies were all represented; below [presumably on the floor or on some kind of table] the geography of the earth was depicted.

In the mid-1980s, the tomb of this emperor was located, but it has not yet been opened and its investigation has been postponed for some time. Traces of mercury have been found in the soil around what is thought to be the entrance. Archeologists have already suggested that this is the

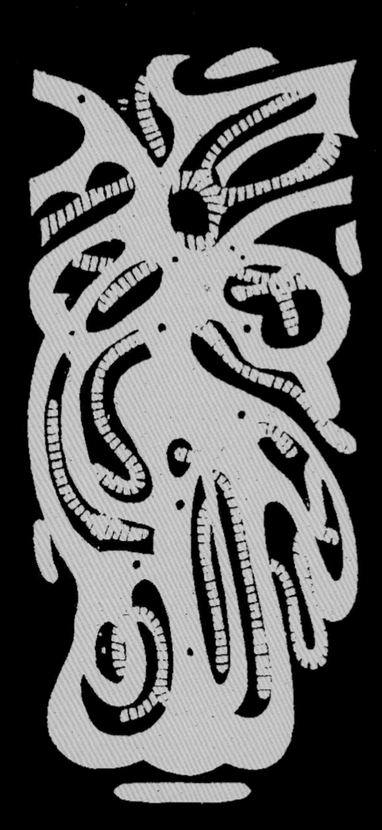

LEFT (132) Relief maps led to two-dimensional contour maps like the one on the left. This is a representation of the sacred T'ai Shan mountain range from a seventeenth-century edition of a much older book, *Map of the True Topography of the Five Sacred Mountains*. To far left, for comparison, is a modern contour map.

very mercury which circulated in the relief map described above. If this be so, we may be on the verge of one of the most exciting archeological discoveries ever made, for the ancient relief map will presumably still be largely intact. The descriptions which survive state that it was carved out of solid rock by the labour of 700,000 convicts. It was therefore absolutely enormous. Automatically triggered crossbows were set up to fire at intruders, and if the strings have survived these would still be in working order.

The Chinese had a curious and persistent legend about an island mountain paradise, named P'eng-Lai, in the Eastern Seas (that is, the Pacific). It was said to be an abode of sages and a heaven upon earth, frequented by the mysterious immortals who, on drinking elixirs of life, had etherealized their bodies, become 'feathered', and were able to soar into the sky. Expeditions were sometimes sent to P'eng-Lai, and at least one full-scale naval task force went in search of the paradise but failed to return. We may never know for certain where this legend originated; perhaps P'eng-Lai was Tahiti, or Hawaii, or even America, but it was often depicted in relief on Han Dynasty pots and censers, dating from the third century BC. One such 'relief map' jar may be seen in Plate 131 (page 195). These jars had a powerful influence in developing relief-map techniques.

A description survives of strategic relief maps made by the general Ma Yüan in 32 AD. The valleys and mountains were represented by modeling in rice. Such military relief maps were so useful that Chiang Fang wrote a special book entirely on the subject, entitled *Essay on the Art of Constructing Mountains with Rice* (c. 845).

Relief maps were also carved in wood. This led to what appears to be the invention of the jigsaw puzzle. We have this description of a 'jigsaw map' in the official history of the Liu Sung Dynasty:

> **Hsieh Chuang (421–66) made a wooden map 10 feet square, on which mountains, water-courses and the configuration of the earth were all well shown. When one separated the parts of the map then all the districts were divided and the provinces isolated; when one put them together again, the whole empire then once more formed a unity.**

The great scientist Shen Kua wrote the following interesting account in his book, *Dream Pool Essays*, of 1086:

> **When I went to a government official to inspect the frontier, I made for the first time**

> a wooden map upon which I represented the mountains, rivers and roads. After having explored personally the mountains and rivers [of the region], I mixed sawdust with wheatflour paste [modelling it] to represent the configuration of the terrain upon a kind of wooden base. But afterwards when the weather grew cold, the sawdust and paste froze and was no longer usable, so I employed melted wax instead. The choice of these materials was dictated by the necessity of making something light which would not be difficult to transport. When I got back to my office [in the capital] I caused [the relief map] to be carved in wood, and then presented it to the emperor. The emperor invited all the high officials to come and see it, and later gave orders that similar wooden maps should be prepared by all prefects of frontier regions. These were sent up to the capital and conserved in the imperial archives.

A wooden relief map was also made by Huang Shang in 1130, which later attracted the attention of one of China's greatest Confucian philosophers, Chu Hsi, who was born in that year and died in 1200. He attempted to find the relief map so that he could study it, and made his own ones, using clay as well as wood. The quaintly named book *Jade Dew from the Forest of Cranes* preserves this description of one of his maps:

> Chu Hsi also made a wooden map of the countries of the Chinese and Barbarians, upon which the convexities and concavities of mountains and rivers were carved. Eight pieces of wood were used, with hinges to connect them together. The map could be folded and one person could carry it. Whenever he travelled, he took this along with him. But it was never really completed.

It is quite likely that the idea of making relief maps was transmitted to the Arabs from the Chinese, and thence eventually to Europe. The earliest certain relief map in Europe was made in 1510 by Paul Dox, showing the neighborhood of Kufstein in Austria. The Arab Ibn Battutah (1304–77) gave an account of a relief or raised map which he saw at Gibraltar. Nothing earlier than this is known outside China.

THE FIRST CONTOUR TRANSPORT CANAL

THIRD CENTURY BC

The Chinese built the world's first contour transport canal (that is, exploiting and following the contour of the land as a way around or over hills) in the third century BC. There is no question that the Chinese were the greatest civil hydraulic engineers in history until modern times. A few facts about the Grand Canal, that gigantic artery for transport by barge, are sufficient to make the point. The Grand Canal was built over many centuries. It extends nearly 10 degrees of latitude on the globe, attaining a length of nearly 1100 miles, and achieves a summit height of 138 feet above sea level. A good way to envisage it is to imagine a broad canal extending from London to Tangiers or from New York to Florida. This great engineering feat was complete by the year 1327.

Nothing remotely approaching the Chinese canal systems existed in Europe until the four great seventeenth-century canals in France, the last of which was not finished until 1775. But none of these

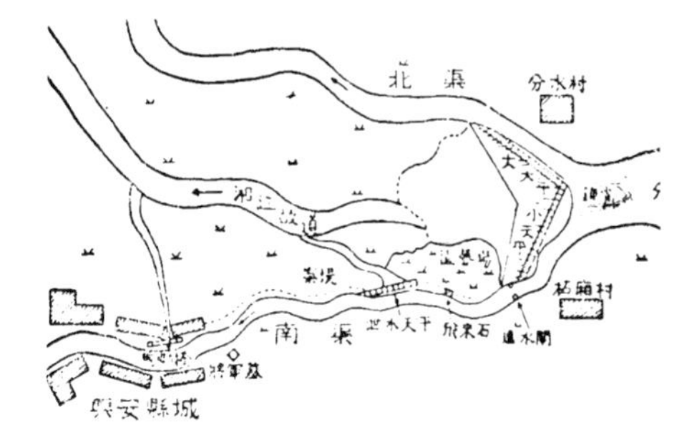

ABOVE (133) The headworks of the Magic Canal. North is at the top left-hand corner. The waters of the Hsiang River flow in from the right, and meet the 'Spade Snout', which divides the waters into two channels. The 'Spade Snout' is flanked on either side by large spillways, to drain away some of the water and reduce the force of the current. The Hsiang continues along the northwards course, and the Magic Canal commences with the channel running along the bottom of the diagram. Massive embankments and walls of dressed stone line the length of the canal. Further spillways keep the water level under control. The canal is 15 feet wide and the water is 3 feet deep. The canal's length is just over 20 miles. It links two rivers flowing in opposite directions, thus making possible continuous inland navigation for a distance of 1250 miles, from the latitude of Peking to Canton. This canal was thus a major economic and military development for China, foreshadowing the later Grand Canal, which runs for nearly 1100 miles.

was longer than 150 miles. There were only 630 miles of canals in all France by the end of the eighteenth century, and even by 1893 the total mileage of French canals had only reached three times the length of China's Grand Canal alone in 1300. The Grand Canal is between 10 and 30 feet deep and often 100 feet broad, whereas the early nineteenth-century canals of England were only 5 feet deep and 45 feet broad. Needham estimates that 'the canals of all Europe probably still fall short of the Chinese artificial navigable waterways in mileage.'

There is no doubt that the ancient Babylonians were highly advanced in irrigation canals, but we are uncertain to what extent they had progressed in transport canals, sluice gates, and so on. Evidence for the use of sluice-gates by the Phoenicians goes back to the second millennium BC in the harbour at Sidon. As regards non-Babylonian canals, an Egyptian canal connecting the Nile with the Red Sea was commenced by Necho (610–595 BC), but was only completed under Ptolemy Philadelphus about 280 BC. There was a slipway across the Isthmus of Corinth by the early sixth century BC which remained in use until the ninth century AD; it was not a canal, and ships were pulled on slipways at either end and carried on wheeled cradles along a masonry road for most of the length. Ancient Egyptian waterworks connected with the Nile were seasonal and intended to help with both irrigation and silt deposition, rather than to act as transport arteries.

The greatest canal of Middle Eastern antiquity was apparently the Nahrawan Canal of about the third century AD, which ran for about 250 miles with an impressive breadth of 400 feet, emanating from and then rejoining the Tigris river. The Babylonian area had much earlier seen vast networks of irrigation canals radiating outwards from the rivers attempting to ensure perennial water supply and to transform the river valleys into artificial deltas. There may have been canals built specifically for transport, but they would not have been contour ones, rather glorified irrigation canals in a flat landscape. We must also not ignore India, and must acknowledge that Nandivardhana constructed a transport waterway in the fifth century BC. Unfortunately, the history of these ancient non-Chinese constructions is still buried in considerable obscurity because of the lack of surviving evidence.

In China, the Hung Kou transport canal is thought to have been the first to be constructed. It was apparently built in the sixth century BC. However, what is of particular interest to us is the world's first contour transport canal, of the third century BC, the Magic Canal (Ling Ch'ü). This was indeed a most impressive pioneering achievement. It was constructed by the engineer Shih Lu on the orders of the Emperor Ch'in Shih Huang Ti, the tyrannical and superstitious monster who first unified China. The impetus for this innovative type of canal was to assist in supplying the emperor's armies sent south in 219 BC to conquer the people of Yüeh. We are told by the great historian Ssuma-Ch'ien that:

> **[the emperor] sent the Commanders (Chao) T'o and T'u Chu to lead forces of fighting-men on boats with deck-castles to the south to conquer the countries of the hundred tribes of Yüeh. He also ordered the Superintendent (Shih) Lu to cut a canal so that supplies of grain could be sent forward far into the region of Yüeh.**

The Magic Canal, which is still used today, is just over 20 miles long. Its chief interest is thus not its length, which is unexceptional. The construction of the Magic Canal, linking as it did two rivers flowing in opposite directions, made possible the continuous

RIGHT (134) A view of the Magic Canal, with one of the boats currently in use on it.

LEFT (135) A spillway of the world's oldest contour transport canal, the Ling Ch'u, or Magic Canal, which connects the River Li and the River Hsiang and was constructed in the third century BC (though since restored and improved).

inland navigation of barge transport for a distance of 1250 miles in a direct line, from the fortieth to the twenty-second parallel. One could thus sail inland from the latitude of Peking in the north as far as Canton and the sea – to what is today Hong Kong. The Magic Canal was the final link in the chain.

The problem that had to be overcome was that the River Hsiang, with its source at Mount Haiyangshan, flowed northwards, while the nearby River Li flowed southwards. If only one could get a boat from one to the other! – for the Hsiang led eventually to the Yangtze and the Li joined a tributary of the West River and led to Canton. Near the little village of Hsing-an, the Hsiang and the Li, in a landscape of limestone hills, are only three miles apart. Simply joining them was not sufficient. Another solution had to be found.

There was a saddle in the hills at this point along which a canal could be dug. The rivers themselves were unruly and a lateral transport canal had to be dug alongside the Hsiang river for 1½ miles at a more even gradient than the river itself had. At the other end, 17½ miles of the Li river had to be canalized in order to regulate it and make navigation possible. Only with the two rivers 'tamed' at either end like this could a 3-mile canal then be dug to join them. A mound shaped like a snout was constructed in the middle of the swiftly running Hsiang to divide its flow, and lead off much of the rushing water. It was backed up by two spillways, and further spillways were made lower down. Several bridges at Hsing-an were constructed to cross the canal, which was 3 feet deep and 15 feet wide. The system of division of the waters and spillways resulted in only about three-tenths of the water from the Hsiang entering the connecting canal, so that it was not overwhelmed.

By being built along the contours of the saddle in the hills, the canal was nearly level. Eighteen flash-lock gates were there by the ninth century at the latest, reducing the number of towers needed for barges by regulating the level and the flow. They were changed to pound-locks by the tenth or eleventh century. The Magic Canal came to be considered a sacred waterway, with a dragon as its governing spirit. The dragon's emissaries were said to be blue snakes called 'dragon colts', which coiled playfully in the hands of visitors. A modern railway bridge goes right over the old Magic Canal, which is still used. Unless one knows its importance, it does not seem particularly impressive, and can easily be missed. But no comparable canal seems to have been built in Europe until the beginning of the thirteenth century.

THE PARACHUTE

SECOND CENTURY BC

Most people know that Leonardo da Vinci left sketches of the parachute, which was the first appearance of the idea in Europe. However, the Chinese seem to have invented the parachute and actually used it well over 1500 years before Leonardo.

The first textual evidence we have for this is in the famous *Historical Records* of China's greatest historian, Ssuma Ch'ien, which was completed about 90 BC. We can therefore safely consider the parachute as dating from at least the second century BC. Ssuma Ch'ien had access to vast archives, and the fact that he attributed the parachute to such remote antiquity means that its origins may well have been some centuries before this time.

As the story goes, the legendary hero, Emperor Shun, was fleeing from his father, who wanted to kill him. He took refuge in a large granary tower, and his father set fire to it, hoping to burn him to death.

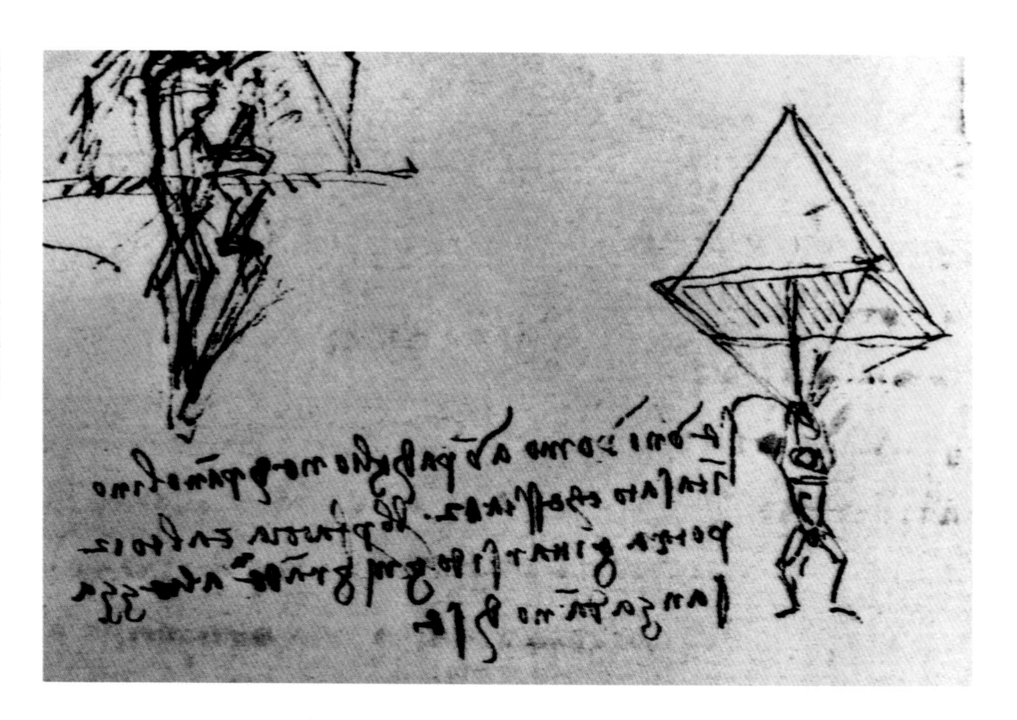

ABOVE (136) This is a famous sketch of a man descending on a parachute, drawn by Leonardo da Vinci (1452–1519) in his notebooks. Until the Chinese invention of the parachute 1,500 years earlier was recognised, people thought that the original inventor must have been da Vinci himself. It is conceivable that da Vinci had heard of parachutes in China. Marco Polo (1254–1324) had not only visited China, along with other Italians, two centuries earlier, but had left a detailed description of a man-flying ki.e in his own memoirs so that such things were 'in the air'.

But Shun tied a number of large conical straw hats together and jumped, using them as a parachute. From this we can assume that there was indeed a jump by someone and that over the years the tale became attached to a legendary episode in the life of Shun. There was a commentary on the story in the eighth century AD by Ssuma Chen (a different person from the historian just mentioned), who remarked that the hats acted like the wings of a bird, making Shun's body light and bringing him safely to the ground.

Needham brings forward a medieval mention of the use of the parachute, from a book called *Lacquer Table History* by Yo K'o. This book was published in 1214 and recounts events witnessed at Canton in 1192. In Canton at that time there was a large Arab community of merchants, who had their own mosques, one of which had a 'grey cloud-piercing minaret like a pointed silver pen' with a spiral staircase inside. At the very top was a huge golden cock, which was missing one leg. The leg had been stolen in 1180 by a cunning thief who had escaped by parachute. The robber's own account is preserved, for he seems to have been something of a local hero. He describes the escape as follows: 'I descended by holding onto two umbrellas without handles. After I jumped into the air the high wind kept them fully open, making them like wings for me, and so I reached the ground without any injury.'

We have documentary evidence that the first construction and use of a parachute in Europe was due to a report of a visitor to Thailand, who witnessed its use by Chinese and Siamese acrobats. The account was written by Simon de la Loubère, appointed Ambassador to Siam by King Louis XIV of France from 1687 to 1688. In his *Historical Relation* he wrote:

> **There dyed [died] one, some Years since, who leap'd from the Hoop, supporting himself only by two Umbrella's, the hands of which were firmly fix'd to his Girdle; the Wind carry'd him accidentally sometimes to the Ground, sometimes on Trees or Houses, and sometimes into the River. He so exceedingly diverted the King of Siam, that this Prince had made him a great Lord; he had lodged him in the Palace, and had given him a great Title; or, as they say, a great Name.**

The historian J. Duhem has established that L.S. Lenormand read this passage a century later, was stimulated to make trials, jumping from the tops of trees and buildings, and was quite successful. In 1783, Lenormand gave the invention its name of 'parachute'.

Lenormand told the brothers Joseph and Etienne Montgolfier about it, and the famous pioneering balloonists were then responsible for A.J. Garnerin's jump from a balloon with a parachute in 1797. This was a direct result of a Westerner having witnessed Chinese parachutes. Needham remarks, aptly: 'There are not many cases in which so clear a line of transmission is detectable.'

It should be noted that the parachutes used in Siam in 1687 were just the same as the one that had been used by the thief at Canton 500 years earlier, namely, a pair of umbrellas. It is quite possible that the Cantonese thief was himself a professional acrobat, and therefore knew how to use the parachute successfully – and even more important, had the confidence to jump. Probably the pair of umbrellas was a stock-in-trade of the highest calibre acrobats for hundreds of years.

In June 1985 a woman named Yang You-hsiang of Liao-chiao in Hopei Province in China was carried 550 yards in the air by a tornado. She landed safely because she had been carrying an open umbrella, which acted as a parachute. Her only injuries were from hailstones. (London *Daily Telegraph*, 22 June 1985.) Clearly very much in the national tradition!

MINIATURE HOT-AIR BALLOONS

SECOND CENTURY BC

By the second century BC, the Chinese were making miniature hot-air balloons using eggshells. A book that was written at that time, *The Ten Thousand Infallible Arts of the Prince of Huai-Nan*, mentions this pastime: 'Eggs can be made to fly in the air by the aid of burning tinder.'

An ancient commentary added to the text explains further: 'Take an egg and remove the contents from the shell, then ignite a little mugwort tinder inside the hole so as to cause a strong air current. The egg will of itself rise in the air and fly away.' Mugwort (*Artemisia vulgaris*) is a very common weed, the long, dried stalks of which were used in China as tinder for lighting fires, and powdered as a flammable element in incense sticks.

Few references are found in Chinese writings to the use of the hot-air balloon principle. Perhaps it was for a long time not thought worthy of much attention, but by medieval times the military possibilities were being exploited. There are several references in European chronicles to the use of hot-air balloons, shaped like dragons, either for signaling or as standards by the Mongol Army at the Battle of Liegnitz in 1241. The principle was in all probability obtained from the Chinese; the Mongol Dynasty finally established full sway over all of China only 19 years after this.

Needham has pointed out that paper was available in China so many centuries before anywhere else (from the second century BC) that 'the development of the classical globular lanterns would have encouraged experimentation. When their upper openings were too small and the source of light and heat unusually strong, they must sometimes have shown a tendency to rise and float free of support.'

A vivid eyewitness description of the use of hot-air balloons in the form of paper lanterns is provided by Peter Goullart, who lived between 1939 and 1949 in the Lichiang region of Yunnan Province in the south of China:

> July, which was the critical month before the rainy season, had several festivals. With the rice already planted, the people did not have much to do and the evenings were devoted by the younger set to dancing and to flying the *kounmingtengs* – the lighted balloons. During the day one could see the young men ... pasting together the oiled sheets of rough paper to form the structure of a balloon. These balloons were then dried in the sun and were ready for use in the evening. Crowds gathered to watch. A bunch of burning *mingtzes* was tied underneath; the balloon swelled and quickly rose into the air to the shouts of excited spectators. The higher it rose the more good luck it promised to its owner. Some went up very high indeed and floated in the sky like red stars for several minutes. At the end they burst into flames and fell, sometimes causing fires by setting light to straw in unwatched farm-houses. Sometimes there were as many as twenty of these balloons floating through the dark sky. Balloon flying lasted for about a couple of weeks and it was great fun.

ABOVE (137) Chinese globe lanterns made of paper, like the one shown here, were used as miniature hot-air balloons in China for centuries.

The invention of paper came at about the same time as the first balloons were tested – the second century BC.

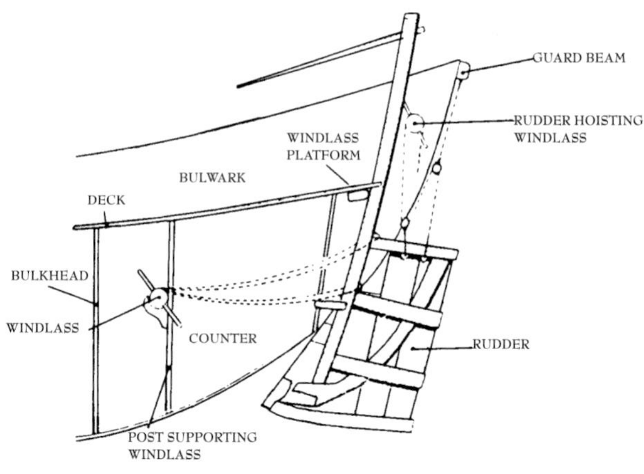

ABOVE (138) The stern of a Hong Kong fishing boat in dry dock, showing the 'fenestrated' rudder. The holes make it easier to turn the rudder through the water, but do not diminish the steering function. This Chinese invention was introduced to Europe in 1901. (Waters Collection, National Maritime Museum, London.)

ABOVE RIGHT (139) The rudder and tiller of a Hangchow Bay freighter. This is a 'balanced' rudder, that is, it curves round somewhat in front of the post (to the left in the diagram). It is also 'slung' by a tackle, which can be raised or lowered by the hoisting windlass. It is thus possible to pull the rudder entirely up above water level and pass over shallows without damaging the rudder.

LEFT (140) The world's oldest depiction of a ship's rudder, on a pottery model of a ship dating from the first century AD. (Kuangchow Historical Museum, Canton.)

THE RUDDER

FIRST CENTURY AD

Until Europeans adopted the rudder from the Chinese, Western ships had to make do with steering oars. This meant that long voyages of discovery by Europeans were impossible. The famous voyages of Christopher Columbus, Vasco da Gama, and others of their time were only made possible by the adoption of Chinese nautical technology. The oldest Western evidence for rudders is found in church carvings of about the year 1180. This is within a few years of the first European evidence for the ship's compass. Rudders and compasses thus seem to have reached Europe at about the same time, which is hardly surprising, since they were so closely associated. The rudder enables one to steer a ship properly, and the compass helped one to decide where to steer it.

The world's oldest representation of a rudder may be seen in Plate 140 (left). This is a pottery model of a Chinese ship excavated from a tomb dated to the first century AD. The model is about 1 foot 10 inches long, and its rope slinging tackle has long since rotted away. But a slung axial rudder may clearly be seen. (A slung rudder is one which can be raised and lowered by rope tackle or chains; when entering shallows, it is often desirable to pull the rudder up so that it will not be snapped off.)

Chinese seagoing rudders grew to many times the size of a man. Huge ships with enormous rudders were used on the Chinese voyages of discovery which preceded the European ones. The Chinese sailed round the Cape of Good Hope in the opposite direction to that taken by the Europeans and at an earlier time. They were also first to discover Australia, landing at the site now called Port Darwin. Chinese trade with the Philippines and Indonesia was common; and trade with the eastern coast of Africa was so extensive that pieces of broken Chinese porcelain are to be found scattered all up and down the beaches of Tanzania and Mozambique, dating back for centuries. The Chinese also made voyages to the American continents, though it is questionable whether they were return voyages. Many Asian influences have been identified in ancient America by Needham and others. But the Chinese who arrived were quite possibly stranded, unable to return home, owing to the greater difficulty of sailing westwards across the Pacific.

Another traditional Chinese invention was the 'fenestrated rudder', which put simply is a rudder with holes made in it (see Plate 138 (opposite)). The Chinese soon discovered that while easing the task of turning the rudder through the water, the holes did not appreciably diminish its steering function. However, it was not until 1901 that fenestrated rudders were introduced to the West. Until that time, a coal-fired torpedo boat travelling at 30 knots was unable to turn its rudder at speed. Fenestration made this possible.

The earliest rudders in China were what is called 'balanced' rudders. This means that part of the blade projected in front of the post. Such rudders are easier to use, but Europeans did not adopt them until the nineteenth century. One of the earliest ships to use such a rudder was the *Great Britain* of 1843. The British were in the forefront when it came to adopting Chinese inventions for naval use, with this as well as the square-pallet chain pump as a bilge pump (see page 63) and watertight compartments in hulls (see page 211). It is no exaggeration to say that the superiority of the British Navy was to a large extent due to its readiness to adopt Chinese inventions more rapidly than other European powers.

MASTS AND SAILING

SECOND CENTURY AD

It could probably be safely said that the Chinese were the greatest sailors in history. For nearly two millennia they had ships and sailing techniques so far in advance of the rest of the world that comparisons are embarrassing. When the West finally did catch up with them, it was only by adapting their inventions in one way or another. For most of history, Europeans used ships which were drastically inferior to Chinese ships in every respect imaginable. They had no rudders, no leeboards, no watertight compartments, single masts, and square sails,

RIGHT (141) A Yangchiang fishing junk becalmed. Note that the masts are staggered thwartwise, with the mast in the front being further to the left, and the masts behind being successively further to the right. This ingenious Chinese technique of preventing the sails from becalming each other was never adopted in the West. (Waters Collection, National Maritime Museum, London.)

ABOVE (142) A three-masted sea-going Chinese junk of modern times, though few using only sails now remain, as engines have been widely adopted for convenience. This photo shows particularly clearly how the lug sails extend in front of the masts (that is, the front edge of the sail, known as 'the luff', is not behind the mast), which was a Chinese innovation, and that the sails are aligned parallel with the boat's long axis (not its keel, because Chinese junks do not have keels), rather than at right angles to it, as was the inefficient Western tradition. This fore and aft rigging enables a ship to tack into the wind, which Western ships were incapable of doing with their square sails at right angles to the boat's long axis, until the adoption of Chinese sailing techniques in the fifteenth century (which then enabled such explorers as Columbus to make their long voyages of discovery).

which left them at the mercy of the winds to an extent which today we would consider ludicrous. This continued to be the case even into the nineteenth century. As Needham says: 'As late as 1800 they sometimes had to wait as long as three months at Hamoaze in order to get into Plymouth Sound, and this was long after the introduction of a lateen sail on the mizzen mast.'

Chinese sails were innately superior to Western ones. Apparently commencing with sailing rafts, the Chinese had recourse to their native bamboo to aid sail construction. This led to the use of sails consisting of bamboo battens with matting stretched between them. Such sails could be hauled up and down like a set of open venetian blinds covered with cloth. Sailing in this way was easier than with Western canvas sails, for it was not necessary for sailors to climb along yard-arms to furl or unfurl the sails when the wind changed: everything could be done from the deck with windlasses and halyards. Furthermore, a sail with battens could be used with as many battens exposed as one wished; in a gale, one could let two battens' worth of sail be exposed, but in a light breeze, the whole sail could be exposed to catch it. This added greatly to control of the vessel. Battens were not used in Western sails until modern racing yachts partially adopted them.

Bamboo and mat sails have another advantage: a ship can sail even if half of the sail consists of holes which have been torn in it, or which have appeared from rotting or other deterioration. A Western canvas sail could not work with so many holes. Also, the battens held Chinese sails taut, which was more efficient aerodynamically. There is unnecessary wind turbulence caused by sails which belly too much in the wind: this reduces speed. Even today, modern racing yachts are thought by some to have insufficiently taut sails and to allow far too much bellying. It is likely that there is still more to be learnt from Chinese sails amongst the yacht-racing fraternity.

However, the greatest advance of Chinese sails was to pass from the basic square sail to the fore-and-aft rig (which enables one to sail into the wind) using a lug sail. This is the sail which everyone has seen in pictures of Chinese junks (see Plates 141 and 142 (pages 205 and opposite)). It has been said that the batten-strengthened lug sail of the Chinese junk is the finest sail ever invented. A modern version of it is the gaff sail; the forward edge (called 'luff' by mariners) has been pushed back to commence at the mast rather than slightly in front of it.

Modern yachts have a gaff and a 'leg of mutton' or 'Bermuda rig' sail, the latter being more or less the hind half of a lug sail – an upright triangle. It is by no means certain that a modern yacht could outperform a comparable Chinese junk built for racing and using lug sails with battens in the classic manner. A race between two such vessels would be interesting. The junk is the direct ancestor of the yacht, but may still be better.

Fore-and-aft rigs with lug sails must have existed in the second-century AD China: by the third century they are clearly described in the book *Strange Things of the South* by Wan Chen, on ships capable of carrying the staggering amount of 700 people and 260 tons of cargo! Wan Chen describes these ships as having four masts, and says:

> The four sails do not face directly forwards, but are set obliquely, and so arranged that they can all be fixed in the same direction (parallel to each other), to receive the wind and to spill it. Those sails which are behind the most windward one receiving the pressure of the wind, throw it from one to the other, so that they all profit from its force. If it is violent, [the sailors] diminish or augment the surface of the sails according to the conditions. This oblique rig, which permits the sails to receive from one another the breath of the wind, obviates the anxiety attendant upon having high masts. Thus these ships sail without avoiding strong winds and dashing waves, by the aid of which they can make great speed.

Another book of 260 AD by K'ang T'ai describes ships with as many as seven masts, used for sailing to Syria. The use of multiple masts was rendered easy and natural in Chinese ships due to the bulkhead construction of their hulls, which have a whole series of obvious cross-timbers capable of providing bases for masts. Wan Chen makes clear that as early as the third century, the Chinese were already well aware, in the region around Canton, of the best way to avoid one sail being becalmed by another behind it. They positioned the masts not directly in a row along the centre of the ship lengthwise, but staggered thwartwise from one side to the other. This brilliant idea was never adopted in the West; even the fastest Yankee Clippers, of which Europe was once so proud, had the pedantic feature of the masts marching along straight above the keel, so the rear sails becalmed the fore sails.

The only way to sail into the wind is by freeing oneself from the fixed notion that a sail must be the traditional square sail whose basic position is at right angles to the boat. It is true that there is a tedious process known as 'wearing', by which square-sailed boats loop round over and over again and slowly creep into the wind. But the only efficient method is what is called 'tacking', and this cannot be done with a square sail.

What is needed in order to tack into the wind is a fore-and-aft rig: a sail whose axis is essentially *along* the long axis of the boat, rather than at right angles to it. On such a rig, the mast is no longer just a long pole holding up the square sail, but becomes a pivot for the sail to swing to one side and then to the other to catch the wind on alternate sides as the ship tacks. This the Chinese did.

When sailing into the wind, a tacking ship can experience a great deal of leeward drift, blowing the ship too much sideways and making too little progress forward. The Chinese therefore invented the leeboard to prevent this. This is basically a board lowered into the water on the lee side of the ship (opposite the direction of the wind) to exert pressure on the water and prevent drift in that direction; it also helps to hold the ship upright. Sometimes the leeboard would be lowered from a slot in the centre of the ship, in which case it can be called a centre-board.

Leeboards existed in China from at least the eighth century. In a book by Li Ch'üan of 759 AD entitled *Manual of the White and Gloomy Planet [of War; Venus]*, we are told of leeboards on certain warships that they 'held the ships, so that even when wind and wave arise in fury, they are neither driven sideways, nor overturn'. Leeboards did not appear in Europe before about 1570, however, when they were adopted by the Dutch and Portuguese who were then trading with China.

When Europeans eventually made their way to China and saw the Chinese ships, with their many masts

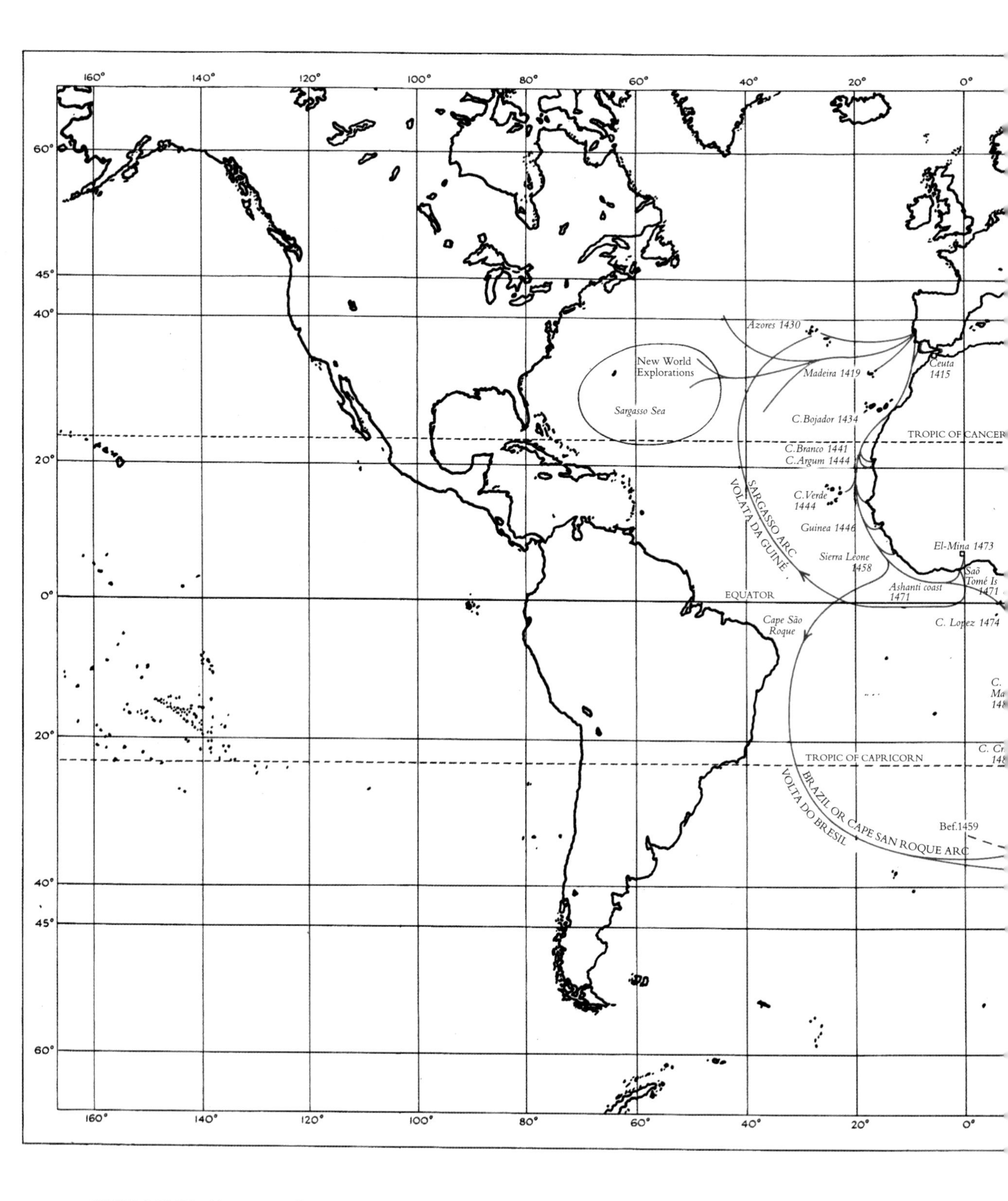

ABOVE (143) This elaborate map of sea voyages was prepared under the personal supervision of Joseph Needham, on the basis of many years of research. The red lines show fifteenth-century Portuguese voyages (extending as far east as Goa); the blue lines show fifteenth-century Chinese voyages for comparison. When the lines are dotted, it indicates that the voyages are conjectural and not proven. Chinese dates of earliest proven visits to the locations are bracketed if they are taken from textual evidence prior to the fifteenth century. The two blue lines emanating out into the Pacific Ocean and ending in question marks represent the unknown fates of convoy voyages heading in the direction of America, which never returned. Needham wrote an entire book on the subject of trans-Pacific sea contacts, and believed that it was possible that the Chinese had reached America more than once but were unable to return home and were thus essentially marooned, leading to possible cultural influence on ancient American civilisations.

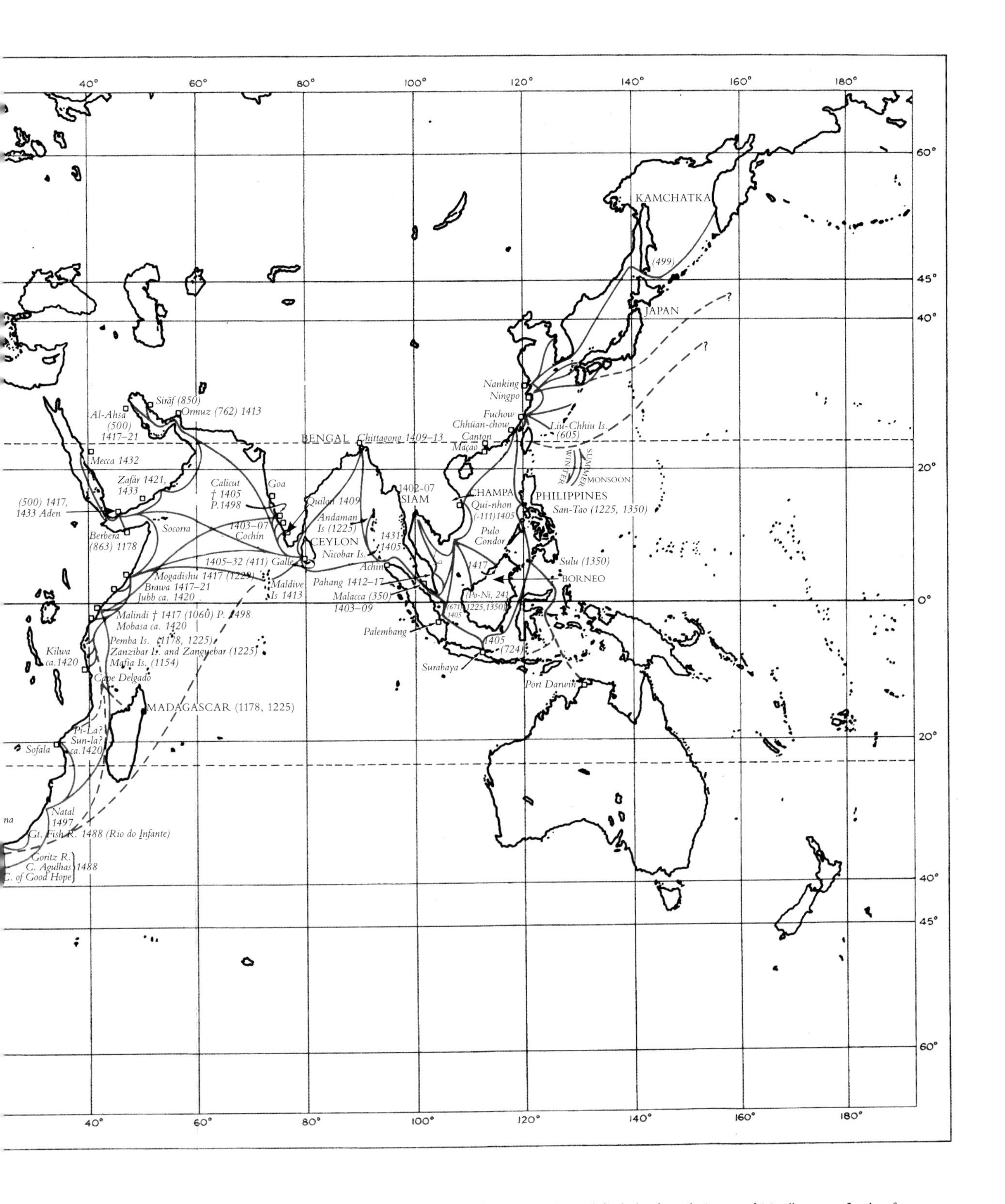

Although the pre-European Chinese voyage to Port Darwin in Australia is shown as conjectural for lack of conclusive proof, Needham was firmly of the opinion that it had occurred and that it was the Chinese who had 'discovered Australia'. Such vast quantities of Chinese pottery have been found littering the East African coast that East Africa has been proven to be a frequent and routine trade destination for the Chinese, and African giraffes were even transported back to the Emperor for his private zoo, and Chinese paintings of them survive. The Chinese sailed as far north-west as Arabia and many centuries earlier even had an ambassador resident at Rome (on which the sinologist Homer Dubs wrote an entire book). Chinese trade with the whole of South-east Asia, India, Ceylon, and the Philippines was frequent and routine, supplying the Chinese with exotic products of the south, such as spices, colourful feathers, tropical wood, Burmese jade, strange fruits, vegetables and plants.

and their fore-and-aft rigs, they were immensely impressed. The first attempt to copy what they saw was the adoption in Europe of multiple masts on ships, with square sails on the fore masts but a fore-and-aft rig for the mizzen mast at the rear of the ship. Since the Arab lateen sail was ready to hand, it was the lateen which was fitted to the mizzen of the European three-masters commencing about 1304, not long after Marco Polo's time in China. But the idea was slow to be assimilated, and only a couple of centuries later did such ships really come into their own.

In 1492 Christopher Columbus sailed such a ship to America. All the Western three-masters of the sixteenth and seventeenth centuries carried lateen sails aft. After that, the more efficient gaff sail replaced the lateen on the mizzen mast of European vessels. The actual Chinese lug sail apparently did not make an appearance there before the sixteenth century in the Adriatic. Certain traditional Venetian boats with lug sails appear to be direct copies of Chinese vessels in other respects as well, having nearly flat bottoms and enormous Chinese-style rudders.

The nautical historian H. Warington Smyth said that these Venetian craft are 'one of the finest forms of sea-going lugger in the world'. They are thought to have derived from designs of Chinese ships brought back by Marco Polo and his Italian peers. But the luggers remained regional, the main line of European development taking the idea of the Chinese ship rather than the *design*. This was undoubtedly a mistake, for the great European sailing ships were truly inefficient compared to sea-going junks, of which Smyth had this to say: 'As an engine for carrying man and his commerce upon the high and stormy seas as well as on vast inland waterways, it is doubtful if any class of vessel is more suited or better adapted to its purpose than the Chinese junk, and it is certain that for flatness of sail and for handiness, the Chinese rig is unsurpassed.' Needham adds: 'The Chinese balance lug ranks indeed among the foremost achievements in man's use of wind power.'

It is probably a lack of bamboo for battens which held back the adoption of the lug sail in Europe except in small localities like the Adriatic, or amongst the Portuguese on their *lorchas* of the sixteenth century. The Chinese stimulus to European sailing was therefore not as powerful as it might have been if direct copying had taken place in rigs, and with masts. But in summing up the situation, it might be well to quote the views of another nautical historian, G.S. Laird Clowes:

> **In 1400 AD northern ships were entirely dependent on a fair wind, and were quite unable, indeed never attempted, to make headway against an adverse one. Before 1500 AD European ships had been able to make the long ocean voyages which had resulted in Columbus' discovery of America, Dias' doubling of the Cape of Good Hope, and the opening of the Indian Ocean trade route by Vasco da Gama. Other scientific advances, such as the introduction of the mariner's compass from China, bore their part in making such voyages possible, but without the far-reaching improvements in masts and sails the great discoverers could never have accomplished their work.**

RIGHT (144) An illustration of a small Chinese fighting ship of the Ming Dynasty or earlier, published about 1700. The wood block for the plate probably dates from at least 1500, and must have appeared in earlier printings. Mid-battle, one man is shooting an arrow as others fight on deck with swords or spears. Particularly clear are the bamboo mat sails with their battens, which could be raised or lowered like blinds according to the wind, work unimpeded by holes, and were so aerodynamically efficient that the little attack boats could manoeuvre round the bigger ships like swooping swallows. It was for lack of such battens that Western ships were so inferior, partially because of the unavailability of bamboo in the West (no suitable material for battens was ready to hand, and battens were therefore never thought of). (Collection of Robert Temple.)

LEFT (145) A junk under repair in dry dock in Hong Kong. Four or five transverse bulkheads can be seen inside the ship and various ribs or frames are also visible. The habit of building hulls like this meant that watertight compartments were possible. If a leak occurred in one section, it could be sealed off from the rest and the ship remained afloat. Chinese ships had watertight compartments from at least the second century AD, but the idea only caught on in the West at the end of the eighteenth century. (Waters Collection, National Maritime Museum, London.)

After a gap of at least 1300 years, Europeans thus adopted Chinese concepts of rigging. In every way, whether for navigation, propulsion, or steering, Europeans were dependent upon Chinese ideas in order to be capable of the Great Voyages of Discovery. The great colonial powers of the West, the Portuguese, British, French, and other Empires, were the direct result of the adoption by Europeans of Chinese technology on the high seas.

WATERTIGHT COMPARTMENTS IN SHIPS

SECOND CENTURY AD

From at least the second century AD, if a traditional Chinese ship received a hole of any kind in its hull, the ship would not sink. This is because the Chinese constructed hulls on the bulkhead principle. A bulkhead is an upright partition separating compartments. A typical medium-sized freighter had fifteen bulk-heads and thirty-seven rib frames. What would happen would merely be that one of the sixteen bulkhead compartments would flood, but the other fifteen, being sealed off from it, would remain dry and the ship would continue to float. There is no way of ascertaining how many thousands of lives would have been saved if Western ships had been built on such sensible principles; but they were not.

Just how novel this technique was only 200 years ago may be seen from a letter written in 1787 by Benjamin Franklin about the mail packets which were envisaged between the United States and France: 'As these vessels are not to be laden with goods, their holds may without inconvenience be divided into separate apartments, after the Chinese manner, and each of these apartments caulked tight so as to keep out water.'

The idea of watertight compartments for ships' hulls was brought to Europe from China by Sir Samuel Bentham (1757–1831), long-time chief engineer and architect of the British Navy. As a young man in 1782, he travelled through Siberia to China and studied Chinese ship construction. He campaigned in Europe for bulkheads, so that by 1795 the Lords Commissioners of the Admiralty asked him to design and build six sailing ships of a new design with 'partitions contributing to strength, and securing the ship against foundering, as practised by the Chinese of the present day'.

As late as 1824, naval writers were still breathlessly expostulating upon the miracle of this simple technique. In that year, a report in Mechanics' Magazine stated: 'There is a method of making it almost impossible to sink ships which ... is now employed by the Chinese. The hold is divided into a number of compartments, so that if the ship spring a leak, or her sides are stove in, she will remain afloat.'

What is surprising is that these ideas were clearly described by Marco Polo in 1295, but no one paid any attention. They were repeated in 1444 by Nicolo de Conti in his own *Travels*, where he said: 'These Shippes are made with Chambers, after such a sorte, that if one of them should breake, the others may goe and finish the Voyage.' But so conservative were European shipbuilders and sailors that it took 500 years after the principles of watertight compartments were made known to the West for them to be adopted.

A small exception to this must be mentioned. In about 1712, some English fishermen adopted the Chinese principle of a sealed hull compartment which could be free-flooded under controlled conditions (recorded in a Chinese text of the fifth century AD and doubtless much earlier than that). The purpose of this 'wet-well' in a ship called a 'well-smack' was to enable fishermen to bring their fish to port fresh, still swimming in water. This is a traditional Chinese practice. In China, such free-flooding compartments were also used to raise and lower the level of a boat when shooting rapids in rivers, giving the captain control over the amount of water-resistance of his ship by acquiring and discharging water ballast. The flooded compartments would be emptied by bilge pumps using the square-pallet chain-pump design, which the British Navy also adopted directly from the Chinese.

The bulkhead construction of ship hulls was a natural notion which the Chinese derived from copying the bamboo, which has transverse septa dividing the stem into a series of compartments. Since Europe did not have bamboo, this inspiration was not to hand.

The use of bulkheads afforded an obvious multiplicity of strong cross-timbers in the hull capable of receiving masts. This not surprisingly resulted in the adoption of multiple masts in Chinese ships, which caused such amazement to Westerners in medieval times and led to the adoption of multiple masts in Europe.

A highly conservative estimate for the adoption of transverse bulkheads in Chinese ship hulls would be the second century AD. Although rafts and coracles were in profuse use at that time, there is no reason to believe that proper ships did not exist before then, whose bulkheads were effectively watertight (if not by design, then at least as a result of the way the hulls were constructed). The bulkhead construction principle may therefore be much older.

THE HELICOPTER ROTOR AND THE PROPELLER

FOURTH CENTURY AD

A description of a helicopter top dating from the fourth century AD was mentioned by the philosopher and alchemist Ko Hung (see the account of manned flight with kites, page 191). By then, helicopter tops seem to have been common toys in China, one of the names most generally used for them being 'bamboo dragonfly'. The top was an axis with a cord wound round it, and with blades sticking out from the axis and set at an angle. One pulled the cord, and the top went climbing up into the air. This simple but fundamental toy had a very important effect on the European pioneers of aviation.

Sir George Cayley, the father of modern aeronautics, studied the Chinese helicopter top in 1809. It had two sets of rotor blades (using feathers, in this case), and a winding device using a spring. This Chinese 'bamboo dragonfly' could rise 20–25 feet in the air. Cayley set about trying to make an improved version. His drawings of 1853 portray his own helicopter top, which could mount 90 feet in the air. Encountering the Chinese helicopter top in 1792 (by way of two Frenchmen, Launoy and Bienvenu) had stimulated Cayley to take an interest in the possibilities of aviation in the first place.

The helicopter top was the model for what became the propeller of modern aircraft. But the Chinese had also preceded Europe in vertically mounting these rotors, as was necessary for use in airplanes. Early in the seventeenth century, kite-flying was temporarily banned in China. Liu T'ung tells us in his book of that time, *Descriptions of Things and Customs at the Imperial Capital*, that people turned instead to vertically mounted 'wind-wheels', which amused people by their pretty red and green colours, which flashed rapidly as the wheels turned in the wind. The wind-wheel could be either set stationary or carried in the hand on sticks. There were also pinwheel varieties, which were able to perform work by depressing a lever and beating a drum. Some of the Chinese wind-wheels may be seen in a painting made in 1310 by Wang Chen-P'eng. We also know of wind-wheels being attached to cambered kites, set vertically mounted, but spinning in the wind 'just for joy'. The Chinese were thus amusing themselves with actual miniature airplanes with proper airfoil wings and propellers, but merely letting them flit about in the wind. There is no record of their having attempted to harness a power source to the propeller and make the kite fly by becoming a real airplane. Thus, the Chinese invention of the helicopter top led in China to nothing but amusement and pleasure. But its influence in the West 1400 years later was to be one of the key elements in the birth of modern aeronautics and manned flight.

THE PADDLE-WHEEL BOAT
FIFTH CENTURY AD

The idea of the paddle-wheel boat did apparently occur to one early European inventor, though it seems never to have been constructed, and involved a very unwieldy power source. An anonymous manuscript entitled *De Rebus Bellicus*, which is thought to date from the late fourth century, contains the suggested design for a ship with three pairs of paddle wheels powered by six oxen walking round and round on the deck, as in grinding a mill. It is considered unlikely that the vessel existed, but we must acknowledge that as far as the paddle-wheel boat is concerned, the original idea seems to have been European, but the first execution Chinese, with no connection between the two. In Europe, therefore, the idea led to nothing, while in China it led to the building of hundreds or thousands of ships.

The first record of the existence of paddle-wheel boats occurs in a Chinese account of a naval action under the command of Wang Chen-O, one of the admirals of the Liu Sung Dynasty, in 418 AD. This is described in *The History of the Southern Dynasties* (compiled in 670):

LEFT (146) This picture of a Ming Dynasty paddle-wheel battleship dates from about 1500, but this printing comes from the 1883 Beijing publication *Strategy for Governing the Country*, which specifically states that it re-used ancient woodblocks that had been carefully stored by the imperial authorities. If the Chinese thought that these old paddle-wheelers would be a match for the steam-powered ones of the British Navy, it is no wonder that they suffered such humiliation in battle. In any case, these no longer even existed, but were merely a centuries-old memory, and the only paddle-wheeler the Chinese were still building by that time was one in stone (see Plate 147 (page 214-15)). (Collection of Robert Temple.)

ABOVE (147) The Empress-Dowager Tz'u-Hsi built this folly in 1889 in the pleasure-gardens of the Summer Palace near Peking. It is a marble boat with paddle wheels on each side. Paddle-wheel boats were invented in China in the fifth century AD. By the twelfth century they had reached lengths of 300 feet, and could carry eight hundred men. Some at that time had as many as ten separate decks.

Wang Chen-O's forces sailed in covered swooping assault craft and small war-junks. The men propelling the boats were all hidden inside the vessels. The Ch'iang [barbarians] saw the ships advancing up the Wei [river] but could not see anyone on board making them move. As the northerners had never encountered such boats before, every one of them was sore afraid, and thought that it was the work of spirits.

An improved version was made by Tsu Ch'ung-Chih between 494 and 497. His boat was called the 'thousand-league boat', and was tested on the Hsin-T'ing river, south of modern Nanking. It could travel enormous distances in a single day without the aid of wind, and must have represented a considerable refinement of the earlier designs.

An admiral of the Liang Dynasty, Hsü Shih-P'u, used a number of paddle-wheel boats in his campaign against the rebel Hou Ching in 552. His paddle-wheel boats were called 'water-wheel boats'. And in another campaign against the same rebel, the admiral Wang Seng-Pien is described as having in his fleet 'ships which had two dragons on the sides to enable them to go very fast'. The text is thought to have become confused, and originally to have stated 'two wheels'. And at the siege of Li-Yang in 573, another admiral, Huang Fa-Ch'iu, who was also a distinguished military engineer, built and used a number of 'foot-boats', which were obviously paddle-wheelers operated by foot treadles.

Between 782 and 785, Li Kao, prince of T'ang, was Governor of Hungchow. The official history of the time tells us that:

Li Kao, always eager about ingenious machines, caused naval vessels to be constructed, each of which had two wheels attached to the side of the boat, and made to revolve by treadmills. These ships moved like the wind, raising waves as if sails were set. As for the method of construction it was simple and robust so that the boats did not wear out.

We are also told of these ships that their speed was 'faster than a charging horse'.

During the medieval Sung Dynasty the paddle-wheel warships really came into their own. They were often rudderless and manoeuvred swiftly and with breathtaking agility by means of a complex system of using varying combinations of their paddle wheels – perhaps three out of six on one side, then two on the other, and so on. They were able to dart in and out between other ships and wreak great havoc. Some of them had rammers fitted to their prows.

The admiral Shih Cheng-Chih in 1168 is recorded as having constructed a 100-ton warship propelled by a single twelve-bladed wheel. This means that some of the paddle-wheel boats were stern-wheelers. And when boats are described, as they often were, as having odd numbers of wheels, one would have been at the stern and the others in opposite pair formation along the sides. However, even though the wheels would have been opposite each other, they would almost certainly have worked quite independently, so that one could stop and the opposite one rotate. This seems to have been the method of steering.

We have an account from 1130 of the construction of some naval paddle-wheel ships:

Kao Hsüan, who had formerly been Chief Carpenter of the Yellow River Naval Guard Force, and of the Pai-p'o Vehicular Transport Bureau of the Directorate of Waterways, submitted a specification for wheeled ships which he claimed could cope with the enemy.... He first built an eight-wheel boat as a model, completing it in a few days. Men were ordered to pedal the wheels of this boat up and down the river; it proved speedy and easy to handle whether going forwards or backwards. It had planks on both sides to protect the wheels so that they themselves were not visible. Seeing the boat move by itself like a dragon, onlookers thought it miraculous.

Gradually the number and size of the wheels were increased until large ships were built which had twenty to twenty-three wheels and could carry two or three hundred men. The pirate boats, being small, could not withstand them.

By this time, the paddle-wheelers came to be called 'wheel ships', and the terminological confusion which had existed for several centuries, when a wide variety of colourful names were used for them, settled down to a standard term. Technological improvements on the 'wheel ships' went on apace. Not long afterwards, Ch'eng Ch'ang-Yü constructed 'wheel ships' up to 300 feet long, capable of carrying between seven and eight hundred men!

The rebels ('enemies') mentioned above captured some of the 'wheel ships', together with the engineer Kao Hsüan. Thus commenced an arms race, with the rebels for a time building bigger and better 'wheel ships' than the Southern Sung Dynasty against which they were in revolt. As we are told in a contemporary history, 'Within two months the pirate bases had over ten many-decked wheel ships that were stronger and better constructed than the government ships.'

At the height of the conflict, the rebel fleet had several hundred 'wheel ships' of this type in operation. It must be emphasized that these ships were quite unsuitable for sea operations, and all the naval engagements took place on rivers and lakes. At a later period of history, under the Mongol Dynasty, when naval operations concerned the sea rather than inland waters, the use of paddle-wheelers went into serious decline. But at the culmination of the Sung period's conflict with the rebels in the twelfth century, paddle-wheelers reached what was probably their most extreme form.

The imperial forces in their turn imitated the paddle-wheel ships of the rebels but made them larger – as much as 360 feet in length, 41 feet in the beam, and with masts 72½ feet high. The largest number of men on record for working treadles on a 'wheel boat' is two hundred. Needham believes that the medieval paddle-wheelers of the Chinese could generate 50 horsepower and would have averaged a speed between 3½ and 4 knots. After this time, paddle-wheelers diminished in size, and also in importance. But they were still around until modern times, and saw naval action against the British in 1841 during the Opium Wars. The British believed them to be rapidly constructed copies of the British Navy's own paddle-wheelers, unaware that the Chinese had been using such ships for sixteen centuries by that time. Indeed, as recently as 1929, paddle-wheelers were still in use for carrying passengers up and down the Pearl River near Canton, though none is known to survive.

LAND SAILING

SIXTH CENTURY AD

Land sailing, particularly on beaches, has become a popular sport today in the West. It originated in sixth-century China. About 550 AD, the Liang Emperor Luan (552-4), a Taoist scholar, wrote his *Book of the Golden Hall Master*. In it he recorded:

> **Kaots'ang Wu-Shu succeeded in making a wind-driven carriage which could carry thirty men, and in a single day could travel several hundred *li* [hundreds of miles].**

These developments continued, and we have the following account, from a book called *Continuation of the New Discourses on the Talk of the Times*, of a very large vehicle built for the Emperor Yang of the Sui Dynasty about 610. It is thought that the account, though obviously exaggerated, describes a genuine sailing vehicle, or at least a vehicle whose progress was significantly assisted by sails, whatever other form of propulsion or haulage it may also have had:

> **Yuwen K'ai built for Sui Yang Ti a 'Mobile Wind-Facing Palace'; it carried guards upon its upper deck, and there was room for several hundred persons to circulate in it. Below there were wheels and axles, and when pushed along it moved quite easily as if by the help of spirits. Among those who saw it there was no one who was not amazed.**

One notices that the vehicle was *pushed along* by some normal means, and sails fitted to the vehicle rendered it astonishingly light, so that this was a semi-sailing carriage, or sail-aided carriage. Such assistance from sails in later times even came to be applied to ploughs. It must have made quite a sight to have farmers going up and down their fields with their ploughs pulled by wind power. But this always seems to have been something of a rarity, unlike the commonplace nature of sailing wheelbarrows.

The Chinese invented the wheelbarrow by the first century BC (page 95). At some unknown time after the first application of the sail to land travel, as just described, sails were put onto wheelbarrows. A.E. van Braam Houckgeest inspected a native Chinese sailing wheelbarrow closely and did a careful drawing of one, which he published in 1798 in his book *An Authentic Account of the Embassy of the Dutch East-India Company to the Court of the Emperor of China in the Years 1794 and 1795*. It has the batten sails and multiple sheets which were so characteristic of Chinese ships.

Sailing wheelbarrows, known as sailing carriages to Western visitors, caused a sensation in the West when they became known to the public, from the sixteenth century. From that date for about two centuries, an imaginary view of a sailing carriage

ABOVE (148) 'Flora's Wagon of Idiots' by Hendrik Gerritszoon Pot (1585–1657). This painting is loosely copied from an earlier engraving by Chrispijn van de Passe, but there were many such pictures at this time showing the sailing wagons introduced into Europe from China by Simon Stevin. This painting is satirical in intent, attacking the wave of mass hysteria which swept over Holland at this time and which has gone down in history as 'tulipomania' – one of the strangest phenomena in European history, when there was such excitement over tulips that people would commit murder to get the bulbs. (Franz Halsmuseum, Haarlem.)

was drawn on nearly every atlas map which showed China. Milton even mentioned them in *Paradise Lost* in 1665, speaking 'Of Sericana, where Chineses drive With Sails and Wind their canie Waggons light' (III, 431).

Jan Huyghen van Linschoten in his *Itinerario* of 1598 wrote:

> The men of China are great and cunning workemen, as may well bee scene by the Workmanship that commeth from thence. They make and use (waggons or) cartes with sayles (like Boates) and with wheeles so subtilly made, that being in the Fielde they goe and are driven forwards by the Winde as if they were in the Water....

Less frequently seen, since they were in less accessible localities, were the Chinese ice-yachts, which ran along the ice on wheels by sail power. Contemporary ice-yachts were in use in 1935 on the Liao River, near Ying-k'ou in the cold region of Manchuria.

The great Dutch scientist and mathematician Simon Stevin actually constructed a Chinese sailing carriage in the autumn of 1600. Stevin was deeply influenced by Chinese science, and helped to transmit to the West the Chinese inventions of decimal fractions and equal temperament in music (see pages 157 and 234). Stevin's land-sailer carried Prince Maurice of Nassau, the young scholar Grotius and many dignitaries along a beach at great speed. It covered a distance of nearly 60 miles in less than 2 hours, so that its average speed was 30 miles per hour, and it must have reached 40. This gave Europeans their very first taste of real speed, and is thought to have exerted a profound psychological effect which lasted in various subtle ways and helped prepare for the 'modern age'. We are fortunate to have a vivid description, written six years later, by Gassendi:

RIGHT (149) Land sailing was being practised in China from the sixth century AD. The most useful form of it was in the attachment of a sail to a wheelbarrow, so that the wind took part of the load. Here we see an engraving from A.E. van Braam Houckgeest's book of 1797, *An Authentic Account of the Embassy of the Dutch East-India Company to … China.* The sail is a batten sail with multiple sheets (rigging), of the sort used on Chinese junks. The wheelbarrow is of the typical Chinese form with a central wheel, which can carry far greater loads than Western forms.

Also he stept aside to Scheveling, to make triall of the carriage and swiftnesse of the waggon, which some yeers before was made with such Art, that it would run swiftly with sails upon the land, as a ship does in the sea. For he had heard how Grave *Maurice*, after the victory at Nieuport, for triall sake, got up into it, with Don *Francisco Mendoza* taken in the fight, and within two hours was carried to Putten which was 54 miles from Scheveling. He therefore would needs try the same, and was wont to tell us, how he was amazed, when being driven by a very strong gale of wind, yet he perceived it not (for he went as quick as the wind), and when he saw how they flew over the ditches he met with, and skimmed along upon the surface only, of standing waters which were frequently in the way; how men which ran before seemed to run backwards, and how places which seemed an huge way off, were passed by almost in a moment; and some other such like passages.

THE CANAL POUND-LOCK

TENTH CENTURY AD

Everyone who has seen a modern canal will be familiar with the pound-lock. It has two gates. The boat enters by one, which closes behind it. The water level in the lock then either rises or falls to match the level of the water towards which the boat is sailing. When it has done so, the further gate opens and lets the boat through. The first certain date for a pound-lock in Europe is 1373.

The canal pound-lock was invented in China in 984 AD. The inventor was Ch'iao Wei-Yo, who, in 983, had been appointed Assistant Commissioner of Transport for Huainan. The impetus for his invention was a concern over the enormous amounts of grain which were being stolen during canal transport at that time. Grain was the normal tax payment throughout China's history. Movement of the grain to central repositories and warehouses was the life-blood of the Empire, and therefore any substantial interruption of this process was a very serious social and political problem.

Until 984, boats could only move between lower and higher water levels in canals over double slipways. Chinese boats had no keels, and were nearly flat-bottomed. A form of portage had been developed in China, therefore, whereby spillways originally designed to regulate water flow were elongated in gentle ramps both front and back, leading into the water. A boat would come along and be attached to ropes turned by ox-powered capstans. Within two or three minutes, the boat would be hauled up a ramp to the higher level, and for a moment would balance precariously in the air. Then it would shoot forward like an arrow out of a bow and scud along the canal at a level several feet higher than it had started. Passengers and crew had to lash themselves tightly to the boat to avoid being hurled into the air and injured. The great disadvantage of this ingenious technique was that boats often split apart or were seriously damaged by the wear and tear of being dragged up the stone ramps. Whenever a boat broke up on a ramp, the contents would promptly be stolen by organized gangs – including corrupt officials – who waited for just such an occurrence. Sometimes, apparently, the ships were roughly handled on purpose, or were artificially weakened or had even been chosen for their weaknesses so that an 'accident' of this kind could be brought about intentionally.

Ch'iao Wei-Yo was determined to wipe out this practice. He therefore invented the pound-lock so that double slipways would not be needed. Here is how the official history of the time relates the story:

> **Ch'iao Wei-Yo therefore first ordered the construction of two gates at the third dam along the West River (near Huai-yin). The distance between the two gates was rather more than 50 paces [250 feet], and the whole space was covered over with a great roof like a shed. The gates were 'hanging gates'; when they were closed the water accumulated like a tide until the required level was reached, and then when the time came it was allowed to flow out.**
>
> **He also built a horizontal bridge between the banks, and added dykes of earth with stone revetments to protect their foundations. After this was done to all the double slipways the previous corruption was completely eliminated, and the passage of the boats went on without the slightest impediment.**

Pound-locks made true summit canals possible. Water levels could differ by 4 or 5 feet at each lock without any problem at all. Over a stretch of territory, therefore, a canal could rise more than 100 feet above sea level, as was the case with the Grand Canal for instance (rising 138 feet above sea level). This made possible a vast extension of the canal network and freed hydraulic engineers from many awkward topographical restrictions. Pound-locks were fitted into the Magic Canal (page 198) by the tenth or eleventh century.

The use of pound-locks meant that precious water could be saved. Many of the Chinese canals went dry in the summer, and days might have to pass before enough water had accumulated from the length of a canal for a flash-gate to be opened to let boats through. There could be hundreds of boats waiting at any given gate. But with the pound-lock, only the one lockful of water was used each time. This meant that the canals, by saving water, could extend the months of their usefulness. Also, Shen Kua tells us in his *Dream Pool Essays* of 1086 of a single canal:

> **It was found that the work of 500 labourers was saved each year, and miscellaneous expenditure amounting to 1,250,000 cash as well. With the old method of hauling the boats over, burdens of not more than 21 tons of rice per vessel could be transported, but after the double gates were completed, boats carrying 28 tons were brought into use, and later on the cargo weights increased more and more. Nowadays Government boats carry up to 49 tons, and private boats as much as 800 bags weighing 113 tons.**

Pound-lock gates later went largely out of use in China, owing to social changes. The Mongols shifted the capital to Peking, making it impossible for the whole of the imperial grain tribute to be transported by canal. Much of the transport of grain was shifted to the sea routes – sometimes more than half of the entire annual grain supplies. As time went on, the need for large barges on the canals diminished, a larger number of smaller vessels came into use, and the pound-locks were allowed to fall into disrepair and eventual decay. By the time the Chinese pound-lock found its way to Europe, it was on the wane in its country of origin, and the fact that it came from China was increasingly easy for not only Europeans but for the Chinese themselves to forget.

LEFT (150) A photograph taken by Cecil Beaton during the Second World War, showing two Chinese sampans in a modern pound-lock. The pound-lock was invented by Ch'iao Wei-Yo in 984 AD, but the first European one was not constructed until 1373 at Vreeswijk in Holland, at a point where the canal from Utrecht joined the River Lek. Shortly afterwards, another was built at Spaarndam. The idea must have been brought from China to Holland by Dutch traders.

Part 10
SOUND AND MUSIC

THE LARGE TUNED BELL
SIXTH CENTURY BC

The Chinese invented, developed and perfected tuned bells long before anyone else. Indeed, they went on to base their entire system of measurements – length, width, weight, volume – on musical pitches of tuned bells. At first, bells were just noise-making instruments. But by the sixth century BC at the latest, and probably long before, finely tuned bells were being manufactured, which when struck emitted precise notes.

The bell seems to have originated from metal grain measures. It evolved into two forms of hand bell – the *to*, which faced upwards, and the *chung*, which faced downwards and the latter became the basis for Chinese measurements. The ancient book, *Kuo Yü*, dating from several centuries BC, states that:

> **The ancient kings made as their standard the *chung* vessel, and decreed that the magnitude of its pitch should not exceed that produced also by the string of the *chün* [7-foot tuner], and that its weight should not exceed 120 catties. The measures of pitch, length, capacity, and weight originate in this standard vessel**

It is doubtful if any other civilization in history based all its measurements on the musical pitches of bells. This system certainly gave an impetus to the development and perfection of the tuned bell in China, since if perfectly tuned bells did not exist, then all would fall into confusion. No standards of measure would exist, cheating and corruption would spread, commercial disagreements could not be avoided, the disruption of trade would ensue, and riots and rebellions might follow. These were real, not imagined dangers. It was imperative that the authorities kept the measures in order by regulating the pitches of the bells. This was one of the most urgent concerns of all ancient Chinese governments. The very safety of the state might have been at stake.

The way in which pitches of bells were transformed into standard measures of length was by way of the *chün*, a stringed tuner 7 feet long. The pitch of the bell would be matched by tuning of the string, which would then be measured, giving a specific length. (Strings composed of identical material and under the same tension give out different notes when plucked according to their different lengths. A sliding bridge may be moved along under a string to alter its effective length when plucked, the resulting length being precisely measurable.) This process could then be reversed, so that a standard length given by a bell at the capital could give the pitch of a string of a *chün* in a distant city, which could then be matched by a bell cast there. The Chinese divided the octave into twelve notes. Consequently, there came to be official sets of twelve bells giving out these fundamental notes, and from them all other instruments would be tuned. Also, at the beginning of any piece of music, the appropriate bell would be sounded to give the key.

OPPOSITE (151) The casting of a giant tuned bell. The man in the foreground is working on one half of the mould, the other half lying bottom left. The bell which has emerged from this mould is being burnished in the background. This illustration comes from *The Creations of Nature and Man* (*T'ien-kung K'ai-Wu*) of 1637. Its text says: 'For casting iron bells without the expenditure of too much fat and wax, an outer layer of the mould is first made from earth, which is then cut into two sections either longitudinally or crosswise. Dowel pins are used to insure the positive alignment of these two sections when the mould is closed.... After the words are engraved, the surface is spread over with a thin layer of ox fat, so that later the bell casting will not stick to the mould. The cope is then laid over the core, and the separate sections are sealed together with mud. The mould is then ready for casting.' (Chinese text translated by Sun E-Tu Zen and Sun Shiou-chuan.)

塑鐘模圖
牛脂受鏤
鋼刻鐵鐘外模

At first the bells did not have clappers and had to be struck to give out their notes. Chinese clapperless bells date from about the sixth century BC. These would have been struck to give the note which would then be the keynote of the musical scale for an orchestra. In music, ancient Chinese bells seem to have been used to stop and start performances, rather than as actual instruments. They essentially served as ancient tuning-forks.

One of the extraordinary features of most ancient Chinese bells is that they were constructed so as to give out not only one note, but two. If struck in the centre they would give out one note, and if struck to one side, near the edge, they would give out another. The intervals between the notes included major and minor seconds, thirds and fourths, and minor sixths.

After some time, the set of twelve bells came often to be replaced by a set of twelve pitch-pipes, which served the same purpose and were much easier to construct. Using pitch-pipes, it was also possible to extend the measures of capacity and volume. The bells had always given measures of grain, having, after all, originated as grain-scoops. But now pitch-pipes were closely studied and the precise number of grains of millet which would fit into them were counted. An imperial history for the first century BC reveals how 'The basis of the linear measure is the length of the *Huang-chung* pitch-pipe [which gave the fundamental note]. Using grains of medium-sized black millet the length of *Huang-chung* is 90 *fen*, one *fen* being equal to the width of a grain of millet.... Using grains of medium-sized millet twelve hundred grains fill its tube....The contents of one *Huang-chung* tube, i.e., twelve hundred grains of millet, weigh 12 *chu* [half an ounce].'

As Needham has remarked, this was a clever system, for though 'millet-grains might vary individually, when large numbers of one given species were used a fairly consistent average would be struck (large and small grains aberrant in size being rejected). This method of ensuring against the loss of standard measures was probably as practical as any that could have been devised.'

The achievement of the ancient Chinese with regard to bells was a truly remarkable feat of early technology. Even today, the proper tuning of bells is considered difficult and highly intricate. To manufacture a bell giving

BELOW (152) The largest set of ancient chime bells ever found in China. It was excavated in 1978 at Sui Hsian in northern Hupei Province, from the tomb of Marquess Yi of the State of Cheng, who was buried in the late fifth century BC. Sixty-four bronze bells make up the chime set, which is mounted on wooden beams with bronze supports and fittings, and stood intact in the tomb for 2400 years. The bases and warrior figures included in the supports together weigh over 150 pounds. The largest of the bells alone weighs over 100 pounds. It is believed that five performers were needed to play the set. As for the notes of the bells, forty-seven of them produce two notes with minor third intervals, and sixteen produce two notes with major third intervals. Each bell bears a lengthy and meticulous inscription describing precisely what notes it gives forth, how they fit into a scale, and how that scale relates to other scales employed by other feudal states of the period. (Hupei Provincial Museum, Wuhan.)

RIGHT (153) A large bell dating from the fifth century BC. It was cast to give two precise musical notes; it would give out one if struck in the centre, and another if struck on one side, near the edge. Bells like this are generally about 8 feet high, and some are much larger. However, as late as 1000 AD, no bell more than 2 feet high had been seen or heard in Europe. Western church bells are thus essentially an idea imported from China. (Victoria and Albert Museum, London.)

out a particular sound, one must consider the proportions of metals in the alloy, the elasticity and thickness of the material, the specific gravity, the diameters at different points, the contours of the curves, the temperature at which the alloy stands when it is poured, the rate at which it cools, and so on. Even when all these things are considered, the resulting sound still may not be right. It is common to have to file bells down in places in order to improve their sounds. Apparently, few ancient Chinese bells show signs of filing. Such was the mastery of bronze and iron casting of the ancient Chinese, that perfect bells were produced at the first attempt.

As late as the year 1000 AD, no bell more than 2 feet high had been seen or heard in Europe. But a cast iron bell precisely dated to 1079 still exists at P'ing-ting in Shansi, in China, which is four or five times this size. Chinese bronze bells have been excavated which date back to at least the fourteenth century BC. While bells of the sixth century BC were just under 2 feet high, no bell higher than 8 inches seems to be known from any ancient Western civilization.

TUNED DRUMS

SECOND CENTURY BC

Drums have existed from time immemorial, and are common among most tribal peoples. But the tribal drums are not tuned. The Chinese appear to have been the first to develop tuned drums. For this they used stretched leather hides, and they were doing so by the second century BC at the latest. An ancient treatise on drums survives, entitled *Chieh Ku Lu* ('On the History and Use of Drums'), written by Nan Cho in the year 848 AD, but it has not yet been studied by any scholar.

In the most ancient Chinese music, a prime concern was the regulating of ritual dances and mimes, which by sympathetic magic were thought

RIGHT (154) A previously unpublished photograph by Ernst Boerschmann, taken about 1900, showing a huge, ancient *chung* bell with a scalloped rim. This bell, resting on its side on the earth, bears lengthy inscriptions, but nothing is recorded of them or of the location of the bell, which appears to have been discarded. It is probably one of the countless old bells which now no longer exist, having been melted down as scrap.

to affect the weather, communicate and resonate with the spirit world of the ancestors, and regulate the mood and behavior of the people. Bells were not sufficient to control the movements, and a combination of drums and bells was common. An ancient book, *The Music Record*, written in the first century BC, describes how:

> **In the ancient times the dancers advanced in ranks, keeping together with perfect precision, like a military unit. The strings, gourds, and reed-tongued organs all waited together for the sound of the tambourines and drums. The music was started by a note from a pacific instrument [drum]. Conclusions were marked by a note from a martial instrument [bell]. These interruptions were controlled by means of the *hsiang* drum. The pace was regulated by means of the *ya* drum.**

We can see from this that various kinds of drum served different purposes. And the one which 'started the music' must have been a tuned drum, serving, as a bell had earlier done, to set the key. The tuning must have been very precise in order to satisfy the requirements of the Chinese. However, drums have an appalling timbre. They lack resonant higher harmonics, and it is just as well that their notes cannot linger long, for if they did, only noise and uncomfortable dissonance would be heard. A drum is an instrument fit only for momentary hearing. Perhaps the rapid dying away of the initial drum sound was found more convenient for the beginning of a piece of music than the lingering resonance of the bell, which may have caused some dissonance with the proceedings of the orchestra.

The drum additionally had a solemn and sacred role at one of the great religious state ceremonies of the year. Every 21 December, the drum was used to announce at the winter solstice that the Sun had 'turned round' and begun its advance once more.

Chinese knowledge and understanding of the drum was so advanced that it was used as an analogy of the human eardrum. The Chinese understood that sound was vibration and that it was carried as waves through the air to the ear. As the Taoist T'an Ch'iao wrote in the tenth century: 'An ear is a small hollow.... It is not the ear which listens to sound but sound which of itself makes its way into the ear.' But much earlier, about 742 AD, the earlier Taoist T'ien T'ung-Hsiu wrote about hearing: 'It is like striking a drum with a drumstick. The shape of the drum is possessed in

RIGHT (155) A previously unpublished photograph by Ernst Boerschmann, taken about 1900. This tuned drum stands outside a Taoist temple and is protected by an ornamental canopy.

ABOVE (156) A detail of a rubbing taken from a Han Dynasty tomb carving of approximately 2000 years ago, showing a man beating a ceremonial or ritual tuned drum. The drum rests on a stand with a tiger base, and is covered by a parasol. (Collection of Robert Temple.)

my person in the form of the ear. The sound of a drum is a matter of my responding to it.'

Needham says of this: 'To expand the analogue slightly, it seems that he believed that sounds strike the inner ear, in fact the eardrum, just as drumsticks strike an actual drum; that is to say, they exert pressure. Nevertheless, it is the response of a sentient being which enables one to describe this process as sound.'

T'an Ch'iao even anticipated in a vague way the possibility of sound amplification, for he envisaged this in principle, saying that under certain imagined circumstances 'even the tiny noises of mosquitoes and flies would be able to reach everywhere.' And his description of the cause of the sound of thunder also came very close to the truth. All this as Needham says, 'is rather remarkable for the tenth century AD'.

ABOVE (157) Large tuned drums hang suspended in a monastery temple, ornamented with swirling patterns based upon the Taoist *yin–yang* symbol. The temple is at the Derge Gonchen Gompa Monastery of the Sakya school of Tibetan Buddhism (a form of Buddhism especially popular with the Manchu emperors of the Ch'ing Dynasty) in Szechuan Province.

HERMETICALLY SEALED RESEARCH LABORATORIES

FIRST CENTURY BC

The Chinese preoccupation with musical pitch led to what is probably the strangest of all the proto-scientific activities of the ancient Chinese. There is possibly no more bizarre story in China's history than how the world's first attempt at a hermetically sealed research laboratory came about, and what the rationale – or rather irrationale – behind this endeavour was.

In order to explain this properly, we have to understand the concept of *ch'i*, a word which is essentially untranslatable. Various meanings of *ch'i* may be conveyed by the following English words: subtle matter, matter-energy, emanation, ether, subtle fluid, rarefied air, aerial or atmospheric disturbance or perturbation, subtle pervasive force, material emanation (such as the 'earthly *ch'i*' which rose from the Earth), energy present in organized form. The Greek word *pneuma* bears some resemblance in its meaning to the Chinese concept of *ch'i*. Similarly, certain mystical uses of the Greek word *logos*, 'word', resemble the meaning of *ch'i*, as in the Bible where *logos* refers to Christ in the sentence: 'The Word was made flesh, and dwelt among us,' in the Gospel of St John.

All living beings inherently possess *ch'i* in addition to their other properties. The Taoists even developed a highly elaborate theory and practice for 'circulating the *ch'i* throughout the human body, which was related to concepts of acupuncture, and was the basis of health and longevity. In the human body, there were particular forms of *ch'i*; there was, for instance the '*ch'i* of the blood'.

It would be possible to continue at great length, but this is sufficient to show that the concept of *ch'i* was fundamental to the ancient Chinese understanding of nature, life and the cosmos. Different forms of *ch'i* were conceived of as emanating from just about everyone and everything. If a person influenced another, his *ch'i* was predominant. Intimately associated with concepts of *ch'i* were the phenomena of pitch, timbre and resonance. Armies were conceived of as having a peculiar group-*ch'i*, which was thought of as an energy field floating over their heads, and was sometimes visible as a cloud or haze as they marched into battle. There were no strict cut-off points or boundaries between the spiritual *ch'i* and physical matter; to the ancient Chinese, not only did the two overlap, they intermingled, which gave explanations for countless natural phenomena.

To detect the nature of the group-*ch'i* of an army, a kind of musical shaman would blow on special pitch pipes, and from the character of the resulting sound would pronounce his conclusions: a note which was weak and did not have sufficient timbre would indicate a weak and vacillating *ch'i*, and would thus foretell defeat or disaster for the army concerned. Pitch-pipe soundings by shamans were taken for the 'home' army upon its marching off on campaign, and for the opposing army as it stood arrayed for battle in the distance. Sometimes, depending on the musical enquiries of the shaman, a battle would be called off, a retreat ordered, or an immediate attack launched. Music, therefore, was a serious business upon which tens of thousands of lives regularly depended.

The Chinese preoccupation with matters of pitch and timbre became a full-fledged obsession. A dreadful anxiety persisted that somehow the true fundamental pitches might become altered in passage of time or through the loss of certain bells and pipes during civil and military disturbances (which were so frequent throughout China's history); so a custom was instituted to safeguard the pitches which was one of the strangest rituals in history. In order to introduce this peculiar tale, a fairly lengthy quotation from the ancient Taoist book of the philosopher Chuang Tzu (c. 290 BC) sets the scene:

> **'The breath of the Universe', said Tzu-Ch'i, 'is called wind. At times it is inactive. But when it rises, then from a myriad apertures there issues its excited noise. Have you never listened to its deafening roar? On a bluff in a mountain forest, in the huge trees, 100 spans round, the apertures and orifices are like nostrils, mouths or ears, like beam-sockets, cups, mortars, or pools and puddles. And the wind goes rushing through them, like swirling torrents or singing arrows, bellowing, sousing, trilling, wailing, roaring, purling, whistling in front and echoing behind, now soft with the cool breeze, now shrill with the whirlwind, till the tempest and the apertures are all empty and still. Have you never observed how the trees and branches shake and quiver, twist and twirl?'**
>
> **Tzu-Yu said, 'The notes of Earth then are simply those which come from its myriad apertures, and the notes of man may be compared**

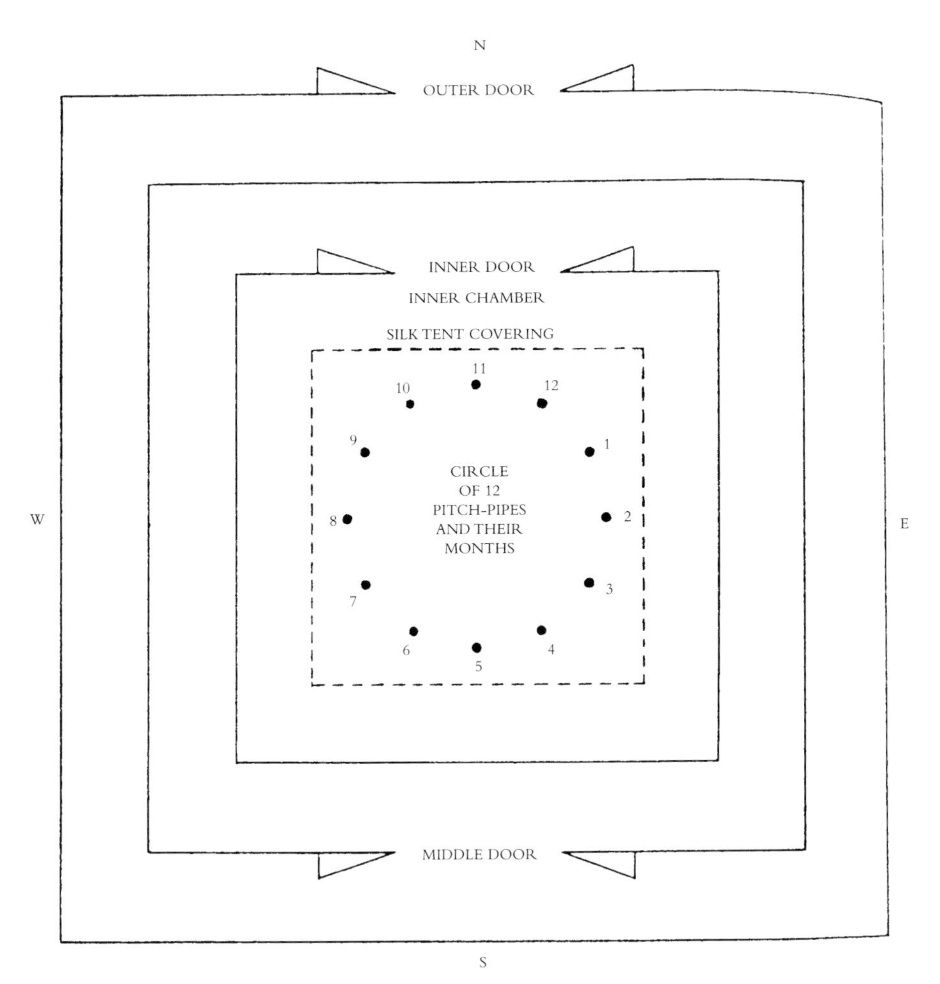

LEFT (FIG. 7) A diagram by Professor Derk Bodde of one of the 'hermetically sealed research laboratories' used from the first century BC onwards to watch for 'blowing of the ashes' under the influence of imaginary earth forces. In order to protect the interior from any possible light, draft or other influence of the elements, an arrangement of imbricated corridors was made, resembling a modern photographic darkroom. One came in by the outer door, walked to the far end of the structure to enter the middle door, and walked to the other end to enter the inner door. Then one entered a tent, inside which were further protective coverings.

to those which issue from tubes of bamboo – allow me to ask about the notes of heaven?'

Tzu-Ch'i replied, 'When the wind blows, the sounds from the myriad apertures are each different, and its cessation makes them stop of themselves. Both these things arise from themselves – what other agency could there be exciting them?'

This early passage shows us with poetic fullness the way in which all of nature was conceived of as emitting musical notes analogous to those of the pitch-pipes. It will help us to understand how the following strange procedure was instituted, and how such a thing could possibly occur to any sensible human being.

In order to verify the correct lengths of the standard pitch-pipes, and thus confirm the pitches emitted by them, which formed the basis for all measurement, by the first century BC the procedure known as 'observing the *ch'i*', or 'the blowing of the ashes', was instituted. As Needham says: 'Some ancient nature-philosophers set out to trap the *ch'i* ... which rose up from the earth combining with the *ch'i* which descended from heaven to produce the different types of wind that blew at different seasons of the year.'

There were, as we have seen previously, twelve basic pitches, and a set of twelve standard pitch-pipes to emit them. But since there were twelve months, the Chinese, who loved to correlate things, assigned one pitch-pipe to each month of the year. As the official history of the first century BC tells us: 'The *ch'i* of heaven and earth combine and produce wind. The windy *ch'i* of heaven and earth correct the twelve pitch fixations.' How did they do this?

An obscure author known as 'Humble Tsan', whose real name is thought to be Yu Tsan, He Tsan or Fu Tsan, writing at some date prior to the end of the fourth century AD, gives the traditional explanation: 'The *ch'i* associated with wind being correct, the *ch'i* for each of the twelve months causes a sympathetic reaction in the pitch-pipes; the pitch-pipes related serially to the months never go astray in their serial order.'

This makes clear that experimental standard pitch-pipes, twelve in number, were somehow 'activated' by subtle *ch'i* forces month by month, the appropriate pitch-pipe being 'activated' in its corresponding month. But what was the technique for accomplishing this, and how was the 'experiment' set up? 'Ts'ai Yung (c. 178 AD) gives the details:

> The standard practice is to make a single-roomed building with three layers [concentric draught-proof walls]. The doors can be closed and barred off from the world outside, and the walls are carefully plastered so as to leave no cracks. In the inner chamber curtains of orange-coloured silk are spread out forming a tent over the pitch-pipes, and certain stands are made out of wood. Each pitch-pipe has its own particular stand, set slanting so that the inner side is low and the outer side high, all the pipes being arranged round the circle of compass-points in their proper corresponding positions. The upper ends of the pitch-pipes are stuffed with the ashes of reeds, and a watch kept upon them according to the calendar. When the emanation [*ch'i*] for a given month arrives, the ashes of the appropriate pitch-pipe fly out and the tube is cleared.

The official history for the first and second centuries AD adds:

> They rely on calendrical calculation and so await the coming of the emanation [*ch'i*]; when it arrives the ashes are dispelled; that it is the emanation [*ch'i*] which does this is shown by the fact that its ashes are scattered. If blown by human breath or ordinary wind its ashes would remain together.

Thus do we see the full lunacy of the 'observing the *ch'i*'. This procedure was practised for at least 1700 years, from the first century BC or earlier, up to the sixteenth century AD, when it was finally discredited as ridiculous. As Needham says: 'We are tempted to feel that there must have been some genuine natural phenomenon, even if only once observed, which sufficed to keep this strange technique living for a dozen centuries. However that may be, no rational basis for the system can be suggested...

But as a by-product of the obsessive need to take precautions to safeguard the correct 'observation of the *ch'i*' and its effects on the pitch-pipes (which in the palace were all made of jade), we find that there arose what was genuinely the world's first attempt at a hermetically sealed research laboratory. Needham has summarized the matter thus:

> The most interesting part of this strange experiment is the care which seems to have been taken to ensure that no ordinary wind could enter the sealed chamber. This sealing is perhaps really the most significant technical feature of the whole story. The precautions against chance breaths of wind and other disturbances reached their greatest degree of elaboration by the middle of the sixth century AD. Besides the tent of orange silk, gauze covers were fitted for each pitch-pipe individually. According to the descriptions, the stands or holders for the pipes were rather like our retort-stands. And the walls were so arranged that the doors of the inner and outer walls were at the south, while the door of the middle wall was at the north. Thus there were imbricated corridors exactly as in modern photographic dark-room practice. These remarkable details of the pursuit of an essentially unreal phenomenon may be found in the commentary of the *Yüeh Ling* written by Hsiung An-Sheng about 570 AD, and in the *Yo Shu Chu T'u Fa (Commentary and Illustrations for the Book of Acoustics and Music)*, due to the eminent mathematician and astronomer Hsintu Fang, his older contemporary.

Never until modern times can such elaborate procedures have been adopted, foreshadowing as they did the modern hermetically sealed research laboratory, as in the pursuit of the fantastic illusion of the supposed 'blowing of the ashes' in ancient China – an imaginary phenomenon of imaginary forces. It all goes to show that in the pursuit of the most irrational aims, the most rational of means may be developed and adopted.

Before leaving this bizarre subject, it is perhaps worth while taking note of some attempts to explain it made by Chinese skeptics of the sixteenth century. These have been studied by Professor Derk Bodde, who in his essay on the subject speaks of 'watching for the ethers' rather than 'observing the *ch'i*'. Sometimes fans were meant to revolve with the *ch'i* when the ashes were blown out of the successive pitch-pipes. Bodde quotes a scathing attack by Hsing Yün-lu in 1580:

> **I understand it! The movements of the fans and flying of the ashes were all based on mechanisms ... with men operating these mechanisms, the fans would move and the ashes would fly without fail, at the appointed times. These false things were secretly done in order to deceive the [emperor]. Unto the present day, officials of the Imperial Board of Astronomy ... using mechanisms, manufacture a false watching for the ethers with ashes.**

Another skeptic pointed out that if the pitch-pipe in the north of the laboratory were meant to respond with an appropriate blowing of the ashes because it was north, one could move the laboratory a few yards further north until the exact same pitch-pipe location had become the south. Would a pitch-pipe in precisely the same patch of ground then not be activated? And if so, then clearly the *ch'i* was not rising from the Earth in the way claimed.

In the end, the curious saga of the hermetically sealed research laboratory helped to bring about a rise in rational thought in China by offering the opportunity for sharper minds to explode what was either a superstitious absurdity or a long-standing case of fraud by the emperor's scientists.

THE FIRST UNDERSTANDING OF MUSICAL TIMBRE

THIRD CENTURY AD

We have already mentioned timbre. It is sometimes called the 'quality' of a musical note. It may be defined as the original fundamental note sounded, together with the overtones present with it. A tuning-fork has no timbre, because it merely sounds the fundamental note free of overtones; but the rich and resonant sounds produced by a modern piano are a triumph of the art of production of harmonious timbre. The only sources of harmonious timbre possible in non-electronic musical instruments are vibrating strings and vibrating columns of air (as in pipes or flutes). Bells can be acceptably harmonious in their timbre, but percussion instruments have bad timbre.

The Chinese achieved a deep and profound understanding of timbre, and went further than any other culture, before or since, in exploiting it for musical purposes. The ancient Chinese zither known as the *ku ch'in*, or just simply *ch'in* (sometimes called the classical Chinese lute, though it is in reality a half-tube zither, usually of seven strings but originally of five and occasionally of nine) is, as Needham tells us, 'the only

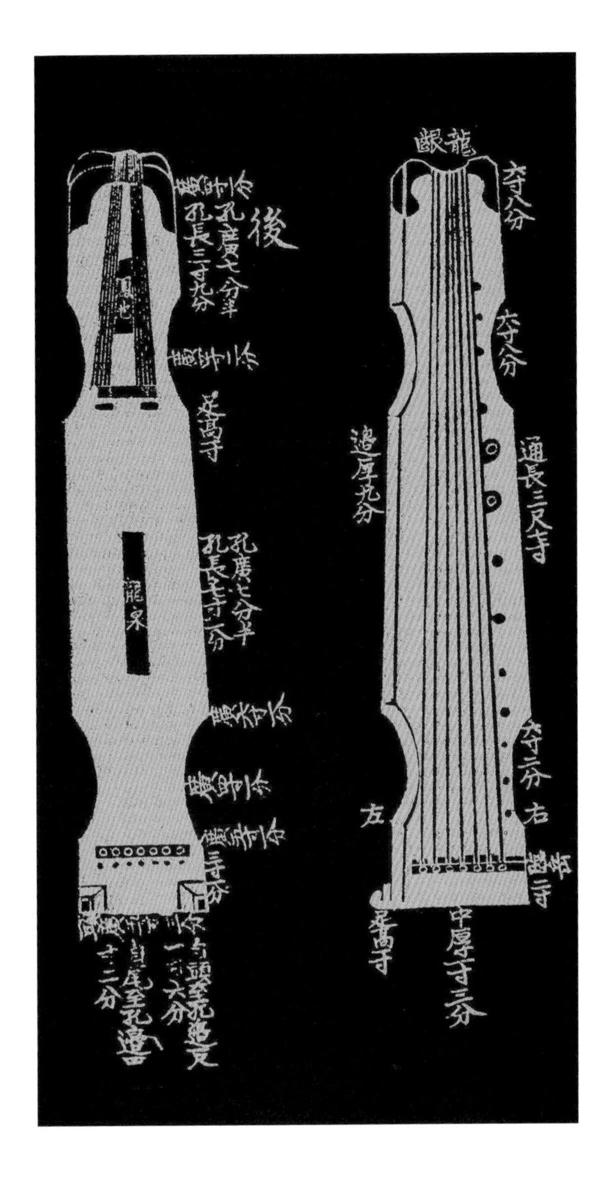

ABOVE (158) A woodcut from the Korean encyclopedia, *Patterns in Musicology*, published in 1493, showing the seven-stringed Chinese zither called the *ku ch'in*, the world's only stringed instrument with no frets. The underside, showing the attachments of the silk strings, is on the left. The *ku ch'in*, or often just *ch'in*, relies for its music upon changes in timbre of its seven notes. These are both highly subtle and surprisingly varied.

ABOVE (159) A Chinese scholar sits reflectively at his table with his *ku ch'in*, which he is strumming with his left hand. Because the instrument has no frets, two hands are not absolutely necessary. His attendant seems riveted by the scene. Watercolour on paper, eighteenth century. (Nottingham Castle Art Gallery; Bridgeman Art Library.)

musical instrument in any culture which has no frets and actually marks the nodes of vibration on the board'. The very fact that there can be such a thing as a stringed instrument whose only musical purpose is the production of timbre variations is astonishing in itself.

The point of this instrument was that the strings were not played in such a way as ever to change their pitch, as happens when a string is pressed onto a fret in other instruments, such as the guitar or the violin. The *ch'in* was played with each string remaining at the same pitch at all times, and the art of playing was the production of different timbres of each string at constant pitch. This is so subtle that many Westerners might balk and question whether the result would strictly be music at all. But it is in fact a wonderfully rich and satisfying music.

The strings of the *ch'in* are all of silk. The traditional manner of playing them involved more than twenty-six different 'touches' or means of plucking or stroking them, for vibrato alone. R.H. van Gulik describes one vibrato technique in *The Lore of the Chinese Lute*, published in 1940, as follows:

> **Remarkable is the *ting yin*, where the vacillating movement of the finger should be so subtle as to be hardly noticeable. Some handbooks say that one should not move the finger at all, but let the timbre be influenced by the pulsation of the blood in the fingertip, pressing the string down on the board a little more fully and heavily than usual.**

Needham comments on 'the infinite subtlety with which any given note could be played,' and remarks: 'Indeed, even today an expert *ch'in* player will himself remain intently listening long after a note has become inaudible to other listeners. As Taoist thought put it (in the Book of Lao Tzu): "The greatest music has the most tenuous notes."'

The subtleties of the *ch'in* caused it to appeal to the Chinese literati, who ended by monopolizing it. At first it was the only musical instrument which a gentleman considered worthy of his playing. Then, so fashionable did it become that every educated man thought it best to have one hanging on his wall, whether he used it or not. And finally, only gentlemen were supposed to play the *ch'in*. This led to a decline in musical standards, since the educated classes naively believed that the qualification for playing the instrument was not musical ability or skill but whether one knew classical literature and poetry and were 'refined'. One direct result was that *ch'in* music became less musical, and was less employed for accompanying singing. The emphasis shifted to the production of the different sorts of timbre for their own sake, rather than as part of musical compositions. Literati would sit and listen to series of subtle timbre variations almost without melodic line. So little did melody matter that the great compilation of *ch'in* music containing 468 tunes was actually lost, at the very time when the numbers of instruments was increasing dramatically. But whereas the musicality declined, the scientific understanding of timbre increased. The adoption of this instrument by the literati therefore both preserved the instrument into an era (the seventeenth century onwards) when it might otherwise have become extinct like the Great Lute, and at the same time advanced musical theory considerably.

The playing of the *ch'in* obviously required a musical notation different from usual. Musical notation for this instrument took the form of instructions, not only on which strings were to be used, note by note, but on the 'touch' by which the fingers were to play each note. Many pieces of music for this instrument also survive which additionally indicate that the rising or falling intonation of the words sung by the singer were to be accompanied by the zither; the change in voice intonation was matched by the variations in timbre of the instrument – a particularly subtle musical technique.

Chinese understanding of the nature of sound as vibration was much increased by studying the production of timbre on the strings of the *ch'in*. When a string vibrates, a wave is manifested in the string, and there are points called 'nodes' where the string remains stationary. Players of the *ch'in* would have become aware that touching a vibrating string at a node has no effect on the vibration, but touching it anywhere else causes the vibration to cease. And the many 'touches' used in playing the *ch'in* exploited these phenomena to the full, in the manipulation of timbre. Since timbre is the sum total of the fundamental note and its higher harmonics, which are of varying consonance and dissonance, the elimination of various of the overtones while leaving others could have a substantial effect.

We can reproduce the essentials of *ch'in* playing on common stringed instruments, and skilled violinists often do this. But if we turn to the piano, let us play the note middle C. If while the string vibrates, we then touch it precisely in the mid-point, we silence the fundamental tone (the note itself) as well as the overtones G of the next octave and E of the octave above that. But the two C notes at one and two octaves removed from the original will continue to sound from the touched string, as will the note G three octaves up.

We can thus 'select out' various overtones and higher harmonics, while leaving others. This is a manipulation of the timbre of a single string such as was practised with elaborate and manifold skill by the *ch'in* players. And in the study and manipulation of timbre, the stretched string, with its utter precision, is the perfect medium. For its higher harmonics are described by simple numerical relations. These overtones theoretically continue to infinity; indeed, in that triumph of resonance, the modern piano, no less than forty-two overtones have been recorded and detected from the vibration of a single string.

The silken strings of the *ch'in* could not be subjected to sufficient tension to be capable of such enormous feats, but they did very well, for silk is extremely strong, and the strings were made of intertwined silken threads sometimes numbering seventy or eighty per string.

The understanding of timbre, overtones and higher harmonics meant that the Chinese were able to contemplate those phenomena of consonance and dissonance which eventually led them to invent the equally tempered musical scales (described in the next account), which most musicians believe to have been invented in the West. One incitement to invent the equal temperament of scales is to avoid certain sounds regarded as dissonant, and knowledge and appreciation of such dissonances partially comes from understanding timbre. For instance, in a vibrating string, the overtones represented mathematically by odd numbers occurring above the fifth are generally considered dissonant sounds to the fundamental note.

The observations of vibration of stretched strings in China were scientifically just about as advanced by the third century AD (when we know for certain that the techniques of the *ch'in* had reached a refined stage) as they were by the nineteenth century in Europe.

EQUAL TEMPERAMENT IN MUSIC

SIXTEENTH CENTURY AD

The inventor of equal temperament in music was Chu Tsai-Yü, who published his invention in 1584. The first mention of Chu Tsai-Yü's invention was in the unpublished papers of the great mathematician Simon Stevin (died 1620). Chu Tsai-Yü was born in 1536, a prince of the Ming Dynasty. But he turned his back on his princely rank and concentrated instead on studying music, mathematics and the science of the calendar. His system of equal temperament appeared in his book *A New Account of the Science of the Pitch-Pipes*, published in 1584. The Chinese did not pay much attention to the new system, but the Europeans quickly saw its advantages.

Two years earlier, the great Chinese scholar of the Jesuits, Matte Ricci, commenced his studies at Macao. From 1580, the Viceroy of the Cantonese province had established biannual 'radio fairs' lasting several weeks, at which Chinese and Westerners exchanged ideas and goods. The interchange between East and West was intense just at the moment when Chu Tsai-Yü went into print with his new theory. It is a case of perfect timing which gives one the feeling that it was 'fated' to happen. We do not know the exact mode of transmission of the idea to Europe; there can be no doubt that Western music was to be totally conquered by the Ming prince, for within 52 years of Chum's publication, his ideas were published by Père Marin Mersenne. The Ming Dynasty ended eight years later, but Ming music today blares from every transistor radio in the world.

The first published reference to the mathematical basis for equal temperament in Europe was by Mersenne in one of his many books of musical theory, entitled *Harmonie Universelle*, issued in 1636. Werkmeister later popularized equal temperament, and Johann Sebastian Bach took up the cudgels on its behalf by composing a series of pieces, collectively entitled *Das Wohl-temperierte Klavier (The Well-Tempered Clavier)*, consisting of 'preludes and fugues in all the tones and semitones ... for the use and practice of young musicians who desire to learn, as well as by way of amusement, for those who are already skilled in this study'.

This epochal work was published in 1722. Yet probably only a few of all those who have ever listened to or played these pieces have had any idea of

ABOVE (160) A painting of a girl playing to an emperor on the lute. His intent stare is not meant to be intimidating: it is evidence of the extreme seriousness with which music was often viewed at court. Such attitudes were partially responsible for the Ming Prince Chu Tsai-Yü's many years spent on the study of musical theory, resulting in his original idea of equal temperament. (British Museum, London.)

新製律準正面小樣

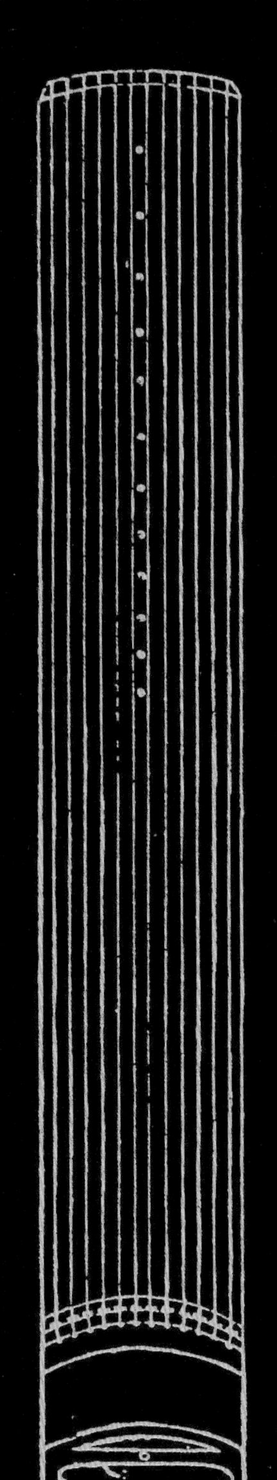

新製律準背面小樣

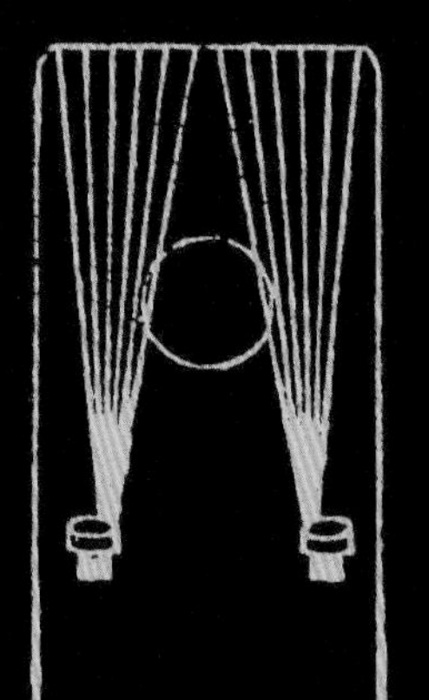

Bach's underlying intentions. The *Well-Tempered Clavier* was a work of propaganda. Bach had adopted equal temperament with a passion later to be matched by that of his fellow-composer Giuseppe Tartini's opposition to it. Equal temperament allows one to modulate fluently from key to key consecutively in the composition and performance of music. It was passionately adopted by those who thought the practical advantages of this were paramount, and was as passionately opposed by those who thought that purity of tone and other substantial factors should not be sacrificed to such base utility. These frenzied debates and disagreements have died down now in the West, at least for the moment. Equal temperament, so hugely aided by the prestige of Bach, triumphed utterly during the nineteenth century, yet many of today's musical experts do not really understand its principles, and no more than a handful of Westerners can now appreciate what music would sound like without it.

The only way for the reader to appreciate the magnitude of this subject is for us to consider the underlying problem which makes perfection of tone in all musical composition impossible. It is one of the most fascinating but least known of the fundamental enigmas of the Universe. There is no music conceivable, on whatever planet, in whatever galaxy, where the dilemma could be escaped.

Let us illustrate the matter in a simple way by discussing a piano. Most people know that a note struck simultaneously with a note eight notes higher (in a scale) is an 'octave'. The low note and the high note are the same (say a C), except that the high note is 'C an octave higher than the lower C'. The higher one is double the frequency of the lower one. This is true whatever note is played; an 'octave' is defined as a note of a particular frequency sounded with a note of twice that frequency. Octaves are quite simple. They are also rather boring, because the same notes sounded an octave apart have no variety. There is neither consonance nor dissonance; there is just uniformity. An octave is like a man standing and staring at a photograph of himself which he holds in his hand: a simple double image. How much more interesting it would be if the man were to stand with his arm around a woman whom he loves, or were to be scowling at one whom he hates. The first would be a consonance, and the second would be a dissonance. The interest of life is in variety and juxtaposition; hence, a C played with any other note than another C takes on greater interest to us than the octave.

The most pleasing consonance of two notes sounded together is what is called a 'fifth'. This is not simply subjective; an analysis of the sound waves involved shows that the higher harmonics of the notes have consonant beats with one another. But there does remain a mystery about the preference human beings find for harmony and the annoyance they experience upon hearing dissonance. Sir James Jeans has written: 'It must be admitted, however, that [there] is a defect of most theories of discord. Innumerable theories are ready to tell us the origin of the annoyance we feel on hearing a discord, but none even attempts to tell us the origin of the pleasure we feel on hearing harmony; indeed, ridiculous though it may seem, this latter remains one of the unsolved problems of music.'

What, then, is this most pleasing and harmonic of all joint sounds, the fifth? For it was the fifth rather than the octave which was the basis for Chinese music. A typical example would be the chord CG. If you then play successive fifths, using the top note of one fifth as the bottom note of a new fifth, you do not come to a C again for quite some time (CGDAEBF♯C♯A♭E♭B♭FC). In fact, you have a succession of twelve fifths before reaching another.

Musical theorists like to speak of the ascending fifths as a 'spiral of fifths', and they draw them in a diagram spiraling upwards. If you count the number of octaves between the first and last note in a spiral of twelve fifths, you will find that there are seven. As the spiral has gone round, it has repeatedly missed the higher octave notes of the original note, until finally after twelve fifths it hits the seventh octave of the original note. Then and only then do the two separate ascending series meet.

An upward (or downward) spiral of fifths and an upward (or downward) spiral of octaves thus only meet up when twelve fifths and seven octaves end on the same note. Until then, on their upward or downward courses, the two means of proceeding

OPPOSITE (161) Prince Chu Tsai-Yü invented equal temperament in music, and published the idea in 1584 in his *New Account of the Science of the Pitch-Pipes*. This woodcut of his personal tuning instrument is from the book, in which he proudly stated: 'I have founded a new system.' Little could he have realized that his system would be universally adopted in the West. Johann Sebastian Bach was its leading promoter.

RIGHT (162) A Sui Dynasty (581–618 AD) figurine of a girl playing the true short lute, which was introduced into China in the second century AD from Central Asia. This instrument was a favourite in China, and is frequently mentioned in literature and poetry.

have been quite separate. It is as if two runners were running the same distance on two separate tracks which went over and under one another repeatedly until they reached the same finishing post, at a point where the separate tracks met for the first time since the start.

This is important, for upon examining this more closely we find that it is not as simple as it seems – there is something curiously wrong. The problem is this: musical tones are very precisely measurable in the laboratory, and an exact number is assigned to every note as its frequency. Now, when one plays a C seven octaves higher than another C, we find that its frequency is 128 times that of the original. (Every octave doubles the frequency, and if you progressively double something seven times you have in fact made it 128 times its original self.) But if you want to have a sequence of fifths which increase by 1.5 times the original frequency (which is what fifths do) twelve times over, you will achieve a final result not exactly equal to the 128 of seven octaves, but equal instead to the slightly different amount of 129.75, which is the value of $(1.5)^{12}$.

The fact is that the mathematics of the fifth is incommensurable with the mathematics of the octave. A note which is a *fifth* higher than another note has a measurable frequency one-and-a-half times that of the lower note. And the number 1.5 is arithmetically incommensurable with the number 2 (which expresses the doubling of the frequency of a note when raised an octave). So the fifth and the octave are out of joint with each other on fundamental arithmetical grounds. The spiral of fifths comes to a stop at a point which is 1.0136 times the sequence of octaves. (Or, the frequency 129.75 is 1.0136th that of 128). This value, 1.0136, is known as 'the comma of Pythagoras', after the Greek philosopher who discussed it.

The different keys are established in Chinese music by the twelve different notes of the spiral of fifths, and these twelve notes are all found within the compass of a single fundamental octave. The spiral of fifths ascends from C to G to D to A to E to B to F$^{\#}$ to C$^{\#}$ to A^{b} to E^{b} to B^{b} to F back to C (except, as just mentioned, this C is not absolutely precise). It will be appreciated that all of these notes fall between one C and the higher or lower C, and give the twelve keys. These twelve notes also give the complete chromatic scale of modern music.

Equal temperament is an artificial system created to get round the fact that the spiral of fifths ends on a note slightly off the end of the sequence of octaves. The tiny fraction of 0.0136 is divided into twelve equal parts, and each part is subtracted from one of the twelve notes into which an octave is divided. This means that the gap between a note and its 'fifth' is no longer precisely 1.5, but is instead a tiny fraction less, namely 1.4983. This 'violence' done to all the fifths squeezes them into the tinier space of a pure octave. All twelve steps in the octave are now precisely equal, and are called semitones. In order to accomplish this, each fifth has been artificially but evenly rendered flat by about one forty-eighth of a semitone. All equally tempered music is thus uniformly and unremittingly 'flat'. But it provides a regular and reliable structure so that one can modulate from key to key, for as much richness and variety of composition as one could desire.

This is not to say that much has not been lost – sacrificed on the altar of utility. Our modern ears have been so debased by hearing only equally tempered music that we no longer know a pure tone. We send for a piano tuner to tune our piano, but in fact he comes and tunes it flat, note by note, as relentlessly as ants march forth from their nest. We are thus subjected, from birth to death, to nothing but flat notes. We never so much as hear a pure tone.

Before equal temperament, in both China and the West, there were various 'modes' of untempered music. For the sake of simplicity, we can speak as well of the ancient Greek ones, which have less difficult names, though they all have Chinese equivalents. There were the Ionian, Dorian, Phrygian, Lydian, Mixo-Lydian, Aeolian and Locrian modes. Instruments could be tuned to only one mode at a time. Some of the modes were happy and carefree, while others were sad and mournful. They represented a vast richness of emotional intensity and experience which has now completely and utterly vanished from the music known to us today. It is impossible to describe them: the colourations and subtleties of the differing modes were somewhat like the difference which we know between major and minor keys, multiplied several times. In the absence of these old modes, not a single person in the entire Western world unafflicted by deafness can avoid hearing at every turn music fashioned from an imported Chinese theory.

Part 11 WARFARE

CHEMICAL WARFARE, POISON GAS, SMOKE BOMBS AND TEAR GAS

FOURTH CENTURY BC

Chemical warfare using poison gas goes back to at least the early fourth century BC in China. It is described in writings of that date by the sect known as the Mohists, founded by a philosopher and social reformer called Mo Ti. In the early Mohist writings, we read of the use of bellows to pump poison gas down into the tunnels of enemies besieging cities. The bellows were made of ox hide and connected to furnaces in which balls of dried mustard and other toxic vegetable matter were burnt – anticipating the mustard gas of the trenches in the First World War by 2300 years.

The use of poison gas was a natural development from the traditional Chinese custom of fumigation of houses, known to be practised in the seventh century BC. Fumigation was also widely practised to kill bookworms; this function was exploited after the introduction of paper (page 92). Even the virtues of steam were appreciated as early as the tenth century AD, when the Chinese seem to have realized that steam could sterilize. Lu Tsan-Ning wrote in his book *Simple Discourses on the Investigation of Things* in 980:

> **When there is an epidemic of a feverish disease, let the clothes of the sick person be collected as soon as possible after the onset of the malady and thoroughly steamed; in this way the rest of the family will escape infection.**

The earliest use of poisons in warfare was obviously the poison-tipped arrow, which was already widespread in many corners of the world, its origins rooted in indefinite antiquity. The next step was the systematic Chinese use of poison gas. The Chinese appreciation of the properties of many gases was perhaps aided by the fact that, unlike Westerners, Chinese did not draw hard and fast distinctions between spirit and matter. For the Chinese, all matter faded into ever subtler gradations of tenuous, ethereal matter. They even believed in a genuine physical immortality of tenuous, ethereal human bodies: if a sage lost weight and appeared to be wasting away before one's eyes (often as the result of poisons in his elixirs), this was looked upon as a marvellous process. He was 'lightening' and was on his way towards physical immortality in the form of a spirit – that spirit being considered *entirely material*, but at the same time, merely tenuous. Belief in this sort of thing obviously led the Chinese to take great interest in tenuous matter of whatever kind, and drew their attention towards gases. The property of evaporation was even seen as an analogy to what humans could look forward to if they were fortunate and took the right potions.

Elsewhere, in the account of the origin of gunpowder (page 250), mention is made of the tradition of incorporating the poison arsenic in many gunpowder mixtures. This may have arisen from an association of arsenic preparation with the first inadvertent discovery of gunpowder. But it was not simply habit and tradition

OPPOSITE (163) The Chinese were using poison gas, including the smoke of burning dried mustard, by the fourth century BC in warfare. Here mustard gas is being used by the Germans against the Allies during the First World War, 2300 years later.

which were involved. Bizarre and terrible poisons were mixed together in a very large proportion of the bombs and grenades of the Chinese (for which, see page 256). One of the most disgusting must certainly have been the medieval excrement bomb, for which we have the recipe:

Ingredients:	*Weight in Ounces:*
Human excrement, dried, powdered and sifted very fine	240 (15 lb)
Wolfsbane	8
Aconite	8
Croton oil	8
Soap-bean pods (to create black smoke)	8
Arsenious oxide	8
Arsenic sulphide	8
Cantharides beetles (*Mylabris cinchorii*)	4
Ashes	16
Tung oil (from either *Aleurites fordii* or *Perilla nankingensis*)	8
Total for one bomb:	316 oz (19.75 lb)

This would have created a stink bomb spreading excrement dust combined with some pretty deadly poisons all over the enemy. All of the above ingredients were either surrounded by or combined with a gunpowder mixture, wrapped in layers of paper which were tied with hemp string and covered with melted resin. This deadly packet would be hurled by a piece of artillery called a trebuchet, which held the bomb on a sling which would be whipped up in the air on an arm. A lighted fuse would cause the bomb to ignite either over the heads of the enemy, or shortly after landing, when it would have burnt fiercely and spread its foul smoke in thick clouds.

Directions also survive as to how to mix poison bomb recipes. For instance, sulphur and saltpetre were pounded together and passed through a sieve. Arsenic and white lead would be ground together. Dried lacquer would be pounded separately into a powder (it was a powerful chemical irritant). Yellow wax, pine resin and various oils would be boiled together into a paste, and the powdered chemicals added to turn it into a thick soupy mass, constantly and carefully stirred. Smaller poison bombs with lighted fuses could be fired from bows or crossbows. In the fourth century BC the Mohists had burnt poisons in a furnace, drawn the deadly smoke off in pipes and blown it with bellows down tunnels constructed by the enemy. But by medieval times, poison bombs were let down and exploded in the tunnels instead. Bamboo fans were used to drive smoke and flame down the tunnels, to stifle and burn the enemy.

The equivalent of our modern tear gas was a blinding smoke, created by bellows blowing out finely powdered lime, which made the eyes run profusely. Yang Wan-Li in his book *Rhapsodic Ode on the Sea-eel Paddle-Wheel Warships* describes a sea fight of 1161 where tear gas was used effectively:

> In the Shao-Hsing reign-period, the rebels of Wanyen Liang came to the north bank of the River in force ... but our fleet was hidden.... Then all of a sudden a thunderclap bomb was let off. It was made with paper carton and filled with lime and sulphur. Launched from trebuchets these thunderclap bombs came dropping down from the air, and upon meeting the water exploded with a noise like thunder, the sulphur bursting into flames. The carton case rebounded and broke, scattering the lime to form a smoky fog which blinded the eyes of men and horses so that they could see nothing. Our ships then went forward to attack theirs, and their men and horses were all drowned, so that they were utterly defeated.

We have better descriptions of these tear gas bombs in an account of the campaign of the Sung general Yo Fei against the bandit chief Yang Yao in 1135:

> ...the army also made 'lime-bombs'. Very thin and brittle earthenware containers were filled with poisonous chemicals, powdered lime, and iron calthrops [sharp pointed objects to injure feet]. In combat they were used to assail the enemy's ships. The lime formed clouds of fog in the air, so that the rebel soldiers could not open their eyes. They wished to make the same kinds of things themselves, but their potters were not able to produce them, so they suffered great defeats.

Chinese tear gas was in fact in use by the second century AD at the latest. The dynastic history relates this suppression of a peasant revolt in the year 178:

> **The bandits were numerous, and Yang's forces very weak, so his men were filled with alarm and despondency. But he organized several horse-drawn vehicles carrying bellows to blow powdered lime strongly forth, he caused incendiary rags to be tied to the tails of a number of horses, and he prepared other vehicles full of bowmen and crossbowmen. The lime chariots went forward first, and as the bellows were plied the smoke was blown forwards according to the wind, then the rags were kindled and the frightened horses rushed forwards throwing the enemy lines into confusion, after which the bowmen and crossbowmen opened fire, and drums and gongs were sounded, and the terrified enemy was utterly destroyed and dispersed. Many were killed and wounded, and their commander beheaded.**

Sometimes the recipes for poison bombs are positively gleeful at the hideous effects caused. One fourteenth-century bomb described in *The Fire-Drake Artillery Manual* of 1412 is the 'bee swarm bomb', producing a burning fire which 'comes forth (and) can stick to the enemy's person and still burn.' The book adds dolefully that (unfortunately) 'it can be extinguished with water.' And from the same book we hear also of the 'flying-sand magic bomb releasing ten thousand fires'. It consisted of a tube of gunpowder put into an earthenware pot containing quicklime, resin, and alcoholic extracts of poisonous plants. It was thrown down from city walls, and its explosion released the deadly poisons.

Poisons, poisonous gases, and tear gas were not only delivered by bombs and grenades. They were fired from guns and proto-guns, as described elsewhere (page 266). An early proto-cannon in the fourteenth century was called the 'poison-fog magic-smoke eruptor', and *The Fire-Drake Artillery Manual* recounts its amazing properties:

> **If blinding gunpowder, flying gunpowder, poison gunpowder, and spurting gunpowder are filled into a shell and fired at the top of a city wall, fire will break out and smoke will spread in all directions as the shell explodes. Enemy soldiers will get their faces and eyes burnt, and the smoke will attack their noses, mouths and eyes. If the right moment is chosen, no defenders can withstand such an attack.**

Among the many sorts of gunpowder whose formulae are given by the early Chinese, several are specified as tear gas mixtures or poison gas producers. A classic tear gas mixture was called 'five-league fog'. It contained a mere 27.8 per cent saltpetre, the same amount of sulphur, and 44.3 per cent carbon. It must have burned slowly, and contained arsenic, sawdust, resin, human hair and chicken, wolf and human excrement. A poison gas bomb called 'soul-hunting fog' contained a very strong explosive with 83 per cent saltpetre, a mere 8.3 per cent sulphur, and 8.4 per cent carbon, together with arsenic sulphides and deadly animal poisons. These bombs are at the two extremes of slow burning and strongly explosive. Practically every animal, plant and mineral poison imaginable was combined in one or another mixture

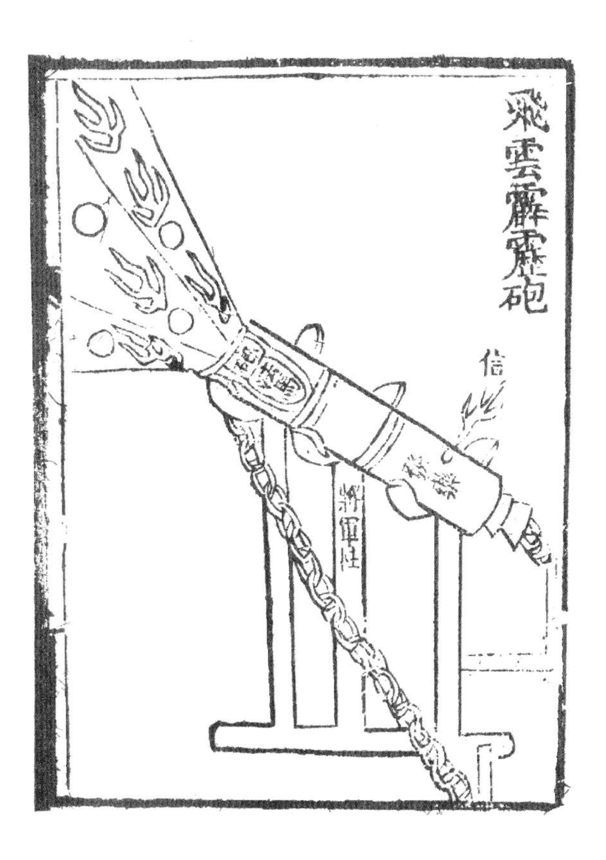

RIGHT (164) A medieval artillery piece, known as 'the heaven-rumbling thunderclap fierce fire eruptor', which fired poison smoke shells at the enemy.

by the Chinese. There hardly seemed to be a deadly substance unknown to them.

They were also informed as to the various effects of the different poisons. A text of 1606 tells us that certain ones will 'cause the enemy's flesh to rot until the bone shows', whereas with dolphin oil, Szechuan varnish and arum poison the enemy will be struck dumb. It adds: 'Smoke from burning wolf excrement, which looks red both in the daytime and on dark nights, can be used for sending warning signals.'

The development of rockets for warfare (see page 262) added a new method of delivering poisonous bombs. The 'free-flying enemy-pounding thunder-crash bomb' used this method. When its rocket charge had exhausted itself the fuse was lit and the bomb fell, releasing a poisonous and irritating smoke even though the bomb was only 5½ inches in diameter.

When did these deadly practices reach Europe? In 1540, Vanoccio Biringuccio in his book *Pirotechnia* described typical Chinese formulae for use in fire-lances. They contained arsenic and the usual poisons. He remarked that when lighted, they sent out 'a very hot tongue of flame more than 2 or 3 yards long, full of explosions and horror'. Even earlier, Leonardo da Vinci had envisaged attacking the enemy with sulphurous smokes, fumes of burnt feathers, sulphur and arsenic, and even toad and tarantula venoms mixed with rabid saliva and conveyed by bombs. That was about 1500. And until 1580 arsenic was a great favourite in Europe. But it was replaced by mercury smoke balls in the seventeenth century. Such, then, is one of China's more baleful gifts to the world.

THE CROSSBOW

FOURTH CENTURY BC

For more than two millennia the crossbow was the standard weapon of Chinese armies. In fact, it was used by the Chinese as recently as 1895, by which time, of course, it was largely ineffective against modern European guns. The crossbow was invented in China in its form as an actual weapon, but the original idea for it came from the primitive bow-trap used by aboriginal peoples to kill game. When a person holds an ordinary bow, he holds it vertically. In setting a bow-trap, however, the bow must be laid horizontal. Furthermore, a stick or long piece of wood must be placed against the bow, at the far end of which the bowstring is pulled taut. Then when an animal comes along and knocks the trip-wire with its foot, the stick is pulled away and the bow fires its arrow at the animal, which is collected by the hunter later as he makes his rounds. Thus, the form of the crossbow was implicit in these primitive beginnings from earliest times.

It later appeared in Europe at two widely separated times; each transmission of the invention is thought to have been from China. Even in neighboring India, though the bow-trap was commonly used in Bengal against tigers, the actual crossbow was only introduced for the first time by the Mogul conquerors of the Middle Ages.

A Chinese text compiled from earlier materials in the second century AD credits the invention of the crossbow to a Mr Ch'in of Ch'u, who was an archery student of Feng Meng. If Feng Meng really existed and was not just a mythical personage, then this date would be the seventh century BC. Unfortunately, we have too few texts from these early periods to be certain of such things. During the Han Dynasty, around the time of Christ, an archery manual by Feng Meng existed – though it has since been lost. This may or may not have been the real name of the author, and he may or may not have been the teacher of Mr Ch'in. This is what we are told about Mr Ch'in himself:

> **Mr Ch'in considered, however, that the bow and arrow was no longer sufficient to keep the world in obedience, for in his time all the feudal lords were fighting against one another with weapons, and could not be controlled by ordinary archery. He therefore added at right-angles to the bow a stock and established a trigger-mechanism within a box or housing, thus increasing its strength. In this way all the feudal lords could be subdued. Mr Ch'in transmitted his invention to the Three Lords of Ch'u ... and it was from them that Ling Wang [reigned 539–527 BC] got it. As he himself said, before their time the men of Ch'u had for several generations guarded their frontiers only with bows of peachwood and arrows of thorn.**

There is additional evidence to indicate that the crossbow was indeed invented in the kingdom of Ch'u, and it would appear that Mr Ch'in may have been a real person who did indeed live in Ch'u in the century before Confucius. However, Needham suggests that what Mr Ch'in may really have invented

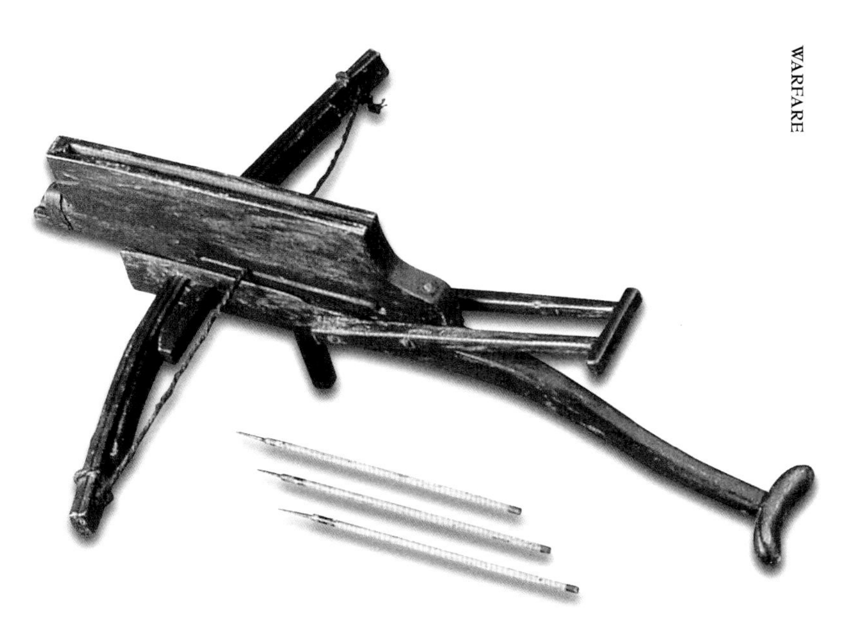

RIGHT (165) A repeating or 'machine-gun' crossbow, capable of firing eleven bolts in fifteen seconds. The magazine resting on top can carry ten bolts. Three of the bolts are displayed beside the weapon. Such repeating crossbows were invented about the eleventh century and were common by the year 1600 in China. They had an effective range of about 80 yards with an extreme range of 200 yards. They provided the earliest form of 'reconnaissance by fire', such as was practised in the Vietnam War (by shooting a spray of bullets). Large numbers of these fired at once could deter most 'human wave' attacks of soldiers. The length of this weapon is 39 inches, with the bolts 14½ inches long. (Simon Archery Foundation, Manchester Museum, University of Manchester.)

was a trigger-mechanism. An earlier form of the crossbow may already have existed. The earliest textual evidence of the crossbow is from a book called *Master Sun's Art of War*, which dates from at least 345 BC and is reputed to date from 498 BC. This book says that 'energy may be likened to the bending of a crossbow, decision to the releasing of the trigger'. The use of such an image can have had little point unless crossbows were sufficiently familiar to the reader for the statement to make sense.

By the fourth century BC the use of crossbows in battle is recorded by Sun Pin, a descendant of Master Sun, who wrote that: 'A very strong body of crossbowmen' was decisive in a victory obtained in that century at the Battle of Ma Ling. Elsewhere, in 336 BC, the official Su Ch'in was recorded as boasting that the Prince of Han has '10,000 suits of armour and the strongest bows and crossbows in the world'.

Dr E.M. Grosser has published a very detailed account of what may be the earliest surviving crossbow. It is a pistol-crossbow, to be held in the hand like a modern handgun. The design on top has been attributed, on stylistic grounds, to the early Chou Dynasty, which might place this excavated bronze pistol-crossbow in the eighth or ninth century BC, or even earlier. However, the earliest certain archeological evidence for crossbows are depictions of them on inlaid bronze vessels, which apparently date from the beginning of the fourth century BC. Further textual evidence from that century exists in the Mohist texts on fortification engineering in *The Book of Master Mo*, which speaks not only of ordinary crossbows but also of large multiple-bolt *arcuballistae* (crossbow-cannons) used for sieges.

By the third century BC textual references to crossbows become abundant. *Master Lu's Spring and Autumn Annals* of 239 BC says: If the mechanism of a crossbow-trigger is out of alignment by no more than the size of a rice-grain, it will not work.' This is perfectly true. Indeed, the perfection of the bronze trigger-mechanisms is the most impressive of the Chinese achievements in developing the crossbow. The trigger-box was inset into the stock like a tenon into mortise, and had a groove on its upper surface for the arrow or bolt. The trigger (called 'hanging knife' by the ancient Chinese, since that is what it resembled) hangs down below the housing case rather like that of a modern gun. Despite the lack of metal lathes, the earliest surviving trigger-mechanisms were manufactured to an incredible degree of machined accuracy, and even more impressive than their quality is the standardization that must have been required for their mass-production. A crossbow trigger-mechanism is a complicated device which contains within its housing three moving pieces on two shafts, each finely cast in bronze and machined to a precision which is difficult to imagine.

In order to appreciate the feat of proto-industrialization which this represented, we must know something of the quantities of crossbows

RIGHT (166) A traditional Chinese crossbow, thought to be of the nineteenth century. The stock is 33½ inches long, constructed of wood, horn and sinew, with an inlay of rosettes and small discs of ivory, and painted paper underneath. The string is of gut. A sight is mounted on the front. (Simon Archery Foundation, Manchester Museum, University of Manchester.)

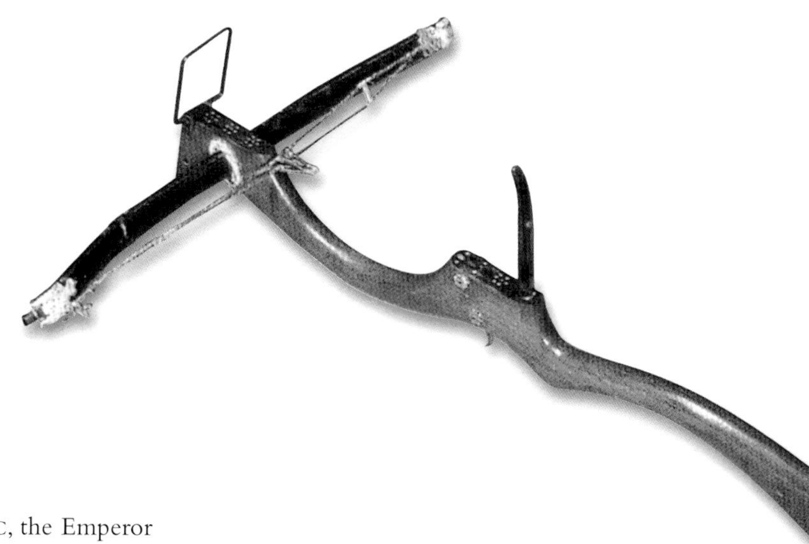

which had to be produced. In 209 BC, the Emperor Erh Shih had crossbow regiments numbering 50,000 men and a similar number of crossbowmen also served under the emperor Wen Ti in 177 BC. But lest we think that this meant that only a few tens of thousands of crossbows existed at this time, we are informed by China's great historian, Ssuma Ch'ien, that in about 157 BC, the prince Liang Hsiao Wang was in charge of arsenals containing *several hundred thousand crossbows*. This means that twenty-one centuries ago, the Chinese were capable of mass-producing complicated mechanisms which were first cast, then machined to a fine standard, and then assembled, in numbers of approximately half a million.

Surely this massive arms production in ancient times is an industrial achievement of the first order. The scholar Homer Dubs observed that the Chinese crossbow trigger-mechanism 'was almost as complicated as that of a modern rifle bolt, and could only be reproduced by very competent mechanics'. So complex were the mechanisms that the Huns were unable to assemble them or copy them. Crossbow arrows were also useless because they were too short for use with a long bow. Therefore, both the weapon and its ammunition were 'capture-proof', in the sense that even if captured they were guaranteed to be useless to the enemy.

The Chinese well appreciated their own advantage in armaments. Here is what Tseng Kung-Liang said in his book *Collection of the Most Important Military Techniques* published in 1044:

> The crossbow is the strongest weapon of China and what the four kinds of barbarians most fear and obey.... The crossbow is the most efficient weapon of any, even at distances as small as 5 feet. The crossbowmen are mustered in separate companies, and when they shoot, nothing can stand in front of them, no enemy formation can keep its order. If attacked by cavalry, the crossbowmen will be as solid as a mountain, shooting off such volleys that nothing can remain alive before them. Although the charge may be impetuous it will not reach them. Therefore the barbarians fear the crossbow. Truly for struggling around strategic points among mountains and rivers and defiles, overcoming men who do not lack bravery, the crossbow is indispensable.
>
> Regarding the method of using the crossbow, it cannot be mixed up with hand-to-hand weapons, and it is most beneficial when shot from high ground facing downwards. It only needs to be used so that the men within the formation are loading while the men in the front line of the formation are shooting. As they come forward they use shields to protect their flanks. Thus each in their turn they draw their crossbows and come up; then as soon as they have shot their bolts they return again into the formation. Thus the sound of the crossbows is incessant and the enemy can hardly even flee. Therefore we have the following drill:

shooting rank
advancing rank
loading rank.

Such praise of the crossbow was traditional in China. Twelve centuries earlier, a eulogy of the weapon was recorded in the imperial official history for 169 BC, part of which says:

> **...the drill of crossbowmen alternately advancing to shoot and retiring to load; this is something which the Huns cannot even face. The troops with crossbows ride forward and shoot off their bolts in one direction; this is something which the leather armour and wooden shields of the Huns cannot resist. Then the horse-archers dismount and fight forward on foot with sword and bill; this is something which the Huns do not know how to do. Such are the merits of the Chinese.**

Manufacture of crossbows was by division of labour. Most crossbow trigger-mechanisms which have been found have been inscribed by their makers, often giving dates. One was excavated in the eleventh century and examined by the writer Shen Kua. He wrote, mournfully, 'We never could find out what dynasty this crossbow mechanism belonged to', but it bore the inscription 'Stock by Yü Shih, rear work of the crossbar by mechanic Chang Jou'.

By the second century BC there was an imperial ban on the exportation of crossbows out of China, and in 125 BC barriers were set up at the frontiers to ensure that this could not happen. But these measures were too late, for crossbows had already been taken out of China. In the first century BC we know that they were to be found in Korea and as far west as Sogdiana in Central Asia. Within China, crossbow manufacture was a state monopoly and it was carried out with all the security precautions familiar to us with modern weapons manufacture today. Candidates for the civil service, in about the year 1030, had to answer a question in their examination, 'What would you do to detect and punish people who kept privately armour and crossbows in their homes?' Presumably only those who gave sufficiently harsh replies were allowed to join the imperial civil service.

Although crossbows came in all sizes, from pistol ones to huge artillery pieces, the problem was always one of arming them. It took a great deal of strength, so that the earlier crossbowmen had to be very strong in order to arm the weapons by brute strength alone, which involved standing on the bows and pulling the strings back with mighty heaves and grunts. Better arming techniques were developed as time went on. Since China was the land where the stirrup was invented (see page 101), it is not so surprising that crossbows were eventually produced with stirrups attached to them, so that one could put one's foot in a stirrup to hold the bow down when rearming. (Standing on a bow was not possible in marshy ground, for instance.) Crossbow stirrups were standard issue to troops by at least the eleventh century. A later improvement was a double-pronged claw for the crossbowman's belt so that he could stand and draw the string of his weapon by the muscular power of leg and back alone, leaving his hands free to hold it, and manipulate the trigger-catch. This was called 'waist-arming'. This technique was thought to have been used in the third century, and then lost for several centuries until revived in the Middle Ages. Rotary arming methods were also used. All of the great crossbow-cannons were armed by winches which pulled back the strings, since no human being alone could possibly have done it. These seem to have been used as early as the eighth century, though relatively rarely, and in connection with sieges. (Regarding winches, it is worth remembering that it was the Chinese who invented the crank handle; see page 49.)

Grid-sights for aiming crossbows were invented in China by the first century AD. These grid-sights were the first in the world, and were similar to the ones used on modern photographic cameras and anti-aircraft guns. Mounted on both rear and forward ends of the stocks, the sights consisted of vertical and horizontal fine wires crossing each other in grids. A third-century text speaks of 'the forester, with his finger on the trigger, and his bow drawn, [who] aims at the target embraced in the graduations of his sights and lets fly'. An even earlier ranging device on the crossbow was a series of graduated notches on the lug of the trigger-mechanism, which goes back to at least the third century BC.

How powerful actually were these Chinese crossbows? A version developed by Li Ting in the eleventh century was made of mulberry wood with the body of brass and the string of hemp. It was presented to the emperor in the year 1068. This remarkable weapon 'could pierce a large elm from a distance of 140 paces.' This was very quickly made standard issue. A crossbow catapult, which took several persons to draw its string and consisted of two bows tied together, could shoot several arrows simultaneously, killing ten persons at a time. The extreme range of a large winch-armed Chinese crossbow was, according to

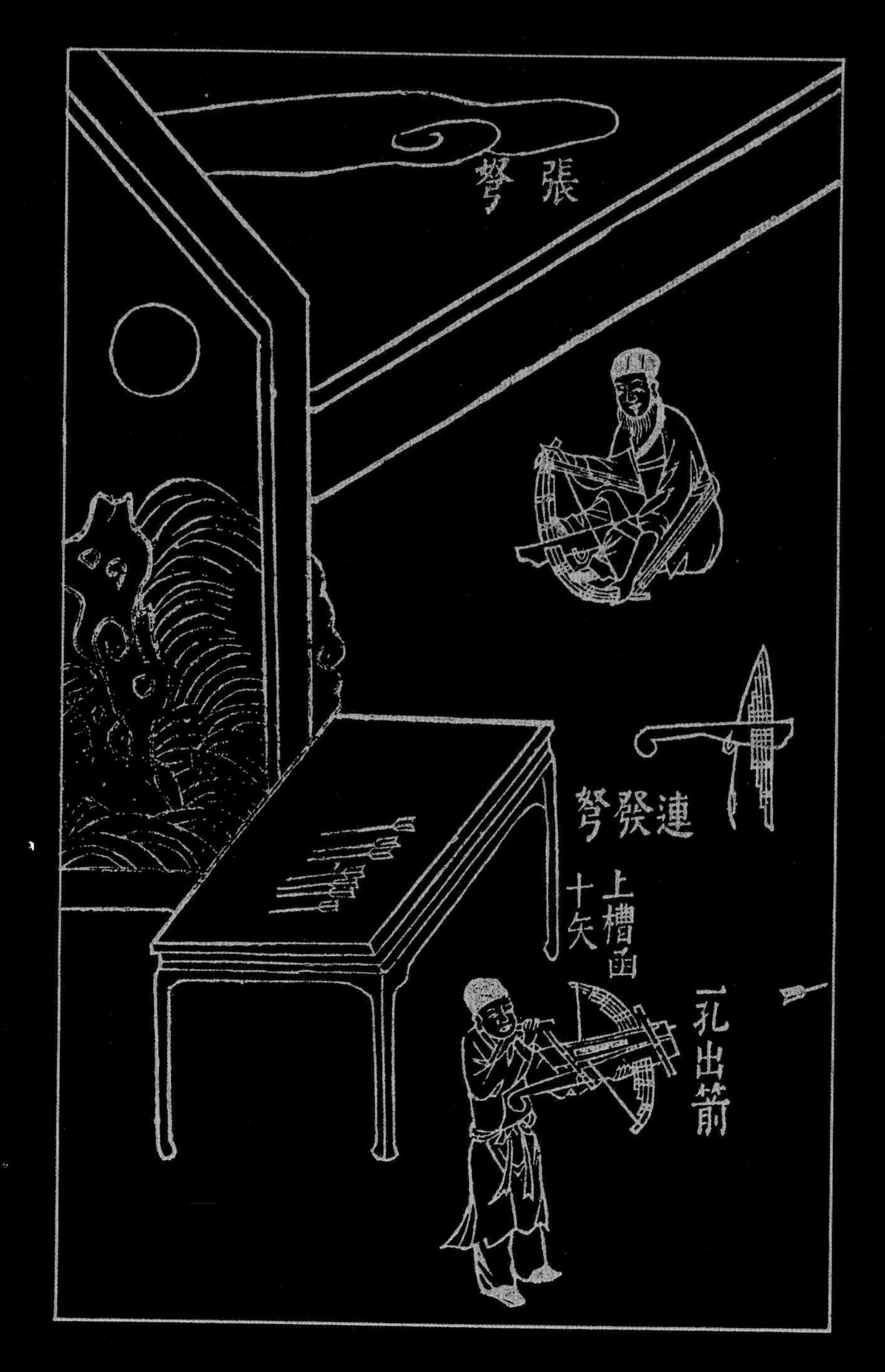

RIGHT (167) Mass manufacture of crossbows during the Ming Dynasty: the original production line, centuries before Henry Ford, and the real beginning of 'the military industrial complex'. As early as 209 BC, 50,000 crossbows were used in a single battle on one side alone. Teams of a hundred archers could each fire 2000 rounds in 15 seconds. This engraving showing crossbows under construction was published in 1637 in *The Creations of Nature and Man* (*T'ien-kung k'ai-wu*).

Needham, 1160 yards. This is only 600 yards short of a mile! It is all the more incredible that the date of this weapon was the eighth century. And at that time, the extreme range of a hand-held crossbow was 500 yards, and that of one fired from horseback was 330 yards.

Although crossbows were very accurate when fired at their extreme ranges from the ground, the necessity of raising the weapon in the air meant that one could not use the sights, and sacrificed accuracy of aim. The Chinese always preferred to use their crossbows from above, firing down upon the enemy. For genuine accuracy at ground level, the range would be only about half the extreme range.

Some crossbows had tubes through which either arrows or bullets were fired. These tubes derived from the bow-tube used by longbowmen for sniping, with the arrow fired through a tube held as an arrow-guide (but this is dangerous, many an experienced bowman having fired an arrow right through his bow thumb by mistake). The word 'musket' is derived from the use of old Italian crossbows with barrels which fired little bullets called *muschettae* ('gadflies'). The word passed from crossbows to barrel guns.

One of the reasons for the lethal effects of the crossbow was that poisoned arrows were commonly used. The poisons used were some of the most deadly in nature, and that meant that if a crossbow bolt scraped past a man's arm, or if he even pricked his finger by picking one up on the battlefield, he quickly fied. But since a properly armed crossbow bolt could pierce through two suits of metal armour with ease, no one was safe.

The desire to enhance the fire-power of the crossbow led to the invention of machine-gun crossbows in about the eleventh or twelfth century. This development overcame the difficulty of arming the crossbow, which proceeded too slowly for rapid refiring. A magazine of bolts was fitted above the arrow-groove in the stock of the crossbow, and as each bolt was fired, another would drop into its place. The mechanism for firing a repeating crossbow was different from usual. The trigger was absent, and a lever was fixed permanently to the stock between the bowstave and the furthest rear position reached by the string when drawn. Pivoted to this lever was the magazine, which usually contained ten or twelve arrows. The lowest arrow rested upon the arrow-groove and against the string. There was a short barrel at the front. Upon pushing the lever forward as far as possible, the string fell into a slot at the rear end of the longitudinal slit along the base of the magazine, remaining caught in it while the lever was drawn backwards to its fullest extent, thus arming the weapon. At this stage the arrow would have dropped fully into the arrow-groove for firing. An 'automatic trigger', in the form of a short vertical hardwood pin moving up and down in the base of the magazine below the string slot, would be forced upwards by contact with the stock, and would release the string and hence the arrow.

This could be repeated so quickly that a dozen arrows could be fired within moments. Replicas have been constructed and tested, and it has been found that using these repeating crossbows, the Chinese could have one hundred men discharge 2000 arrows in 15 seconds. The ranges of repeating crossbows were less, the extreme being 200 yards, but with an effective range of 80 yards. Although the power of the weapon was diluted, such vast, rapid and continuous showers of bolts raining down on an enemy were said to be extremely demoralizing, especially when one considers that the bolts were usually poison-tipped. Machine-gun crossbows were widespread in China by the year 1600 and many examples survive in museums. One may be seen in Plate 165 (page 245).

In Europe, crossbow-type artillery pieces were known to the ancient Greeks, who used them to fire bolts in sieges, and they were used in 397 BC at Syracuse. A Pythagorean named Zopyros of Tarentum in Sicily developed some of these devices at approximately that time, two years after the death of Socrates. It is difficult to know what to conclude about these early crossbow devices in Europe, but Needham believes they were introduced from China. Sicily, which was within easy sailing distance of Carthage, had Carthaginian colonies, and there were at least four prominent Pythagoreans from Carthage who were presumably known to Zopyros. And since the Carthaginians traded continually with the region of what is today Syria and Lebanon, there was certainly a route for transmission. Syria was known to the Chinese as Ta Ch'in, and trade existed between the two lands. Perhaps a Chinese crossbow *arcuballista* was smuggled out of China and sold at a high price as a high-technology secret weapon. Or perhaps just a description of one was passed along the trade routes and copied by the Greeks (what is called 'stimulus-diffusion'). Or the Greeks may even have made the invention entirely independently, drawing inspiration as the Chinese had done from the simple bow-trap.

Heron of Alexandria has left an elaborate description of a crossbow artillery piece, though only the Germans have translated this fascinating work entitled *Belopoika*. Two ancient authors who wrote about these devices were Biton and F. Vegetius Renatus (fourth century AD). But with the decline of Rome as a power the crossbow principle fell into disuse and seems to have been more or less forgotten. The Romans had certainly used hand-held crossbows for hunting, but in the eleventh century the Byzantines were horrified to encounter their first crossbow in the hands of the first Crusaders. The Byzantine historian, the princess Anna Comnena, wrote of this new weapon, called the *tzaggra*:

> **The *tzaggra* is a barbarian bow hitherto quite unknown to the Greeks. [Description omitted.] ... They will pierce the stoutest metal armour and sometimes wholly imbed themselves in a stone wall or other such obstacle when they strike it. In short, the *tzaggra* is a diabolical and murderous instrument, which fells men to the ground with such a shock that they do not even know what hit them.**

The crossbow had made its appearance again in western Europe in the tenth century, after a gap of 500 years, and it is believed to have been used at the Battle of Hastings. So shocked were those who lived in the

Mediterranean region, that the Second Lateran Council of 1139 condemned crossbows under anathema of the Church, except for use against infidels. But 50 years later, Richard I (Lion Heart) took many of them on his Third Crusade. In 1521, Cortes used crossbows as one of his main weapons in subjugating what is now Mexico.

It is curious that the crossbow should have been used, though rarely, at an early period in Europe, and then have disappeared. This perhaps demonstrates something about the disruptions and lack of continuity of the European Dark Ages. The device may have survived as a hunting weapon for five centuries in the far west of Europe, to re-emerge in the tenth century as a weapon of war. Or it may, as Needham suspects, have been reintroduced a second time from China, perhaps through the Central Asian people known as the Khazars.

GUNPOWDER

NINTH CENTURY AD

Gunpowder first came to the attention of the West in the late twelfth century. By that time, the Chinese had carried its development through many stages and even perfected the barrel gun and the cannon. The evolution of gunpowder and its uses was therefore essentially complete before Europe had even heard of it.

Gunpowder was invented in China not by people seeking better weapons or even explosives, but by alchemists seeking the elixir of immortality. What greater irony could there be than that men wishing to find a drug to enable them to live for ever should instead find a simple substance destined to kill millions of people?

Before gunpowder could be developed, first it was necessary to recognize and obtain the most important of its three ingredients. This is saltpetre, the chemical name for which is potassium nitrate. It was completely unknown in the West until the Middle Ages, and because it generally forms natural deposits only in hot climates there is a considerable shortage of it in Europe, while China always had abundant local supplies of it. But it is not something which is just sitting there waiting to be used; it has to be recognized for what it is, differentiated from all other similar-looking chemical salts, and purified.

How is it possible to know when one has real saltpetre, which looks more or less like any number of other chemicals? The potassium flame test is crucial for detecting saltpetre, because it burns with a violet or purple flame. This was being used to test for true saltpetre in China by at least the third century AD (2000 years before the West), the test being performed by putting a sample of the saltpetre on a piece of charcoal and watching it burn. Later, an even more important test was developed, which is described by Sheng Hsüan Tzu in his book of 1150, *Illustrated Manual on the Subduing of Mercury* (as quoted by the later scientist Li Shih-chen of the Ming Dynasty in the chapter on minerals in his famous 52-volume work, *Compendium of Materia Medica*):

> **If you heat a piece of white quartz and then put a drop of the saltpetre on it, it will sink in. The Taoist books say that saltpetre from Wu-Ch'ang (Udyana) can liquefy or dissolve all metals and minerals. If consumed it can prolong life. The places where it is produced have an extremely loathsome smell, so that birds cannot fly over them.... Pieces shaped like little goose quills are the best kind ... some of the saltpetre used nowadays is not natural saltpetre.... At Shang-ch'en and at Huai Chou and Wei Chou, all in Hopei Province [but nowadays in modern Henan Province], as Tsui Fang records in his book *Wai Tan Pen Tsao*, people scrape it up from the salty soil, and make it from the filtered drippings ... it is named 'solve-stone' because it can dissolve and transform all kinds of ores and minerals.**

Here we see its initial attraction to the Chinese. Long before anyone had any idea that saltpetre could be used as an ingredient of the then still unknown gunpowder mixture, it was prized for its ability to liquefy ores and to dissolve otherwise indissoluble minerals, such as cinnabar, into watery solutions. Saltpetre was used for this purpose at least by the second century BC, as well as acting as a flux to promote metallurgical processes. Since the name 'solve-stone' goes back to the fourth century BC, it is likely that saltpetre's uses were appreciated then, though they were not specified in surviving texts until the second century BC.

When saltpetre finally came to the knowledge of the West, the Arabs (who were the first to learn of it) called it 'Chinese snow'. The oldest surviving mention

ABOVE (168) The figure on the right is firing a bombard, and this cave carving is the oldest known depiction of a gun in the world, dating back to 1128. I visited this carving together with Joseph Needham and Lu Gwei-Djen in 1986, in the extensive Buddhist cave complex at Dazu in Szechuan Province. It is in an extremely remote country area reached by ten hours of driving over rough dirt roads.

of it is in an Arab book of 1240 called *The Book of the Assembly of Medical Simples* by Abu Mohammed al-Malaqi Ibn al-Baitar. Although knowledge of gunpowder reached the West in the twelfth century, the formula for it was somewhat delayed and it was the end of the thirteenth century before saltpetre's use in gunpowder was known to the Arabs, and hence the Europeans. Until the eighteenth century, when sources in India were opened up by the British, saltpetre was in such short supply in Europe that the uses of gunpowder were much curtailed.

Another essential constituent of gunpowder is sulphur, and this too the Chinese were able to purify. Pure sulphur is mentioned in the Chinese book of the second century AD, *Pharmacopoeia of the Heavenly Husbandman*. Evidence from the eleventh century indicates that by that time the method of obtaining pure sulphur was by roasting iron pyrites ('fools' gold') piled up with coal briquettes in an earthen furnace. A still-head sent over the sulphur as vapour, after which it solidified and crystallized (from the conversion of the sulphide to oxide). The emperor issued an edict in 1067 banning the sale to foreigners of either sulphur or saltpetre, and banning all private transactions in both commodities altogether. Large private enterprises in those commodities were thus put out of business, and the usual Chinese practice of nationalizing major industries and forming government monopolies was followed.

In the gunpowder mixture, whatever the proportions (which vary widely), it is the sulphur which lowers the ignition temperature to 250°C and, on combustion, raises the temperature to the fusion point of saltpetre (335°C). It also helps to increase the speed of combustion, but the explosive element in gunpowder may be said to be saltpetre, which burns merrily even on its own. This chemical contains much oxygen; whereas ordinary fires burn, that is, oxidize the fuel, by taking oxygen from the air, gunpowder burns by taking oxygen from the saltpetre within it, and which is ready to hand. And the more saltpetre in the gunpowder, the more explosive it becomes. Early Chinese gunpowders tended to contain about 50 per cent saltpetre, and thus were not truly explosive. For a really big explosion, about 75 per cent saltpetre is needed. The Chinese slowly edged their way upwards towards this daring proportion, which would eventually enable them to have all the bombs, grenades, land and sea mines which are described below (page 256).

With saltpetre, sulphur and the readily available carbon of charcoal and other substances all in their hands continually, it was inevitable that the alchemists would eventually put the three together and stumble upon gunpowder. Indeed, by the third century AD we find evidence that saltpetre and sulphur were being combined in the search for artificial gold. The famous alchemist Ko Hung produced a recipe in around 300 AD for mixing half a pound of sulphur with a pound each of saltpetre, mica, hematite, clay, and so forth, and heating them together to form a mysterious 'purple powder' which by projection would turn molten lead into gold, or something like it. And, he says, 'the same purple powder will at once turn heated mercury in an iron vessel into silver.' What the purple powder was remains uncertain, but one thing is certain – sulphur and saltpetre were both ingredients in an alchemist's formula at the beginning of the fourth century.

Even more to the point, Ko Hung gave a recipe for getting elementary metallic arsenic, which was also isolated by the Chinese centuries before its isolation in Europe. In this process, saltpetre, sulphur and carbonaceous material were all combined, getting very close to proto-gunpowder. The saltpetre oxidized the realgar (arsenic disulphide) into arsenious oxide, and the carbonaceous material then reduced this to volatile elementary arsenic vapours which, Ko Hung said, 'will arise like wisps of cloth, and arsenic sublimes as white as ice.' Needham believes that these early associations with arsenic may have been responsible for setting a pattern, whereby arsenic was routinely incorporated into Chinese gunpowder in later centuries, where its presence in bombs acted as a poison adding to the destructive effect of the explosions.

No textual evidence survives of any approximation to gunpowder mixtures for another 350 years. A book called *Methods of the Various Schools of Magical Elixir Preparations* records that around 650 the alchemist Sun Ssu-Mo made some preparations which included sulphur and saltpetre, and which were inflammable. But another 150 years were to pass before more such mixtures appeared. Chao Nai-An compiled a book about the year 808 entitled *Complete Compendium on the Perfect Treasure of Lead, Mercury, Wood and Metal* which mentioned a 'Method of subduing alum by fire' involving the mixing together of two ounces each of sulphur and saltpetre with a third of an ounce of dried birthwort herb (*Aristolochia*). The dried herb would have contained sufficient carbon to make this mixture liable to ignite suddenly and burst into flames, though it would not actually have exploded. This has been confirmed by a modern experiment where the mixture was prepared. It may be that over the centuries since 300 AD, such inflammable mixtures were continually found and just as continually avoided as being too dangerous, and hence did not enter into the texts, which were concerned with useful (or ostensibly useful) recipes. But this is speculation. The fact is that five-and-a-half centuries passed after Ko Hung before Chinese alchemists recorded a formula which can be genuinely described as a gunpowder mixture. And so it is that we set 850 AD as the approximate date for the invention of gunpowder.

The first textual recording of a proto-gunpowder formula is found in a book entitled *Classified Essentials of the Mysterious Tao of the True Origin of Things*, preserved in the great collection of Taoist classics. It was attributed to Cheng Yin, the teacher of Ko Hung, but this is believed to be a fictional authorship, and the book is dated to about 850. The book contains thirty-five elixir formulae which are to be avoided as wrong or dangerous, three of these involving saltpetre. But the crucial passage reads:

> **Some have heated together sulphur, realgar (arsenic disulphide), and saltpetre with honey; smoke and flames result, so that their hands and faces have been burnt, and even the whole house where they were working burned down. Evidently this only brings Taoism into discredit, and Taoist alchemists are thus warned clearly not to do it.**

It should be noted that the honey would have contained carbon, and the drier the honey the more the carbon. We do not know how long this sort of thing had been going on, but Needham takes the date

LEFT (169) A woodcut published in the 1643 edition of the popular novel *Chin P'ing Mei* (named after its three main female characters), showing a public fireworks display. To the left of the pole near the bottom is a spinning catherine-wheel. Fireworks were the first form of Chinese gunpowder to reach the West, in the late twelfth century. The Chinese knew how to make them in a wide variety of different colours.

ABOVE (170) The Chinese passion for fireworks has been continuous for about a thousand years. Here we see a display in modern Hong Kong during the National Day celebrations.

of this book, 850 AD, as the date of the first true appearance of what the Chinese called the 'fire-chemical', that is, gunpowder.

It was around the year 1040 that Tseng Kung-Liang published a true gunpowder formula for the first time in history. Certainly true gunpowder and its uses had been known by this time for at least a century. Tseng gives three gunpowder formulae for three different weapons: a quasi-explosive bomb to be hurled by a kind of catapult, a burning bomb with hooks which could catch onto wooden structures and set them on fire, and a poison-smoke ball for chemical warfare. The gunpowder in these was not explosive but rather what is called 'deflagrative', that is, it burned 'with a sudden and sparkling combustion, producing a "whoosh" like a rocket'. The bombs did not yet contain enough saltpetre to explode. This would come later.

The first use of gunpowder in war was not, however, in anything so dramatic as an incendiary bomb. Around 919, we find it used in the form of a gunpowder-impregnated slow-match fuse for the ignition chamber of the flame-thrower. This hideous device (see page 254) sprayed enemies with burning gasoline. As it passed from the pump it was ignited by the gunpowder-impregnated fuse, so that sheets of flame descended upon them.

When explosive gunpowder did come, truly explosive bombs, rather than merely incendiary bombs, were produced in vast quantities. As the proportion of saltpetre in the mixture is increased, a point is reached where 75 per cent of the gunpowder is saltpetre – more or less the proportions found in 'modern' gunpowder. It is now past the point of mere explosion; one now gets detonation. When this stage was reached in China, barrel guns and cannon became possible (see page 266). Upon ignition, gunpowder of this type suddenly produces 3000 times its bulk in gas, reaching temperatures of the order of 3880°C. The rapidity of the phenomenon is made possible by the fact that the oxygen is in the saltpetre within the mixture, so that no oxygen needs to be sucked in from the surrounding air.

All these stages, from alchemical accidents and singed Taoist beards through to the true barrel gun and cannon, had been reached by the Chinese before anyone in the West had ever even heard of gunpowder, let alone seen it in use. But what Westerners lacked in initial knowledge, they made up for in later enthusiasm.

THE FLAME-THROWER

TENTH CENTURY AD

The flame-thrower was not invented in the twentieth century, as most people would naturally assume. If one considers a flame-thrower to be a device capable of emitting a continuous stream of flame in warfare, it was invented by the Chinese by the tenth century AD. An earlier form, which might more properly be called a 'flame-squirter', was used in 675 by Callinicus in defence of Byzantium. It had a 'siphon' which apparently pumped flame by means of a single-acting force-pump which was rather like a large syringe (invented in the West by Ctesibius in the third century BC). The proto-flame-thrower of the Byzantines was apparently incapable of ejecting a continuous stream of flame; it sent out a burst of flame, was repumped, and then gave another burst.

The reason for the superiority of the Chinese device, and the reason we consider it to be the first genuine flame-thrower, was the continuous stream of flame made possible by the Chinese invention of the double-acting piston bellows (see page 46), which was also used in chemical warfare by the early fourth century BC for spraying soldiers with clouds of poison gas (page 240). The superiority of Chinese metallurgy was also apparent here, as the flame-thrower was made of the very best cartridge-quality brass, containing 70 per cent copper.

What was it that the flame-thrower actually threw? What was burnt to make this stream of fire? According to Needham, it was either gasoline or kerosene, in other words, a 'distilled light fraction of petroleum'. The Chinese had stills for manufacturing this and they certainly used petroleum products (see page 89). But there is no need to think that the Chinese actually

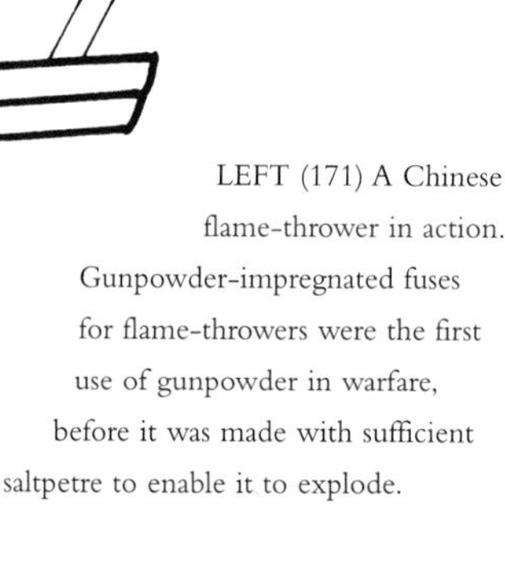

LEFT (171) A Chinese flame-thrower in action. Gunpowder-impregnated fuses for flame-throwers were the first use of gunpowder in warfare, before it was made with sufficient saltpetre to enable it to explode.

invented this, for J.R. Partington in his *History of Greek Fire and Gunpowder* (1960) concluded that the 'Greek Fire' used by the Byzantines in their flame-thrower was the same substance. And Needham points out that 'Greek Fire came into China by about 900 AD.'

The first use of 'Greek Fire' in a flame-thrower in China appears to be in the year 904. Lu Chen, in his *Historical Memoir of the Nine States*, recounts an engagement of that date and says: 'Cheng Fan's men let off flying fire machines, which burnt the Lung-sha Gate...'. We have a far clearer account of flame-throwers in a description of the naval battle on the Yangtze River in 975, preserved by Shih Hsü-Pai in his book *Talks at Fisherman's Rock*:

> **Chu Ling-Pin as Admiral was attacked by the Sung emperor's forces in strength. Chu was in command of a large warship more than ten decks high, with flags flying and drums beating. The imperial ships were smaller but they came down the river attacking fiercely, and the arrows flew so fast that the ships under Admiral Chu were like porcupines. Chu hardly knew what to do. So he quickly projected petrol from flame-throwers to destroy the enemy. The Sung forces could not have withstood this, but all of a sudden a north wind sprang up and swept the smoke and flames over the sky towards his own ships and men. As many as 150,000 soldiers and sailors were caught in this and overwhelmed, whereupon Chu, being overcome with grief, flung himself into the flames and died.**

There is another mention of flame-throwers in the same year, in the *History of the Southern T'ang Dynasty*, where we are told that a commander named Ts'ao Pin 'came down upon Chinling. He had large ships furnished with bundles of reeds saturated with thick oil, with the intention of taking advantage of the wind to start conflagrations.... But in urgent situations, then they used the machines to shoot the fire-oil forwards to resist the enemy.'

In a book of 1137 by K'ang Yü-Chih entitled *Dreaming of the Good Old Days*, we find a description of the storage and use of gasoline for flame-throwers. He speaks of 'a large reservoir more than 10 feet square in order to store "fierce fire oil"', which he says 'should only be stored in real glass vessels.' He recalls:

> **I myself still remember the district commanders coming to it to study and practise water-combat with their troops, and to test the fierce fire oil. The opposite bank of the lake represented the fortified camp of the enemy. Those who were in charge of the oil sprayed it about, and as it was ignited it broke into a sheet of flame, so that the [fictitious] fortifications of the enemy were all in a short time completely destroyed. What is more, the oil had a secondary effect on the water, for all the water-plants were killed, and the fishes and turtles died.**

The mention of ignition in the passage above brings us to the question: how was the ejected gasoline ignited as it left the flame-thrower? Obviously it could not be burning before it left, for then the man holding the machine would be destroyed by flames himself. The answer is that a lighted fuse was held in front of the nozzle, so that when the gasoline was squirted out, it was ignited after it was already on its way through the air. The fuse was impregnated with gunpowder; as mentioned in the account of gunpowder (page 250), this was the first military use of the substance. This gunpowder was so low in saltpetre content that it could not explode, but only sparked and burnt slowly in the fuse.

By 1044, the flame-thrower was standard issue to Chinese armies. A military encyclopedia of that date, *Collection of the Most Important Military Techniques*, gives drawings of a flame-thrower with design details (see Plate 172 (page 256). The text says: If the enemy comes to attack a city, these weapons are placed on the great ramparts, or else in outworks, so that large numbers of assailants cannot get through.' There is a lengthy description of the device, which commences:

> **On the right is the gasoline flamethrower. The tank is made of brass, and supported on four legs. From its upper surface arise four [vertical] tubes attached to a horizontal cylinder above; they are all connected with the tank. The head and tail of the cylinder are large, the middle is of narrow diameter. In the tail end there is a small opening as big as a millet-grain. The head end has two round openings 1½ inches in diameter. At the side of the tank there is a hole with a little tube which is used for filling, and this is fitted with a cover.**

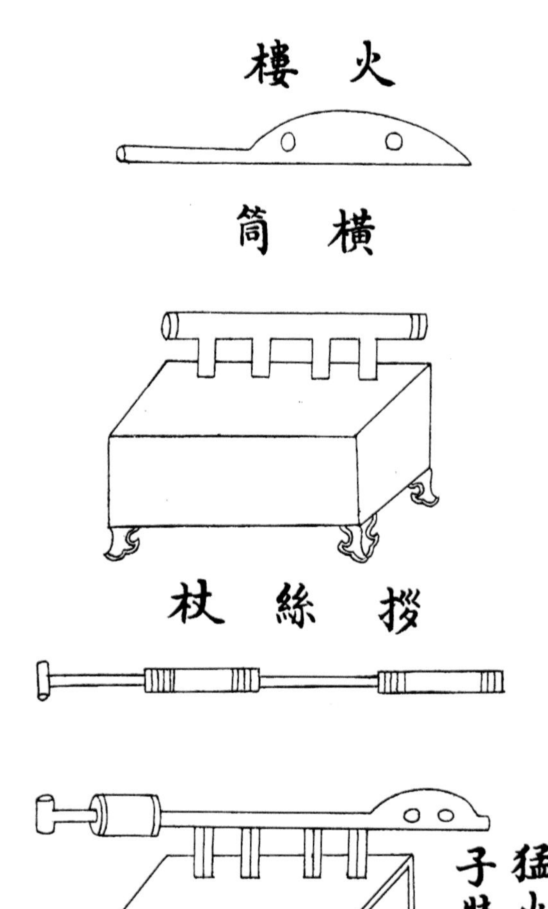

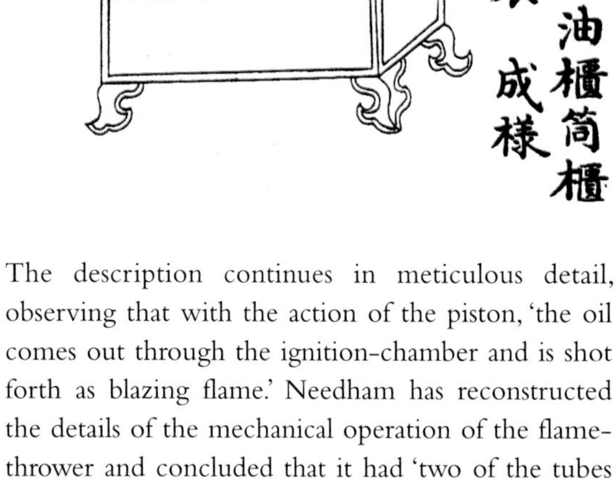

LEFT (172) An illustration of a Chinese flame-thrower, published in 1601, redrawn from an encyclopedia of 1044. The Chinese invented the continuous flame-thrower in the tenth century, having known of the seventh-century spurting flame-thrower of the Byzantines via Central Asia. Here we see the tank standing on four feet, with the pump and ejector above it. Because the Chinese invention of a double-acting piston-bellows was used with this device, a continuous stream of flame could be emitted. The metal used was cartridge-quality brass.

The description continues in meticulous detail, observing that with the action of the piston, 'the oil comes out through the ignition-chamber and is shot forth as blazing flame.' Needham has reconstructed the details of the mechanical operation of the flame-thrower and concluded that it had 'two of the tubes secretly connected within the tank. Such a design is very compatible with the directions in the text that the machine was to be started with the piston-rod pushed fully forward, and it also agrees with the statement that the "two" communicating feed tubes are alternately occluded.' The flame-thrower gave a continuous jet of flame 'just as the double-acting piston bellows gave a continuous blast of air, and the most obvious way of effecting this was to have a pair of internal nozzles one of which was fed from the rear compartment on the backstroke'.

FLARES, FIREWORKS, BOMBS, GRENADES, LAND MINES AND SEA MINES

TENTH CENTURY AD

Fireworks existed in China before gunpowder was invented. Pieces of bamboo were thrown into fires and exploded as fireworks from at least 200 BC. Even after gunpowder fireworks were invented, the exploding bamboo was retained until modern times. Because a fireworks tradition already existed in China, the advent of gunpowder was seen as a natural progression.

The Chinese love to accompany celebrations with explosions. Here is a description of early fireworks from a book entitled *Dreaming of the Capital while the Rice is Cooking*, published in 1275 by Wu Tzu-Mu:

> **Inside the palace the fire-crackers made a glorious noise, which could be heard in the streets outside.... All the boats on the lake were letting off fireworks and fire-crackers, the rumbling and banging of which was really like thunder.**

The Chinese rapidly developed every conceivable type of fireworks, including spinning catherine wheels and rockets (for the invention of the rocket see page 262). They also had coloured explosions, using a wide variety of special materials. They could obtain brilliant sparkling effects by combining steel dust or tiny shavings of cast iron, reduced to a powder, with the gunpowder. They used these effects in their Roman candles, which also preceded any made in the West. Blue-green flashes could be obtained by using indigo, white by using white lead carbonate, red by using red lead tetroxide, purple by using cinnabar, black by using lignite and soap-beans, yellow by using arsenical sulphides, and violet by using – of all things – cotton fibres. No festivity in China ever lacked colour after the invention of gunpowder.

It has already been mentioned (page 251) that the first bombs were incendiary ones, using a gunpowder with insufficient saltpetre to cause a proper explosion. There was a long history of incendiary weapons of all kinds throughout the world, but when gunpowder became available in China, it was incorporated into incendiary weapons to deadly effect. No longer were such things as pitch and flammable resins and oils the main substance of incendiary arrows and bombs. In the tenth century, a whole new wave of incendiaries appeared, using gunpowder. Fire arrows were now tipped with little gunpowder bundles wrapped in paper and sealed with wax, and had gunpowder-impregnated fuses. We have seen how in 919 the flame-thrower used such fuses (page 255).

In 994 a gigantic force of 100,000 besiegers was driven away from the city of Tzu-t'ung using these gunpowder fire-arrows. These gunpowder projectiles would set fire to anything inflammable, whether tents, clothing, besieging engines, trench walls made of wood, stores of hay or food wagons. The commander Chao Hsieh is reported to have had a supply of 250,000 gunpowder-armed arrows in the year 1083; mass-production of these weapons was evidently in full swing.

By the first half of the eleventh century, a new kind of gunpowder bomb came onto the scene. Called a 'thunderclap bomb', this was truly explosive, and had a high percentage of saltpetre. It was enclosed in a weak casing of bamboo or paper and hurled from trebuchets, which had previously been used to hurl great stones at enemy fortifications with slings, and then incendiary bombs. Thunderclap bombs were far more effective in starting fires; they also terrified the enemy's horses with explosive sounds, and many barbarian tribes were seriously demoralized by the noise alone, at first. These bombs either had fuses or were ignited by a red-hot poker just before being hurled.

Here is the eleventh-century description of how to make an explosive bomb:

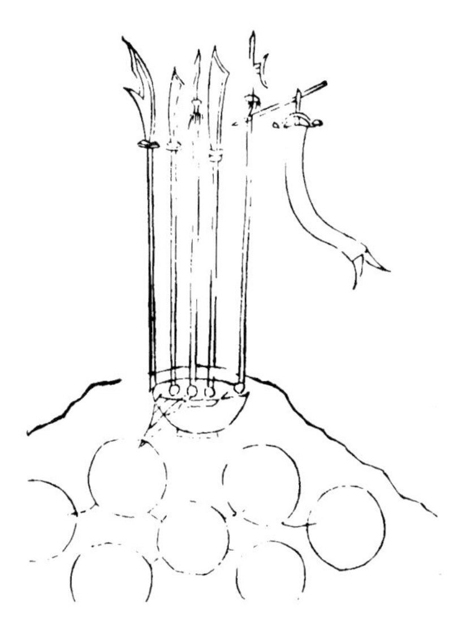

LEFT (173) A fourteenth-century land mine booby-trap called 'the underground sky-soaring thunder'. A group of halberds, pikes and banners is stuck into the ground as a tempting trophy. When the enemy approaches to seize them, he steps on a mechanism which lights a fuse, and he is blown up by the land mines buried underground.

The thunderclap bomb contains a length of two or three internodes of dry bamboo with a diameter of 1.5 inches. There must be no cracks, and the septa are to be retained to avoid any leakage. Thirty pieces of thin broken porcelain the size of iron coins are mixed with 3 or 4 lb of gunpowder, and packed around the bamboo tube. The tube is wrapped within the ball, but with about an inch or so protruding at each end. A gunpowder mixture is then applied all over the outer surface of the ball.... A long red-hot iron brand is used to set off the thunderclap bomb, which produces a noise indeed like thunder.

This extract is from the *Collection of the Most Important Military Techniques*, published in 1044. There is an eye-witness account of the use of thunderclap bombs in fighting off a siege in the late eleventh century, written by the commander Li Kang: 'At night the thunderclap bombs were used, hitting the enemy well, and throwing them into great confusion. Many fled, howling with fright.'

Thunderclap bombs also came in the form of grenades which could be hurled by hand. A scholar, Yuan Hao-Wen, recorded an extraordinary story in a book published in 1187, describing the use of thunderclap grenades by a hunter:

Towards the end of the Ta-Ting reign-period there lived north of T'aiyuan a certain hunter named T'ieh Li. One evening he found a great number of foxes in a certain place. So knowing the path that they followed, he set a trap, and at the second watch of the night he climbed up into a tree carrying at his waist a vessel of gunpowder. The coven of foxes duly came

ABOVE (174) The only known picture of a thirteenth-century bursting bombshell, this woodcut published in 1293 comes from a Japanese book describing the Mongolian invasion of Japan in 1274. Although the Japanese obtained the idea of bombs from China, the bomb shown here, a cast-iron 'thundercrash' type, is probably a Chinese one, since the Mongols were ruling China by this time. It was two-and-a-half centuries before bombs of this type appeared in Europe.

under the tree, whereupon he lit the fuse and threw the vessel down; it burst with a great report, and scared all the foxes. They were so confused that with one accord they rushed into the net which he had prepared for them. Then he climbed down the tree and killed them all for their fur.

This grenade would probably have been housed in a narrow-mouthed pottery vessel. As for lighting the fuse, the hunter probably used a match (see page 110).

The Chinese also invented flares, which they called 'signal bombs'. In 1276, when A-Chu was attacking Yangchow, flares were fired by the Chinese to send messages to distant troop detachments. These were soft-shelled bombs timed to explode in mid-air, possibly coloured like fireworks.

By 1221, we find in use a new type of bomb in China. Instead of being called a thunderclap bomb, it is called a thunder*crash* bomb. These bombs were far more deadly because instead of having soft casings, they were real bombs in the sense we would recognize today, with casings of metal. This meant that they gave off deadly shrapnel, mutilating and blinding the enemy. Also, in order to burst the metal casings asunder, these new bombs were made of a gunpowder with even more saltpetre. No longer did this gunpowder merely explode: it detonated. Gunpowder of this strength was fully comparable with modern gunpowder, with about 75 per cent saltpetre in the mixture.

The thundercrash bombs appear first in the hands of the Chin forces at their successful siege of Ch'i-chou. Here they defeated the Sung Chinese, who were gradually being forced southwards (the Chin themselves were being forced southwards in their turn by the Mongols who were at their back). The new bombs were in cast iron casings (cast iron having been invented in China, see page 44). Chao Yü-Jung records: 'Their shape was like that of a bottle-gourd with a small opening and they were made from cast iron about 2 inches thick.' It is thought that the same bombs were in the hands of the defenders, but because they lost the siege there is less information about them. These thundercrash bombs were said to make a noise like thunder, shaking the walls of houses and killing and wounding countless people. They are not called thundercrash bombs in this account. The first use of that term occurs ten years later, in 1231, when the Chin were themselves besieged by the Mongols in a city in Shansi Province. The Chin general escaped by sea down the Yellow River with three thousand of his men, pursued along the northern bank by the ferocious Mongols who kept up a steady rain of arrows upon him. The Chin dynastic history tells us:

> But the Chin ships had on board a supply of those bombs called 'thunder-crash' missiles and they hurled these at the enemy. Flashes and flames could distinctly be seen ... eventually the Chin fleet broke through, and safely reached Tung-kuan.

In the following year, 1232, the Mongols took the capital in the north, K'aifeng, from the Chin. In the Chin history we read of the siege:

> Among the weapons of the defenders there was the heaven-shaking thunder-crash bomb. It consisted of gunpowder put into an iron container; then when the fuse was lit and the projectile shot off there was a great explosion the noise whereof was like thunder, audible for more than 100 *li* [tens of miles], and the vegetation was scorched and blasted by the heat over an area of more than half a *mou* [many acres]. When hit, even iron armour was quite pierced through. Therefore the Mongol besiegers made cowhide sheets to cover their approach trenches and men beneath the walls, and dug as it were niches each large enough to contain a man, hoping that in this way the Chin troops above would not be able to do anything about it. But someone up there suggested the technique of lowering the thunder-crash bombs on iron chains. When these reached the trenches where the Mongols were making their dug-out, the bombs were set off, with the result that the cowhide and the attacking soldiers were all blown to bits, not even a trace being left behind. Moreover, the defenders had at their disposal flying fire-spears. These were filled with gunpowder, and when ignited, the flames shot forward for a distance of more than ten paces, so that no one durst come near. These thundercrash bombs and flying fire-spears were the only two weapons that the Mongol soldiers were really afraid of.

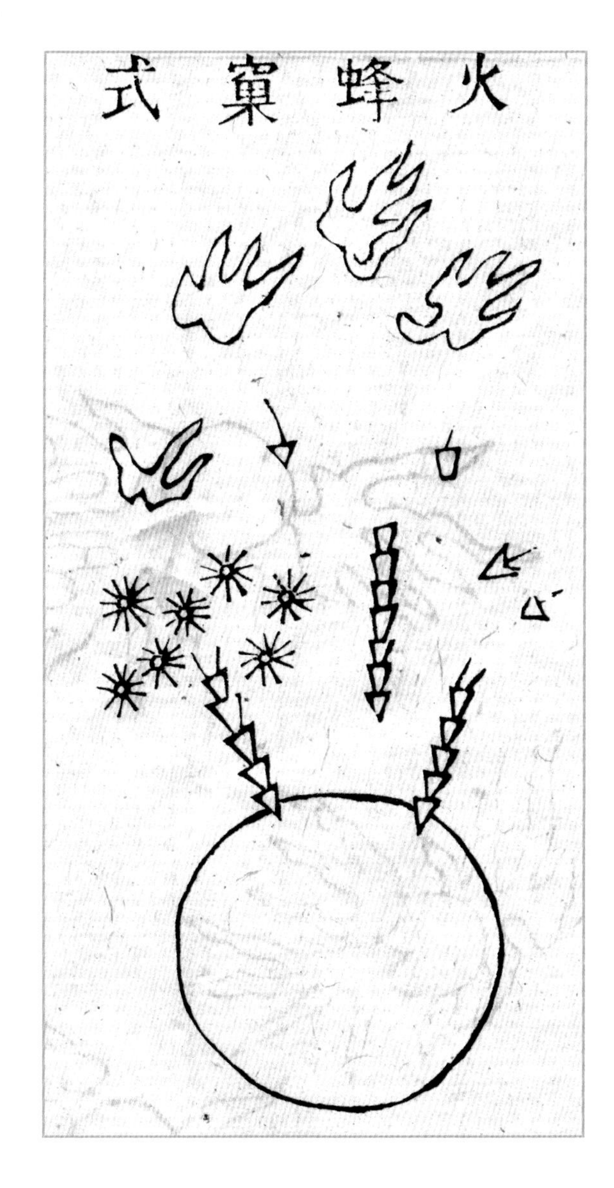

RIGHT (175) A picture of a very nasty Ming Dynasty device called a Wasps' Nest Fire Bomb. It spewed out pellets, barbs, and spikes which 'stung', in addition to the usual fire and explosives. This engraving dates from circa 1500 and was reprinted from the original block in 1883. (Collection of Robert Temple.)

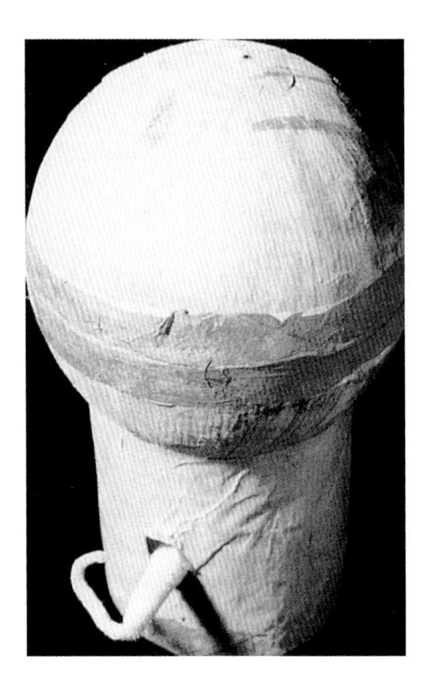

RIGHT (176) A seventeenth-century Chinese bomb with fuse, of a kind used for firing into the clouds in an attempt to precipitate rain. The gunpowder is wrapped in paper and sealed with wax, and the fuse is impregnated with gunpowder.

The Mongols eventually destroyed the Chin power and moved on to attack the Sung Chinese who had long since set up a new capital in the south; this dynasty is called the Southern Sung. However, the Southern Sung were not very well prepared for the Mongol onslaught. A scholar official named Li Tseng-Po bemoaned this in 1257:

As for the weapons for attack by fire, there are or should be several hundred thousand iron bomb-shells available. When I was in Chingchow they were making one or two thousand of them a month, and they used to despatch to Hsiang-yang and Ying-chou ten or twenty thousand a time. Yet now at Chingchiang we have no more than 85 iron bomb-shells, large and small, 95 fire-arrows, and 105 fire-lances. This is not sufficient for a mere hundred men, let alone a thousand, to use against an attack by the Mongol barbarians. The government supposedly wants to make preparations for the defence of its fortified cities, and to furnish them with military supplies against the enemy, yet this is all they give us. What chilling indifference!

The only surviving contemporary picture of a bursting bombshell of the thirteenth century is a Japanese drawing showing a Chinese-style thundercrash bomb in a cast iron casing exploding in mid-air. The drawing was made in 1293. As for cast iron bombshells in Europe, the earliest record of their definite occurrence and use dates from 1467, exactly 246 years later than the earliest recorded use of them in China.

Over the coming years, the Chinese developed a bewildering variety of different bombs with specialist uses, some packed with what military writers today call anti-personnel material, to increase the shrapnel effect. There were also many poison bombs, gaseous bombs, and bombs filled with human excrement. Some of the best-known bombs and grenades were 'the bone-burning and bruising fire-oil magic bomb', the 'magic fire meteoric bomb that goes against the wind', the 'dropping-from-heaven bomb', the 'bee-swarm bomb', the 'match for ten thousand enemies bomb', the 'flying-sand magic bomb releasing ten thousand fires' and the 'wind and dust bomb'.

By 1277 the Chinese had developed bombs to the point of their being used as land mines. From that year we have reports of an 'enormous bomb' being used which was more like a land mine than an actual bomb. And by the middle of the fourteenth century descriptions of how to make land mines were actually being published in books, which is a sure sign that they were no longer secret. Here is one such description:

The mine, made of cast iron, is perfectly spherical in shape. It holds one peck or five pints of black powder, depending on its size. The 'magic gunpowder', 'poison gunpowder' and 'blinding and burning gunpowder' compositions are all suitable for use in this device. Hard wood is used for making the wad, which carries three different fuses in case of defective connection, and they join at the 'touch hole'. The mines are buried in places where the enemy is expected to come. When the enemy is induced to enter the mine-field the mines are exploded at a given signal, emitting flames and fragments and a tremendous noise.

This is from *The Fire-Drake Artillery Manual*, which was published in 1412 and contains some mid-fourteenth-century material. Another mid-fourteenth-century land mine recorded in the same book was actually a network of mines called 'the ground-thunder explosive camp', described as follows:

> These mines are mostly installed at frontier gates and passes. Pieces of bamboo are sawn into sections 9 feet in length, all septa in the bamboo being removed, save only the last; and it is then bandaged round with fresh cow-hide tape. Boiling oil is next poured into the tube and left there for some time before being removed. [This is thought to have been to seal the bamboo against damp from the ground as well as insect attack, to which bamboo is notoriously prone.] The fuse starts from the bottom of the tube, and black powder is compressed into it to form an explosive mine. The gunpowder fills up eight-tenths of the tube, while lead or iron pellets take up the rest of the space; then the open end is sealed with wax. A trench 5 feet in depth is dug for the mines to be concealed. The fuse is connected to a firing device which ignites them when disturbed.

The same also describes a 'self-tripped trespass mine':

> It is made of iron or rock, or even porcelain or earthenware, with a cavity inside, very like the explosive mine mentioned above. Outside, the fuse runs through a series of 'fire-ducts', which connect together several of these devices installed at strategic points. When the enemy ventures onto ground containing one of these mines, all the others are set to explode quickly one after another.

Yet another land mine described in the same book, called 'the Supreme Pole combination mine', featured a battery of eight little guns pointing in different directions, all set off by an automatic trigger mechanism. It was more in the nature of a booby-trap to be laid in mountain passes, dating from the early fourteenth century.

Some of the trigger-mechanisms of Chinese land mines remained secret until the seventeenth century, when descriptions of them were finally published. Some involved arrangements of flint and steel, which when automatically triggered by a cord would strike and send sparks onto tinder, which then lit gunpowder-impregnated fuses leading to the mines. There were at least two steel wheel systems. These land mine triggers would seem to be the ancestors of the flintlock rifle, going back to 1360 at least.

In Europe, the wheel-lock musket first appeared in a drawing by Leonardo da Vinci in 1500, but the

BELOW (177) An articulated barge of the late sixteenth century, used as a minelayer. The front is loaded with sea mines. When the barge approaches the target, the front portion is detached and left with a time fuse burning, while the men in the rear portion quietly row away.

first actual example to be recorded dates from 1526. And the first European flintlock did not appear until 1547. Needham believes that the flint and steel land mine triggers of China were the direct forebears of these European mechanisms, though perhaps transmitted by 'stimulus-diffusion', which is the transmission of an idea or description rather than an actual object. The first appearance of land mines in Europe is thought to have been in the war between Pisa and Florence in 1403. And the first European land mines using triggers for ignition at a distance were developed by Samuel Zimmermann of Augsburg in 1573.

The Chinese did not restrict themselves to land mines. They also invented sea mines. *The Fire-Drake Artillery Manual* gives this description:

> **The sea mine called the 'submarine dragon-king' is made of wrought iron, and carried on a submerged wooden board, appropriately weighted with stones. The mine is enclosed in an ox-bladder. Its subtlety lies in the fact that a thin incense-stick is arranged to float above the mine in a container. The burning of this joss-stick determines the time at which the fuse is ignited, but without air its glowing would of course go out, so the container is connected with the mine by a long piece of goat's intestine through which passes the fuse. At the upper end the joss-stick container is kept floating by an arrangement of goose and wild-duck feathers, so that it moves up and down with the ripples of the water. On a dark night the mine is sent downstream towards the enemy's ships, and when the joss-stick has burnt down to the fuse, there is a great explosion.**

This Chinese sea mine of the fourteenth century was thus two centuries earlier than the oldest known European plan for a sea mine, which was presented by Ralph Rabbards to Queen Elizabeth in 1574. The Chinese were still using sea mines in fighting against the British in 1856 on the Canton River.

The Chinese were thus not slow to exploit all the explosive potentials of gunpowder. We shall shortly see how they developed the barrel gun and the cannon, which were also implicit in the development of their 'fire-chemical'.

THE ROCKET, AND MULTI-STAGED ROCKETS

ELEVENTH AND TWELFTH CENTURIES AD

The invention of the rocket in China seems to have been doubly inspired. On the one hand, incendiary fire-arrows gave way to the idea of an arrow-shaft which could have a rocket mounted on it, so that arrows would no longer need to be fired from bows at all. On the other hand, inspiration came from a type of firework known as the 'ground rat' or the 'earth rat' which sped along the ground spewing flames from behind. We have an amusing story of a 'ground rat' running amok in the imperial palace in the year 1264:

> **When the Emperor Li Tsung retired, he prepared a feast in the Ch'ing-Yen Tien Palace Hall on the 15th day of the first month of the year in honour of his mother, the Empress-Mother Kung Sheng. A display of fireworks was given in the courtyard. One of these, of the 'ground-rat' type, went straight to the steps of the throne of the Empress-Mother, and gave her quite a fright. She stood up in anger, gathered her skirts around her, and stopped the feast. Li Tsung, being very worried, arrested the officials who had been responsible for making the arrangements for the occasion, and awaited orders from the Empress-Mother. At dawn next day, he went to apologise to her, saying that the responsible officials had been careless, and took the blame upon himself. But the Empress-Mother laughed and said, 'That thing seemed to come specially to frighten me, but probably it was an unintentional mistake, and it can be forgiven.' So mother and son were reconciled and just as affectionate as before.**

There were also 'water rats', fireworks of this type which were tied to floats or little skis and went skidding across lakes and ponds in firework festivals. Chou Mi in his book *Customs and Institutions of the Old Capital*, written in the middle of the thirteenth century but describing festivities of the mid-twelfth century, speaks of fireworks of those days:

> **Some of these were like wheels and revolving things, others like comets, and others again shooting along the surface of the water.**

The latter were water rats, but comet-like fireworks were a type of rocket. Around 1150 it crossed someone's mind that an 'earth rat' or 'water rat' tube attached to a feathered stick would constitute a rocket-arrow equipped to fly without being launched from a bow. We even have an early description of how to make one:

> One uses a bamboo stick 4 feet 2 inches long, with an iron or steel arrow-head 4½ inches long smeared with poison; and some smear that on the rocket-tube too. Behind the feathering there is an iron weight four-tenths of an inch long. At the front end there is a carton tube bound onto the stick, where the 'rising gunpowder' is lit and it is oiled to prevent its getting wet. When you want to fire it off, you use a frame shaped like a dragon, or else conveniently a tube of wood or bamboo to contain it or launcher boxes of different kinds.

Another book gives further details about the crucial invention of the balancing weight:

> An iron weight is fixed at the rear end of the rocket-arrow, behind the feathering, of such a mass that the fulcrum of the balance is situated just four finger-breadths away from the mouth of the rocket-tube.

These counter-weights enabled the rockets to travel considerable distances by holding down their rear ends in order to avoid the rocket arrow's tipping down and hitting the ground.

The same early text adds the even more remarkable description of the need to bore a hole in the centre of the gunpowder (a dangerous business; the boring tool needed to be moistened continually with water to avoid explosion from the friction). When the gunpowder burned, there would be equal areas of combustion surface along an internal cavity in the powder, which is a basic requirement for an efficient rocket. The text tells us:

> If the hole is straight-sided [i.e., parallel with the walls of the tube] the arrow will fly straight; if it is slanting the arrow will go off at a tangent. If the hole is too deep the rocket will lose too much flame at the rear, if it is too shallow, it will not have enough strength, so the arrow will fall to the ground too soon. If the rocket-tube is 5 inches long, the cavity must extend into it some 4 inches. The shaft has to be absolutely straight, and the rocket-tube and end-weight of the arrow must balance perfectly when suspended two inches from the neck, or throat [i.e., nozzle], of the rocket-tube, while the feathering should be almost as long as the rocket-tube itself.

There were even diagrams published showing the boring tool, and illustrating the cavity. All of these developments had taken place during the twelfth and thirteenth centuries. By 1300 at the latest, the furthest refinement of rockets had taken place whereby the orifice of the rocket-tube was constricted to increase the flow-velocity of the issuing gases, giving greater power. This choke or nozzle thus represented the 'Venturi-tube effect', which is one of the most elementary principles of aerodynamics since it gives the explanation for the occurrence of lift in connection with airplane wings. The principle was formulated in Europe by G. B. Venturi (1746–1822). The Chinese were therefore about 500 years in advance of him in their use of the principle.

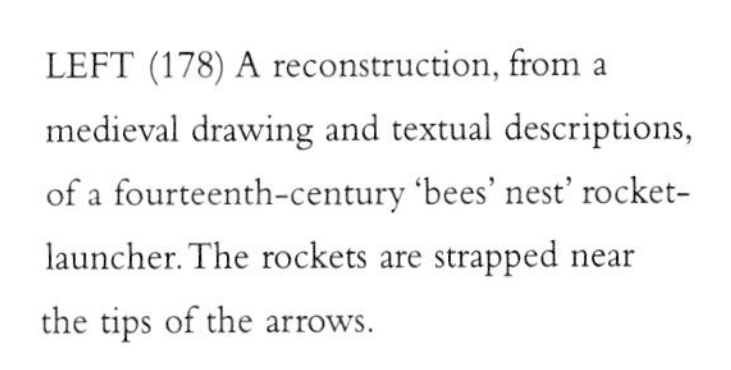

LEFT (178) A reconstruction, from a medieval drawing and textual descriptions, of a fourteenth-century 'bees' nest' rocket-launcher. The rockets are strapped near the tips of the arrows.

RIGHT (179) Rockets were invented in China by the eleventh century, and were commonly strapped to arrows with poisoned tips. Here we see portable basketwork rocket-launchers in use in the fourteenth century. These rockets had a range of 200 yards or more. The launchers and rockets were not heavy, and were frequently carried by large numbers of foot soldiers. Tens of thousands of rockets were often fired during a single battle.

The gunpowder used in these early Chinese rockets is thought to have been about 60 per cent saltpetre. There were various mixtures, of which the names but not the proportions have survived: 'wind-opposing gunpowder', 'flying-in-the-air-gunpowder', 'flying gunpowder', and so on. Needham believes that the proportions were suppressed for the sake of military secrecy.

These rockets could fly between 500 and 1150 yards, though during the fourteenth or fifteenth century a device called the 'flying powder tube' was invented, which is said to have flown by rocket power to the enemy, dropping an explosive charge of poison and being flown back by a second rocket lit by a fuse. This device, which in theory could be re-used indefinitely, probably hardly ever worked. It was difficult enough controlling the flight direction of rockets, without also having to rely on their correct return. But it was certainly a bold and ingenious idea.

It soon became obvious to the Chinese that setting off the occasional rocket on its own was insufficient. They needed batteries of them. By fixing them within frames of rocket-launchers, they could direct them more reliably. And by making the frames take batches of them at once, barrages of rocket fire could and did wreak great havoc in warfare. There were even hand-held portable rocket launchers woven of basketwork. These conical basket launchers would not have outlasted one mass firing, but were easily replaceable; they were really the medieval equivalent of the bazooka so widely used in the Second World War.

The wheelbarrow (invented in China; see page 95) was regularly used to hold portable batteries of rocket launchers. These were called 'fire-frame combat-vehicles'. Wonderfully evocative names were given to the different types of rocket launchers: 'Mr Facing-Both-Ways rocket-arrow firing basket', 'Five-tigers-springing-from-a-cave rocket-arrows', 'Pack of one hundred tigers running together', 'Covey of hawks catching rabbits', 'Leopard pack unexpectedly scattering', 'Bees' nest', and 'Forty-nine simultaneously fired rocket-arrows'. The splayed rocket launchers had internal diaphragms with holes in them to keep the rockets apart, and the launchers widened at the top from a narrower base, to ensure a wide area of dispersion of the points of impact of the rockets. An early text gives the description of one of the smaller portable ones:

> The small bamboo rocket-arrow tube:
>
> Each tube holds ten short rocket-arrows, only 9 inches long, and poison is applied to the head of each. The total weight of the tube and its contents does not exceed 2 pounds, and each soldier can carry four or five of them on its sling easily. The enemy would not know exactly what they were transporting. At a distance of some 170 yards away, the rocket-arrows are all fired as one. These arrows, though small, are fast, and the enemy cannot avoid them; so one soldier can do as much harm with these arrows as several dozen others using more conventional arms. These rocket quivers can be carried by the personal guards of the commander, or by the detachment

of soldiers surrounding the flag, or else by men scattered among ordinary fighting units. The rocket-arrows should be tested to ensure that they can penetrate thin wooden planks. If the bamboo tube is slightly raised at the time of firing, the arrows can reach over 340 yards. This weapon should not be overlooked just because the arrows are so small.

Descriptions of rocket-launchers in medieval Chinese books tell of batteries which could each dispatch 320 rockets at a single moment. These consisted of four 'long-serpent' rocket launchers in rows on wheelbarrows together with two rectangular wooden 'hundred-tiger' rocket launchers, one on either side. During the Ming Dynasty there is a battle account of a hundred of these operating together in an engagement. Such a super-battery was theoretically capable of launching 32,000 rockets at a single time, and during the course of a battle must have used up roughly a million rockets.

Stabilizing devices for rockets in flight were introduced by the Chinese not long after 1300, and are recorded in *The Fire-Drake Artillery Manual* published in 1412 but incorporating material a century older. These took the form of fins and wings, so that the result looked like a bird with rockets strapped under its belly. The text says:

> The 'flying crow with magic fire' winged rocket bomb:
>
> The body of the bird is made of fine bamboo laths or reeds forming an elongated basketwork, in size and shape like a chicken, weighing over 1⅓ pounds. It has paper glued over to strengthen it, and it is filled with explosive gunpowder. All is sealed up using more paper, with head and tail fixed on before and behind, and the two wings nailed firmly on both sides, so that it looks just like a flying crow.
>
> Under each wing there are two slanting rockets. The fourfold branching fuse, connected with the rockets and about a foot long, is put through a hole drilled on the back of the bird. When in use, this main fuse is lit first.
>
> The bird flies away more than 1000 feet, and when it eventually falls to the ground, the explosive gunpowder in the cavity of the bird is automatically lit, and the flash can be seen miles away. This weapon is used against enemy encampments to burn them, but also at sea to set ships on fire. It should never fail to bring victory.

It will immediately be obvious how strong a resemblance this weapon bears to the notorious V-1 rockets of the Second World War. As in the case of the V-1, there was little control over where the bomb might drop and explode when the rockets burned out. This is the world's oldest winged rocket. There were others, including a flying grenade called the 'free-flying enemy-pounding thundercrash bomb', which was a flying bomb in a cast iron casing which also gave off poisonous smoke and spilled out poison-tipped calthrops – sharp pointed objects on which men and horses would tread.

Multi-stage rockets in China go back to the early fourteenth century. We have already come across one which was supposed to fly over the enemy, drop an explosive charge, and fly back on a second rocket. Here is a description from *The Fire-Drake Artillery Manual* of the automatic lighting by a fuse of the

BELOW (180) The first of all multi-stage rockets, the 'fire-dragon issuing from the water', of the early or mid-fourteenth century, which was used in naval engagements. When the rockets near the head burnt out, they lit fuses which ignited the second-stage rockets at the rear. The tube with the dragon's head was 5 feet long, and the fuses ran inside the body. This rocket flew in a flat trajectory, 3 or 4 feet above the water, for over a mile. It was thus an eerie forerunner of the modern Exocet surface-skimming naval rockets. This woodcut is from *The Fire-Drake Artillery Manual*, published in 1412.

second stage of a rocket called 'the fire-dragon issuing from the water':

> A tube of bamboo 5 feet long is taken, the septa removed, and the nodes scraped smooth with an iron knife. A piece of wood is carved into the shape of a dragon's head and fitted on at the front, while a wooden dragon tail is made for the rear end. The mouth must be facing upwards, and in the belly of the dragon there are several 'mysterious mechanism rocket-arrows'. At the dragon head there is an opening through which go all the fuses of the rockets inside.
>
> Beneath the dragon-head on both sides there are two big rocket-tubes.... Their fuses and orifices should face downwards and backwards, and their front ends must face upwards and forwards; and they are fixed tight to the body by bands of hempen cloth secured with skin- and fish-glue. The fuses of the rocket-arrows within the belly lead out from the head of the dragon, and they are divided into two. Oiled paper is used to make them firm, and they are so arranged as to be connected with the front ends of the outside rocket-tubes. And under the tail of the dragon on each side there are also two big rocket-tubes, fastened in the same style. The fuses of the four rockets are twisted into a single one. In a naval battle, the apparatus can fly 3 or 4 feet above the water.
>
> Upon lighting it will fly over the water as far as 1800 yards [just over a mile]. At a distance it really looks like a flying dragon coming out of the water. When the gunpowder in the rocket-tubes is nearly all finished, that in the rocket-arrows within the belly is ignited, so that they fly forth, destroying the enemy and his ships. It can be used either on land or on sea.

This strange multi-stage rocket was a forerunner of the modern Exocet, operating in the same way by skimming the water for a long distance and hitting a ship.

Rockets thus seem to have been invented in China about 1150, and they were used in warfare by at least 1206, in particular by the Sung Chinese in their defence of Hsiangyang against the Chin. As fireworks, they were in use in 1180 at Hangchow, the Southern Sung capital, as we have seen in the account by Chou Mi. 'Earth rats', which were simple random ground firework-rockets, were probably the first rockets actually used in warfare at some time before the siege just mentioned. By about 1280, the Arab Hasan al-Rammah was already describing rockets as 'Chinese arrows'.

When did rockets first appear in the West? They are first mentioned in connection with the Battle of Chioggia between the Genoese and the Venetians in 1380. They must have come to Italy with Italian travellers, in the wake of Marco Polo. By 1405, Konrad Keyser in his *Bellifortis* was able to write that a rocket must have a hollow space inside its charge and an arrow-shaft. The knowledge of rockets thus came to Europe much more rapidly than did most Chinese inventions. The delay was just over two centuries. But it was only in the twentieth century that liquid-fuel rockets were developed, which give the promise to mankind of colonizing outer space. The eventual use of rockets to take man in great numbers beyond his own world leads Needham to believe that the rocket was probably China's most important invention, and its greatest technological contribution to mankind.

GUNS, CANNONS, MORTARS AND REPEATING GUNS

THIRTEENTH CENTURY AD

About 905 AD the first 'proto-gun', called a fire-lance, was invented in China. The world's oldest depiction of a 'gun' is in a detail from a silk banner painted in the middle of the tenth century (see Plate 181 (opposite)). Buddha is meditating, and the demons of the evil goddess Mara the Temptress are trying to distract him. A grenade is about to be thrown at him while a demon with three snakes issuing from its head fires at him with a fire-lance. This incontrovertible physical evidence dates this invention no later than 950, much earlier than its first textual mention.

OPPOSITE (181) The world's oldest pictures of a gun and a grenade, in a painted silk banner of the mid-tenth century found at Tunhuang. The banner shows the assault of Mara the Temptress and her demons on the meditating Buddha; they are seeking to distract him from his attainment of understanding of the nature and mechanism of the universe, and to prevent his enlightenment. To the Buddha's top right one demon holds a proto-gun in the form of a fire-lance, whilst another just below brandishes a weak-casing bomb, from which flames have already begun to emerge. (Musée Guimet, Paris.)

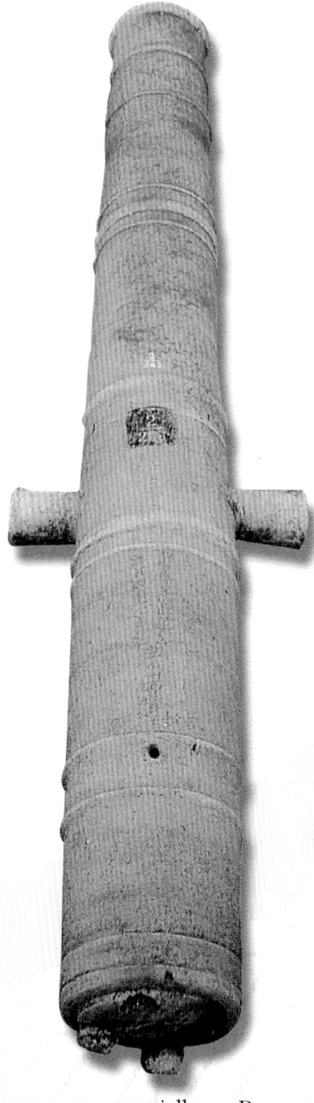

RIGHT (182) A bronze Chinese cannon which, according to its inscription, was cast during the Ming Dynasty (ended 1644). Its calibre is 5½ inches; length 7 feet 10 inches. (Royal Artillery Institution, Woolwich.)

At first the fire-lance was essentially a Roman candle firework tied to a spear, but it was more powerful than a mere firework. In fact it acted as a five-minute flame-thrower, and it could kill, blind, or maim. It was particularly useful for defending city walls against besiegers. Large numbers of fire-lances were wheeled round in trolleys of several layers, so that when the five minutes were up the defending soldier would be passed another fire-lance in order to keep up the fire. A fire-lance would also set on fire the clothes of an enemy, or his besieging engines, or his tents. Even after the invention of the true gun, fire-lances remained popular and in fact were used right up to the middle of the twentieth century in China as protection against pirates who tried to board ships. In Europe, the fire-lance seems to have been used for the last time in the first siege of Bristol during the English Civil War in 1643; by the second siege, in 1645, it had been abandoned.

Fire-lances went through many improvements. Having started as lengths of bamboo tube which simply spurted fire, they progressed to metal barrels. Many projectiles were fired from them, though not true bullets, since the projectiles did not perfectly fit the bore of the barrel (which is necessary for a true gun). Any old bits of broken pottery, stones, and iron or steel chips or balls, would be blown out of a fire-lance in a deadly swarm to splatter the unfortunate enemy. Clouds of projectiles could effectively be shot out 30 or 40 yards, and it was common practice to include poisonous chemicals in the gunpowder, so that enemies were blasted with arsenic and other lethal substances. But of course, this could not be done if the wind blew in the wrong direction! Fire-lances also frequently spewed out arrows with poisoned tips.

In 1233, the Mongols were besieging the Chin, who launched a daring night raid using fire-lances. We are told in the dynastic history:

> **On the fifth day of the fifth month they sacrificed to Heaven, secretly prepared fire-lances, and embarked 450 Chin soldiers outside the south gate, whence they sailed first east and then north. During the night they killed the enemy guards outside the dykes, and reached the Wang family temple.... Kuan-Nu divided his small craft into squadrons of five, seven and ten boats, which came out from behind the defences and caught the Mongols both from front and rear, using the fire-spouting lances. The Mongols could not stand up to this and fled, losing more than 3500 men drowned. Finally their stockades were burnt, and our force returned.**

The fire-lance doubled as a pike in close combat, as we learn from this account of a battle in 1276:

> **At daybreak Chiang Tsai, seeing that Shih Pi's troops were few, pressed an attack, but Shih Pi resisted furiously. Two Sung cavalrymen rushed at him to pierce him with fire-lances, but he so defended himself with his sabre that to left and right every man fell; and he himself personally killed more than one hundred.**

Later fire-lances dropped their spear-points, especially when they had progressed to stronger and longer metal barrels. There were always many kinds to choose from, for different purposes.

An artillery piece evolved from the fire-lance, as well as a gun which fired an ill-fitting shell, and was hence not a true gun. Since there is no specific name for these proto-cannons, Needham has christened them 'eruptors' – and erupt is what they did. Like fire-lances, they spewed forth fire and flame mixed with poisons and assorted bits and pieces. They fired arrows, pellets, stones and cannon balls. Some of them even fired exploding canisters with attached fuses which blew up after landing.

Should anyone think these fire-lances and eruptors were just noisy toys unlikely to cause much damage, it pays to read these remarks from the mid-thirteenth century about a 3-foot-long fire-lance which fired arrows and was called the 'single-flight magic-fire arrow':

> **Use a barrel 3 feet long cast from high-grade bronze, and designed to take only a single arrow. Put 0.3 ounces of 'blinding gunpowder' as charge into the barrel before firing, whereupon the arrow is sent flying like a fiery serpent, with a range of between 200 and 300 paces. It can pierce the heart or the belly when it strikes a man or a horse, and can even transfix several persons at once.**

Fire-lances with several barrels were frequently used, and were built so that when one fire-tube had exhausted itself, a fuse ignited the next, and so on. One triple-barrelled fire-lance was called 'the triple resister', another was called 'the three-eyed lance of the beginning of the dynasty'. Several old woodcuts of these weapons survive, so that we know what they looked like and how they were constructed. One curious weapon was the 'thunder-fire whip' – a fire-lance in the shape of a sword, 3 feet 2 inches long and tapering into a muzzle. It discharged three coin-sized lead balls. An even stranger one was the 'vast-as-heaven enemy exterminating Yin–Yang shovel' which had a broad crescent-shaped blade at its end, and emitted poison as well as flames and lead pellets.

Certainly one of the most ingenious and useful fire-lances was the 'mattock gun'. It was fixed at right angles to a pole and could be hoisted to the top of the battlements while a hidden operator crouched behind and fired at besiegers coming over the wall. Besides flames, it shot six or seven pellets at a time from its metal barrel.

There were also huge batteries of fire-lances which could be fired simultaneously from mobile racks. One such was called 'the ingenious mobile ever-victorious poison-fire-rack', and it seems to have originated in the fourteenth century. A great frame with several wheels would hold many layers of sixteen fire-lances one after the other. It took ten men just to light all the fuses at once. A quaint description of its use in defending a besieged city relates:

BELOW (183) A fifteenth-century cannon of the Ming Dynasty, from an old woodblock of circa 1500, reprinted in 1883. (Collection of Robert Temple.)

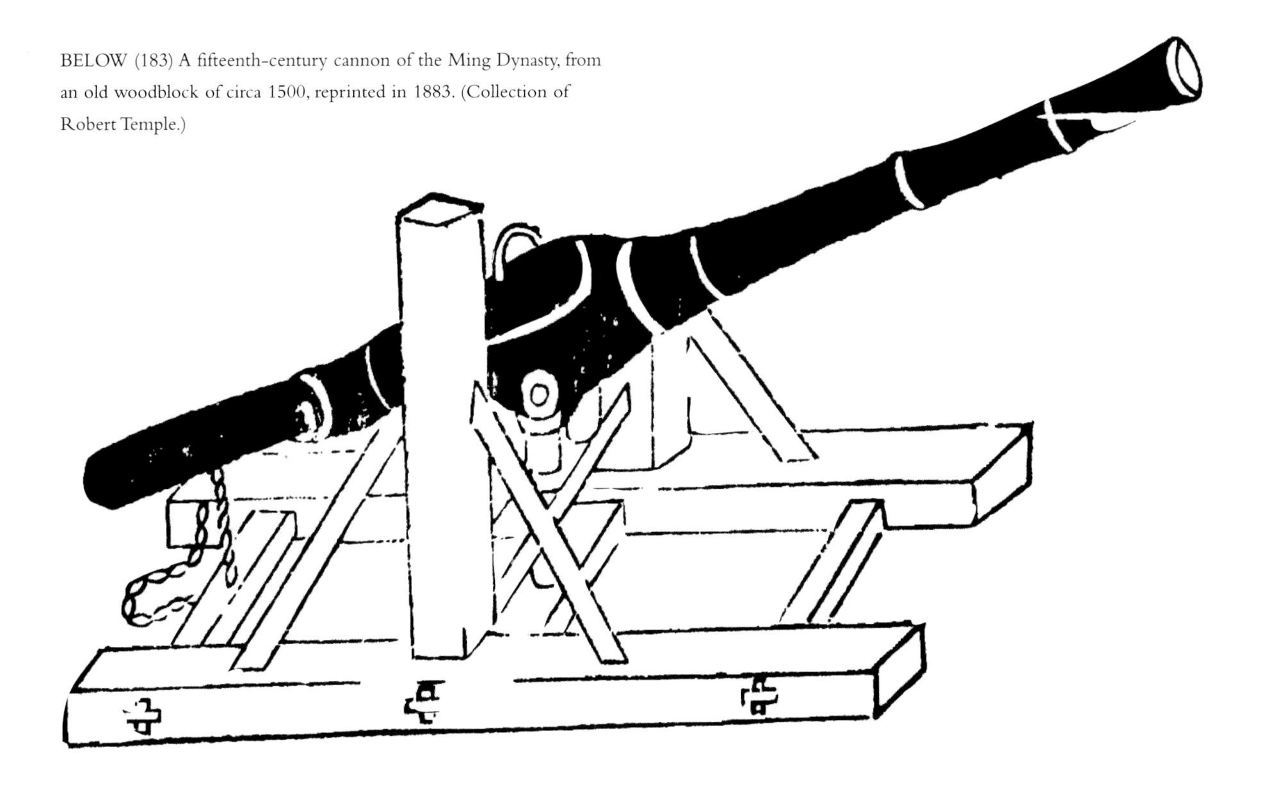

When the enemy approaches the gate, all the weapons are fired at a single moment, giving a noise like a great peal of thunder, so that his men and horses are all blown to pieces. You can then open the city-gate and, relaxing, talk and laugh as if nothing had happened; this is the very best device for the guarding of cities.

Since these fire-racks were highly mobile, they were probably the earliest form of proto-tank, though of course no one was inside them. The blast from a fire-rack containing perhaps 200 fire-lances fired at once is hardly to be imagined, and the men operating them must have gone deaf very soon. The nearest thing to the fire-rack in modern times is perhaps the Russian 'Stalin-Organ', though this is used to launch rockets.

Yet another variation on the fire-lance was the 'one-eyed magically-efficient gun', which was a swivel gun. It had a metal barrel and wooden handle, and was inserted into one of three rings on a pole thrust into the ground on a spike to keep it steady. The swivel or forked rests later used in Europe with early muskets may have originated from this weapon.

The earliest evidence we have of fire-lances in Europe is from a Latin manuscript dated 1396, 450 years later than the Chinese version. It depicts them vividly, and contains a very dramatic picture of a mounted horseman charging ahead with a long, blazing fire-lance in his right hand. The Chinese protocannons – the 'eruptors' – appeared in the thirteenth century. *The Fire-Drake Artillery Manual* of 1412 describes one:

The shells are made of cast iron, as large as a bowl and shaped like a ball. Inside they contain half a pound of 'magic' gunpowder. They are sent flying towards the enemy camp from an eruptor; and when they get there a sound like a thunder-clap is heard, and flashes of light appear. If ten of these shells are fired successively into an enemy camp, the whole place will be set ablaze and his men will be thrown into confusion. You can use any kind of gunpowder in the shells – blinding powder, flying powder, violent powder, poison powder, bruising and burning powder, and smoke-screen powder, according to the circumstances.

Chang Hsien wrote a poem called 'The Iron Cannon Affair' in the year 1341, which describes an eruptor in operation:

The black dragon lobbed over an
egg-shaped thing
Fully the size of a peck measure it was,
And it burst, and a dragon flew out
with peals of thunder rolling
In the air it was like a blazing and
flashing fire.
The first bang was like the dividing
of chaos in two,
As if mountains and rivers were
all turned upside down....

Just as the multi-barrelled fire-lance and the eruptors were ingenious innovations, so was the 'multiple bullets magazine eruptor'. Undoubtedly the idea of a magazine of projectiles had occurred to someone because of the widespread use of magazines on repeating crossbows. The eruptor was 4 feet 5 inches in length and made of cast bronze. A magazine a foot long, containing about a hundred lead balls, fed into it from the side as it spewed out flame. When the

ABOVE (184) A fourteeth-century bomb in the process of exploding. This one contains poisons and iron calthrops (small triangular spiked objects intended to litter the ground and wound the feet of men and horses). Reproduced from *The Fire-Drake Artillery Manual of Gunpowder Weapons*, 1412.

eruptor was at rest the barrel was rotated, enabling the magazine to fall out and be loaded. Then, as firing began, the barrel would be rotated back again and the pellets would start falling down into the flames and be shot forth – the world's first repeating-cannon! It must be stressed, however, that only the projectiles could be termed 'repeating', not the charge.

A naval version of the eruptor fired 'chain-shot'. This consisted of two cannon-balls joined together by a chain or an iron bar. When fired from an eruptor, these rotated at great speed through the air, smashing the spars and rigging of an enemy ship and clearing its upper deck of men.

The gunpowder in these eruptors and fire-lances usually had 60 per cent of saltpetre in its composition, as was the case with rockets (see page 264). For the true gun, higher percentages of saltpetre were necessary – between 70 and 80 per cent. But it is worth pointing out that with this amount of saltpetre, rockets will not function properly because the mixture becomes too explosive. Thus a wide variety of gunpowder mixtures was used for fire-lances, eruptors, rockets and true guns – each mixture tailored specially for each weapon. And, of course, there were different types of poisonous gunpowder for use when the wind was right, and coloured gunpowders for signal flares.

The true gun was eventually developed in China at some time during the thirteenth century. A bronze hand gun from 1288 has been excavated at Heilunchiang Province in Manchuria. It is more than 1 foot long and weighs 8 pounds. The gunpowder chamber has a small touch-hole for ignition; the bore is even. The explosion chamber is strengthened by a bulbous enlargement of the barrel at that point, which was a common feature of early guns to prevent them from splitting under the force of their own internal explosions.

For such a perfect little gun to exist in 1288, how long had guns existed prior to that date? The fact is, new inventions did not at once attract new words; at first a gun was hardly different from a fire-lance, so the same word was used for both weapons. It is safe to presume that guns evolved around 1250, at the latest around 1280.

When did guns reach Europe? The earliest evidence is a picture of a bombard (a small bulbous cannon with a constant bore) which fired arrows, depicted in a 1327 manuscript by Walter de Milamete's *On the Majesty, Wisdom, and Prudence of Kings*, in the Bodleian Library at Oxford. Whereas it took 450 years for the fire-lance to reach Europe from China, the true gun did so in a tenth of that time.

How were these things transmitted? Needham believes it is quite likely that gunpowder in the form of fire-crackers reached Roger Bacon in England in 1265 through his friend the Franciscan William Ruysbroeck, who returned from China in 1256. Alternatively, Chinese evidence exists of a mission to the emperor by Scandinavian trader-envoys who reached China in 1261, several years before Marco Polo, apparently having travelled overland through what was then the independent kingdom of Novgorod.

After 1260, many foreigners were welcomed and employed by the Mongol emperors in China. It was their policy to treat the native Chinese as third-class citizens, hardly even regarding them as human. The Mongols did not trust the Chinese to run things, and tried to create a civil service of Mongols, Europeans and Arabs. With these constant foreign visitations, the transmission of guns to Europe was inevitable. Indeed, the writer Miu Yu-Sun quotes an obscure scholar, Yü Wei, as saying that in the second half of the thirteenth century a Mongol, Ch'i Wu Wen, went to Europe and took with him a complete knowledge of gunpowder technology and gunnery.

The similarities between the earliest European eruptors and cannons and the slightly earlier Chinese ones are so striking that it seems likely that actual guns were transported to

LEFT (185) A pivoting nineteenth-century breech-loading Chinese mortar gun of cast iron, with a wooden stock. The total length of this weapon is 8 feet. This type of gun was used for the defence of fortifications, and was not carried in the field. It was probably seized by British forces during the nineteenth-century conflicts in China and carried home as a specimen. (The Royal Armouries, Tower of London.)

LEFT (186) A fourteenth-century cast iron Chinese cannon. The cannon length is 18.7 inches, the bore length is 16.1 inches, and the calibre 4.15 inches. (Royal Artillery Institution, Woolwich.)

Europe for direct copying. For instance, the earliest piece of field artillery we know of in China was on four wheels and called the 'thousand-ball thunder cannon'. It no longer had a bulbous body round the explosion chamber, and was ringed by bands of iron. A surviving woodcut from an old Chinese book shows that it was virtually identical to an early European field artillery piece portrayed in a German manuscript of about 1450. The German cannon was probably a direct copy, or it might even have been an obsolete Chinese model sold by some corrupt official to a Dutch trader.

Chinese fire-lances and eruptors did not fit the description of true guns as having a constant bore and a smoothly fitting projectile of exactly the right size. But between 1250 and 1280, the Chinese finally did achieve the manufacture of true guns, and by the 1320s these had reached Europe.

The evolution of guns and cannons proceeded apace in China, where perfectly cast iron cannons were being produced before Europe had even learned how to make cast iron (see page 44). The bulbous reinforcing of the barrel was superseded by even barrels along the whole length, while iron bands encircling the barrels provided reinforcement. These grew and flattened out as development proceeded, so that eventually the barrels were more than half covered by large flattened bands. Cannons commonly bore inscriptions, like crossbow trigger mechanisms, giving their precise dates of manufacture. There are literally hundreds of medieval Chinese cannons still in existence today.

Projectiles also survive: rounded stones or carefully cast bronze or iron balls, and exploding cast-iron 'thundercrash' bombs (described at length on page 258). Bags of shot which smoothly fitted the bore could also be fired, as in the shotguns of today. While some of the cannons were laid flat, others were set at an angle and lobbed projectiles over, acting as mortars. As metallurgy developed, cannons became larger, longer and heavier. We read in *The Fire-Drake Artillery Manual* (1412) of a cannon which weighed 159 pounds; it was called the 'long-range awe-inspiring cannon':

> **Each weighs 72 kilograms [159 pounds] and measures 2 feet 8 inches long. The touch-hole is 5 inches from the base and 3.2 inches from where the belly begins. The diameter of the bore at the muzzle is more than 2.2 inches. Above the touch-hole there is a movable lid to protect the priming powder from rain. This cannon does not give a great bang or much recoil. With 8 ounces of gunpowder one uses one large lead ball weighing 1.2 kilograms (2.6 pounds), or a hundred small lead bullets in a bag, each weighing 0.6 ounces. Firing is done very conveniently by hand.**

However, cannons grew in size from this until one weighing 1389 pounds, called 'the great invincible general', was cast. A later giant cannon from the middle of the fifteenth century was the 'great general gun'. We have a description of this and of a later improved model:

> Among the large firearms there is none that is greater than the 'great general gun'. Its barrel used to weigh 80 kilograms (176 pounds), and was attached to a stand made of bronze weighing 600 kilograms (1322 pounds).... Yeh Meng-Hsiung changed the weight of the gun to 150 kilograms (330 pounds) and doubled its length to 6 feet, but eliminated the stand, and now it is placed on a carriage with wheels. When fired it has a range of 800 paces. A large lead shell weighing 3.5 kilograms (7.7 pounds) is called a 'grandfather shell' and the next shell of medium size weighing 1.8 kilograms (3.9 pounds) is a 'son shell', while a smaller shell weighing 0.6 kilograms (1.3 pounds) is a 'grandson shell'. There are also 200 small bullets each weighing 0.3 to 0.2 ounces contained in the same shell and called 'grandchildren bullets', while the saying is that the 'grandfather' leads the way and the 'grandchildren' follow. They are supplemented with iron and porcelain fragments previously boiled in cantharides beetle poison. The total weight of the projectile is some 12 kilograms (26 pounds). A single shot has the power of a thunderbolt, causing several hundred casualties among men and horses. If thousands, or tens of thousands, of this weapon were placed in position along the frontiers, and every one of them manned by soldiers well-trained to use them, then we should be invincible. This weapon is indeed the ultimate among all firearms. At first its heavy weight caused some doubt as to whether or not it was too cumbersome; but if it is transported on its carriage then it is suitable, irrespective of height, distance or difficulty of terrain. During the sixth year of the T'ien-Sun reign-period (1462) 1200 gun carriages were made.... During the first year of the Ch'eng-Hua reign-period (1465) 300 different 'great general guns' were manufactured and 500 carriages for cannons were made. This was an excellent strategy in using Chinese expertise to keep the barbarians under control.

Attempts to develop repeat-firing guns continued. The Chinese came up with a method of doubling the rate of fire with small cannons by developing something called the 'Mr Facing-Both-Ways' gun. It consisted of two small cannons whose rear ends joined together in one long barrel. The first one was

RIGHT (187) A detail from a previously unpublished photograph, taken by Ernst Boerschmann in about 1900, of an unidentified medieval bas-relief. The figure on the right is holding a fire-lance from which an arrow is protruding. Many fire-lances shot arrows with poisoned tips, followed by several minutes' spurting of fire and flames. The earliest proto-guns to reach Europe fired arrows.

BELOW (188) Sixteenth-century Chinese mortars being fired. These vase-shaped mortars have a uniform bore inside, despite the bulge outside. These were called the 'flying, smashing and bursting bomb-cannons', and had fuses running to the touch-holes through bamboo tubes. They fired cast-iron bombshells which, when they exploded, also showered the ground around with calthrops – nasty little metal objects covered in sharp, poison-tipped points, which would be trodden on by the enemy's men and horses.

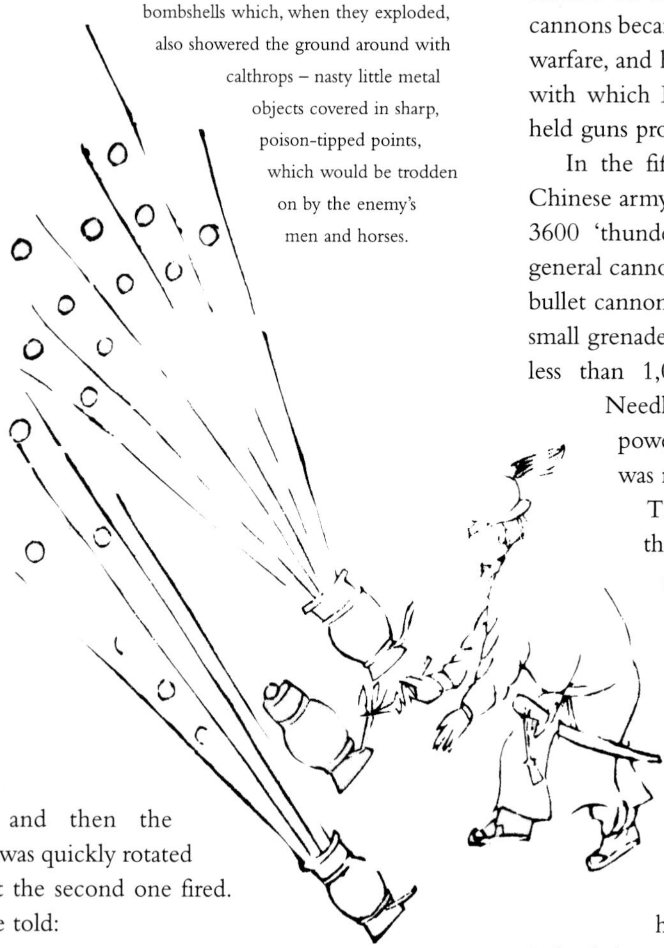

fired, and then the barrel was quickly rotated so that the second one fired. We are told:

> **Immediately after firing the first gun the second is rotated into position and fired, each one being muzzle-loaded with a large stone projectile. If the gun is aimed at the hull of an enemy ship below the water-line, the cannon-balls shoot along the surface and smash its side into splinters. It is a very handy weapon.**

A further step was the development of the 'cartwheel gun'. It had thirty-six barrels radiating from its centre like the spokes of a wheel. These guns were small enough that a single mule could carry two, one on each side of its pack. However, as some of the barrels pointed at the gunner, there must have been some nasty accidents, so these 36-repeaters were never widely adopted.

Batteries of true guns were used, just as batteries of fire-lances had been. For instance, there was a 'nine ox-jar battery' with nine cannons resting together on a frame and fired by one fuse. Guns were also wheeled around on wheelbarrows. As the centuries wore on, cannons became more and more important in Chinese warfare, and huge artillery duels took place of a kind with which Europeans are well familiar. And hand-held guns proliferated even more.

In the fifteenth century a single battalion of a Chinese army was provided with 40 cannon batteries, 3600 'thunder-bolt shells', 160 'wine-cup muzzle general cannons', 200 large and 328 small 'continuous bullet cannons' firing grape-shot, 624 hand guns, 300 small grenades, some 6.97 tons of gunpowder and no less than 1,051,600 bullets, each of 0.8 ounces. Needham remarks: 'This was quite some fire-power, and the total weight of the weaponry was reckoned to be 29.4 tons.'

The vast quantities of weapons used and the enormous numbers of deaths which resulted make European war-death statistics before the nineteenth century seem puny. It is important to realize that Chinese armies commonly numbered hundreds of thousands at a time. By contrast, the English Civil War of the seventeenth century was fought by only a few thousand people, and if casualties for a single battle mounted into the hundreds it was considered horrific. Indeed, the mass-production of armaments in medieval China rivaled the output of twentieth-century Detroit with its assembly lines of automobiles; division of labour and assembly-line techniques of producing crossbows, guns, gunpowder, porcelain, and all manner of commodities, both helpful and harmful, were traditional for centuries. The 'military-industrial complex', entirely state-run, with stringent security and highly classified new weapons projects, is a two-thousand-year-old phenomenon in China. The Chinese were as expert at killing people as they were at inventing beneficial things; they were arms manufacturers on a scale undreamed of until modern times in the West. The Chinese inventive genius certainly did not hold back in pacifist timidity from the design and manufacture of weapons. No nation in the world could match the Chinese expertise in warfare for two millennia.

CHINESE DYNASTIES

	Hsia Kingdom (legendary?)	c.2000–c.1520 BC
	Shang (Yin) Kingdom	c.1520–c.1030 BC
	Chou Dynasty	
	Early Chou period	c.1030–722 BC
	Ch'un Ch'iu period	722–480 BC
	Warring States period	221–207 BC
First Unification	*Ch'in Dynasty*	
	Han Dynasty	
	Ch'ien Han (Earlier or Western)	207 BC–9 AD
	Hsin interregnum	9–23
	Hou Han (Later or Eastern)	25–220
	San Kuo (Three Kingdoms period)	221–265
First Partition	*Shu (Han)*	221–264
	Wei	220–265
	Wu	222-280
Second Unification	*Chin Dynasty*	
	Western	265–317
	Eastern	317–420
	(Liu) Sung Dynasty	420–479
Second Partition	*Northern and Southern Dynasties*	
	Ch'i	479–502
	Liang	502–557
	Ch'en	557–589
	Northern Wei Dynasty	386–535
	Western Wei Dynasty	535–556
	Eastern Wei Dynasty	534–550
	Northern Ch'i Dynasty	550–577
	Northern Chou Dynasty	557–581
Third Unification	*Sui Dynasty*	581–618
	T'ang Dynasty	618–906
Third Partition	*Wu Tai (Five Dynasty Period)*	
	Later T'ang (Turkic), Later Chin (Turkic), Later Han (Turkic), and Later Chou	907–960
	Liao (Ch'itan Tartar) Dynasty	907–1124
	West Liao Dynasty	1124–1211
	Hsi Hsia (Tangut Tibetan) State	986–1227
Fourth Unification	*Northern Sung Dynasty*	960–1126
	Southern Sung Dynasty	1127–1279
	Chin (Jurchen Tartar) Dynasty	1115–1234
	Yuan (Mongol) Dynasty	1260–1368
	Ming Dynasty	1368–1644
	Ch'ing (Manchu) Dynasty	1644–1911
	Republic	1912–
	People's Republic	1949–

95° E
100° E
105° E
40° N
35° N
30° N
25° N
Tunhuang
KANSU
NINGHSIA
Tsaidam Plateau
Ch'ilian Shan
Alashan
Ninghsia
K'unlun Shan
Nan Shan
Paiyu Shan
Ching River
Lanchow
Liup'an Shan
CH'INHAI
Huang-ho
KANSU
SHENSI
Ch'in-ling Shan
Hanchung
Chialing River
Mekong River
SIKANG
SZECHUAN
Ch'engtu
Tahsüeh Shan
Salween River
Minya Konka
Omei Shan
Yangtze River
Chungking
Taliang Shan
Talou Shan
KWEICHOW
Kweiyang
Tali
YUNNAN
K'unming
Hungshui River
Chiang River
95° E
0
300 km
0
200 miles
International boundaries
Provincial boundaries
100° E
105° E

115° E
120° E
125° E
40° N
35° N
30° N
25° N
Liaotung
Peking
Hêng Shan
T'aihang Shan
T'ienching
(Tientsin)
HOPEI
Thaiyaun
(Yellow River)
T'ai Shan
Ch'ingtao
(Tsingtao)
Anyang
SHANTUNG
Grand Canal
Huang-ho
K'aifeng
Sung
Shan
HONAN
CHIANGSU
Huai River
Yangtze River
Nanking
Shanghai
HUPEI
Tapieh Shan
ANHUI
I-ch'ang
Huaiyuan Shan
Hankow
Hangchow
EAST CHINA SEA
CHEKIANG
Poyang Lake
Tungting
Lake
Ch'ang-sha
CHIANGSI
HUNAN
Hêng Shan
Min River
Fuchow
Wu-i-Shan
FUKIEN
T'aipei
Hsiamen
(Amoy)
Nan-ling
TAIWAN
KUANGTUNG
Kuangchow
(Canton)
Shant'ou
(Swatow)
SOUTH CHINA SEA
Hong Kong
GENERAL MAP OF
CHINA
Shan = Mountain(s)
115° E
120° E

TIME LAGS BETWEEN CHINESE INVENTIONS AND DISCOVERIES AND THEIR

AGRICULTURE	Row cultivation of crops and intensive hoeing	6th century BC
	The iron plough	6th century BC
	Efficient horse harnesses – trace	4th century BC
	– collar	3rd century BC
	The rotary winnowing fan	2nd century BC
	The multi-tube ('modern') seed drill	2nd century BC
ASTRONOMY AND CARTOGRAPHY	Recognition of sunspots as solar phenomena	4th century BC
	Quantitative cartography	
	Discovery of the solar wind	
	The Mercator map-projection	
	(Mounted) Equatorial astronomical instruments	
ENGINEERING	Spouting bowls and standing waves	5th century BC
	Cast iron	4th century BC
	The double-acting piston bellows: ⋆ air; # liquid	4th century BC
	The crank handle	2nd century BC
	The 'Cardan suspension', or gimbals	2nd century BC
	Manufacture of steel from cast iron	2nd century BC
	Deep drilling for natural gas	1st century BC
	The belt drive (or driving-belt)	1st century BC
	Water power	
	The chain pump	
	The suspension bridge	
	The first cybernetic machine	
	Essentials of the steam engine	
	'Magic mirrors'	
	The 'Siemens' steel process	
	The segmental arch bridge	
	The chain-drive	
	Underwater salvage operations	
DOMESTIC AND INDUSTRIAL TECHNOLOGY	Lacquer: the first plastic	13th century BC
	Strong beer (*sake*)	11th century BC
	Petroleum and natural gas as fuel	4th century BC
	Paper	2nd century BC
	The wheelbarrow	1st century BC
	Sliding callipers	1st century BC
	The magic lantern	
	The fishing reel	
	The stirrup	
	Porcelain	
	Biological pest control	
	The umbrella	
	Matches	
	Chess	
	Brandy and whisky	
	The mechanical clock	
	Printing – block printing	
	– movable type	
	Playing-cards	
	Paper money	
	'Permanent' lamps	
	The spinning-wheel	
MEDICINE AND HEALTH	Circulation of the blood	6th century BC
	Circadian rhythms in the human body	2nd century BC
	The science of endocrinology	2nd century BC
	Deficiency diseases	
	Diabetes discovered by urine analysis	
	Use of thyroid hormone	
	Immunology – inoculation	

2200 years
2200 years
00 years
1000 years
2000 years
1800 years
2000 years
2nd century AD 1300 years
6th century AD 1400 years
10th century AD 600 years
13th century AD 600 years
NEVER
1700 years
* 1900 years; # 2100 years
1100 years
1100 years
2000 years
1900 years
1400 (signif. realised in West 1800 years)
st century AD 1200 years
st century AD 1400 years
st century AD 1800 years (poss. over 2200 years)
3rd century AD 1600 years (poss. 3000 years)
5th century AD 1200 years
5th century AD 1500 years before understood in the West
5th century AD 1300 years
610 AD 500 years
976 AD 800 years
11th century AD 800 years
3200 years
NEVER
2300 years
1400 years
1300 years
1700 years
2nd century AD 1800 years
3rd century AD 1400 years
3rd century AD 300 years
3rd century AD 1700 years
3rd century AD 1600 years
4th century AD 1200 years
577 AD 1000 years
6th century AD 500 years
7th century AD 500 years
725 AD 585 years
8th century AD 700 years
1045 AD 400 years
9th century AD 500 years
9th century AD 850 years
9th century AD NEVER
11th century AD 200 years
1800 years
2150 years
2100 years
3rd century AD 1600 years
7th century AD 1000 years
7th century AD 1250 years
10th century AD 800 years

MATHEMATICS	The decimal system	14th century BC
	A place for zero	4th century BC
	Negative numbers	2nd century BC
	Extraction of higher roots and solutions of higher numerical equations	1st century BC
	Decimal fractions	1st century BC
	Using algebra in geometry	
	A refined value of *pi*	
	'Pascal's' triangle of binomial coefficients	
MAGNETISM	The first compasses	4th century BC
	Dial and pointer devices	
	Magnetic declination of the Earth's magnetic field	
	Magnetic remanence and induction	
THE PHYSICAL SCIENCES	Geo-botanical prospecting	5th century BC
	The First Law of Motion	4th century BC
	The hexagonal structure of snowflakes	2nd century BC
	The seismograph	
	Spontaneous combustion	
	'Modern' geology	
	Phosphorescent paint	
TRANSPORTATION AND EXPLORATION	The kite	5th/4th century BC
	Manned flight with kites	4th century BC
	The first relief maps	3rd century BC
	The first contour transport canal	3rd century BC
	The parachute	2nd century BC
	Miniature hot-air balloons	2nd century BC
	The rudder	
	Masts and sailing: Batten sails – Staggered Masts	
	Multiple masts – Fore and aft rigs	
	Watertight compartments in ships	
	Leeboards	
	The helicopter rotor and the propeller	
	The paddle-wheel boat	
	Land sailing	
	The canal pound-lock	
SOUND AND MUSIC	The large tuned bell	6th century BC
	Tuned drums	2nd century BC
	Hermetically sealed research laboratories	1st century BC
	The first understanding of musical timbre	
	Equal temperament in music	
WARFARE	Chemical warfare: poison gas, smoke bombs and tear gas	4th century BC
	The crossbow	4th century BC 200 years
	Gunpowder	
	The flame-thrower	
	Flares and fireworks	
	Soft bombs and grenades	
	Metal-cased bombs	
	Land mines	
	Sea mines	
	The rocket	
	Multi-staged rockets	
	Guns, cannons and mortars – firelance	
	– true gun	

ADOPTION OR RECOGNITION IN THE WEST

2300 years
1400 years
1700 years
600 years
1600 years
3rd century AD 1000 years
3rd century AD 1200 years
1100 AD 427 years
1500 years
3rd century AD 1200 years
9th century AD 600 years
11th century AD 600 years
2100 years
1300 years (but 2000 to Newton)
1800 years
130 AD 1400 years
2nd century AD 1500 years
2nd century AD 1500 years
10th century AD 750 years
2000 years
1650 years
1600 years
1900 years
2000 years
1400 years
st century AD 1100 years
2nd century AD NEVER
2nd century AD 1200 years
2nd century AD 1700 years
8th century AD 800 years
4th century AD 1500 years
5th century AD 1000 years
550 AD 1050 years
984 AD 400 years
2500 years
UNKNOWN
2000 years
3rd century AD 1600 years
1584 AD 50 years
2300 years
9th century AD 300 years
10th century AD 1000 years
10th century AD 250 years
1000 AD 400 years
1221 AD 246 years
1277 AD 126 years
14th century AD 200 years
11th century AD 200 years
14th century AD 600 years
1120 AD 450 years
1280 AD 50 years

INDEX

The use of **bold** indicates main subject areas and subject main entries.

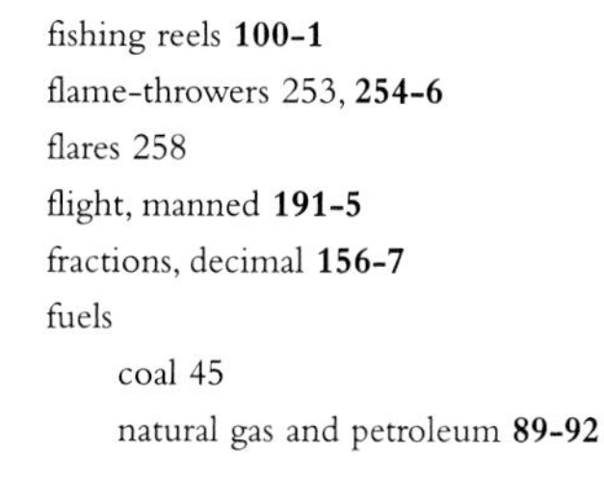

J

K

L

M

N

O

P

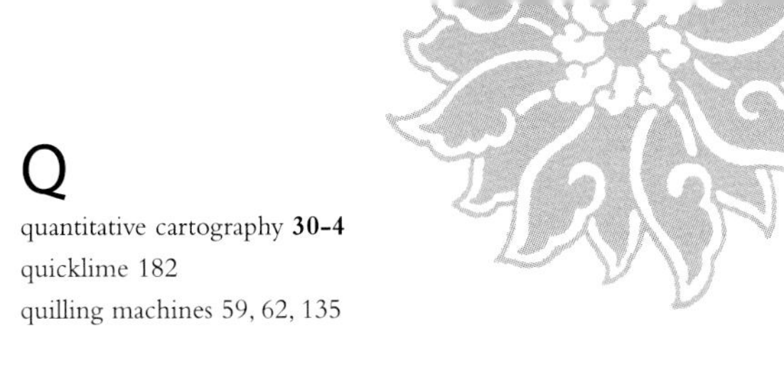

Q

R

S

T

U

V

W

Z

AUTHOR'S ACKNOWLEDGEMENTS

My deepest debt of gratitude is owed to the late Dr Joseph Needham who authorised me to popularize his great work *Science and Civilisation in China*, in its published and unpublished sections, as well as all of his related writings. He threw open to me the magnificent facilities of what was then called the East Asian History of Science Library on Brooklands Avenue in Cambridge, which later moved its premises and became the Needham Research Institute, adjoining the grounds of Robinson College, in Cambridge. He extended to me hospitality, assistance and advice, as well as access to all necessary materials, including manuscripts of his. I am also deeply grateful to Dr Needham's second wife and closest associate, the late Dr Lu Gwei-Djen, who persuaded Joseph to agree to this project by pointing out to him that although he had always dreamed of doing this job himself, and had first publicly announced his intention to do so as early as 1946, he was too old (in his 80s) to carry it out and too burdened with other work. It was with the greatest sadness that Joseph relinquished his cherished dream of doing this book for the ordinary reading public himself, and allowed me to do it for him. His praise at the end of the task was my sufficient reward. Gwei-Djen was enthusiastically friendly and supportive to me at every turn, and her warm friendship and good will can never be forgotten. I was the only Western colleague of Joseph's who ever travelled to China with him. It was to be Joseph's last trip and my first. We had so many remarkable adventures together! Most amazing of all were our forays into the remotest areas of Chinese countryside, which has such overwhelming natural beauty. Anyone who has been to its truly hidden regions is forever touched by the exquisite and haunting mystery of China, which is unfathomable and infinite.

The book would not have been possible without the friendship and assistance of the remarkable Miss Carmen Lee, who was at that time the Librarian of the Library (later the Institute). She worked tirelessly on difficult problems as they arose, calling my attention to, or locating, obscure materials, and being generally supportive by her unflagging enthusiasm.

Dr Simon Mitton of the Cambridge University Press expedited and assisted this project in every way which was within his power, in a most generous manner.

Of Joseph Needham's collaborators who have helped me, I am most grateful of all to Dr H. T. Huang (Huang Hsing-Tsung). He made possible my entry 'Strong Beer' from an oral interview with him and vetted my account in person, allowing it to appear in print before he had even written his own first draft on the subject, which is true scholarly generosity. He also brought to my attention his subject of Biological Pest Control when it was still in proof stage. My next debt of gratitude is owed to Dr James C.Y. Watt, who cleared up many deeply confusing issues for my entry on porcelain (about which Joseph never wrote a word himself), and whose redrafting of my initial two paragraphs I have largely adopted. Without his help, clarity would never have been attained in this highly specialized subject. Mr Edward McEwen was most helpful and generous on information about the crossbow, as was Dr Robin D. S. Yates about the stirrup (he was also the one who sent Joseph, Gwei-Djen and myself to the depths of Szechuan Province to see the oldest depiction of a gun which appears in Plate 168 (page 251), as he was the first to notice it). Others of the collaborators who were of some assistance were Dr Francesca Bray, Mr Kenneth Robinson and Professor T. H. Hsien (Ch'ien Ts'un-Hsün).

I am deeply grateful to Madame Chen Zhili, State Counsellor for Education, Science, Technology and Culture, in China, for writing her introduction to my book, and for taking such a keen interest in it and recommending it to the Chinese secondary school system when she was Minister of Education.

My gratitude to my old friend Dr Song Jian, former State Counsellor, and former Minister of Science and Technology, may be seen from my dedicating this edition of my book to him. His brilliant work as a scientist before he became a minister was entirely within the finest Chinese tradition of invention.

I am grateful to Christer von der Burg and Therese O'Connor for locating books and making helpful suggestions; and to David Goddard for his invitation to sail on and study his Chinese junk, Keying II, and for information and demonstrations about how steer and sail it. I am grateful to Dr Larry Schulz for going out of his way to take a photograph in China, and to Professor Melvin Kranzberg for putting me in touch with Dr Schulz and for helpful comments. I am also grateful to Lindsay Badenoch and Mrs You-Hsien (Isabella) Lewis, and extremely grateful to Graham Hutt of the Chinese Department of the British Library, for his generosity and help.

For many of the photographs I am indebted out of the ordinary to the extraordinary Miss Charlotte Deane, for her enthusiastic and devoted picture research, which on many occasions involved her in situations of unprecedented difficulty. Her tireless efforts on behalf of this project went beyond the call of duty. I am also uniquely indebted to Ulrich P. Hausmann for an enormous amount of help given out of a delightful enthusiasm he has for all matters Chinese. He has provided me with information, loaned me precious books, and resourcefully produced several magnificent photos for this book.

I would like to give special thanks to Michael Lee, who brought his uncanny expertise to bear on scanning illustrations and making the 'impossible' ones reproducible, such as Plate 156 (page 225), and cleverly joining up old Chinese engravings which were split into two halves, so that they now appear to be one image. He also found ways to scan large pictures, hangings, and rubbings, and they came out looking perfect (as may be seen in Plate 143 (page 208–9). For his efforts, he has been given the honorific Chinese name Li Mihai, and elevated to the Taoist pantheon.

I especially wish to acknowledge information which I have obtained from four particularly useful sources: an article on porcelain by Hung Kuang-chu, the article 'The Chinese Cosmic Magic Known as Watching for the Ethers' by Derk Bodde, the book *The Vermilion Bird* by Edward H. Schafer (where additional information about the Magic Canal was found), and the article 'Porcelain and Plutonism' by Cyril S. Smith.

I wish most profusely to thank my secretary in Beijing, Miss Zhang Jing, for her tireless and determined efforts to assist me in this project over a long period, gathering material and making translations. She always shows ingenuity and rapidity in everything she does. I also wish to thank our mutual friend Du Xiansheng for his inspired and intelligent assistance over the years., and my old friend Professor Pan Jixing for valuable discussions and suggestions, and generous exchange of publications.

I wish to thank Miranda West, formerly of Andre Deutsch, and Gareth Jones and Steve Behan of Andre Deutsch for their wonderful editorial and photographic efforts respectively on this edition. I also wish to acknowledge also the magnificent design work of Clare Baggaley and Michelle Pickering..

I wish to thank the People's Educational Press in Beijing for their meticulous attention to detail in gathering from a wide variety of specialist scholars all over China a number of minor corrections to dates and names for the Chinese edition, which have duly been incorporated in this new English edition, in order to perfect its accuracy in every detail. In particular I wish to thank President Han Shaoxiang of P.E.P., Wei Guodong, Editor-in-chief, Chen Chen, Wang Cunzhi and Deng Wenfeng. I also wish to thank all the scholars throughout China who contributed to those corrections, and Professor Wang Cuncheng of Tsinghua University, who was in charge of this difficult and comprehensive project, and did such an excellent of coordinating it. I also wish to thank Professor Chen Yangzheng for his earlier work on my book.

I have left it to last to offer my fondest thanks of all to my wife, Olivia – not only for her general support at all times and in all circumstances, but also for her customary editing of everything written in this book, as in all my other books. She is the most brilliant editor I have ever known, and there is not a paragraph that she has not improved. She advises on and takes an interest in everything that concerns my work, and was especially active on this project.

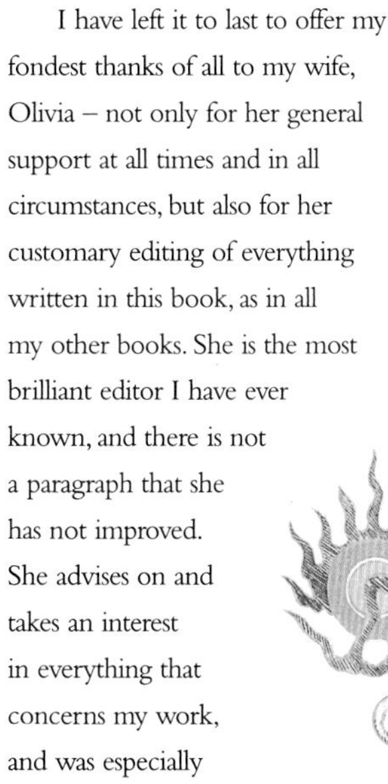

PICTURE CREDITS

The publishers would like to thank the following sources for their kind permission to reproduce the pictures in this book.

Alamy Images: /© Michael Patrick O'Neil: 148–49, /© WizData,inc.: 147; /**Aldus Archive**: 87, 104; /**Ashmolean Museum**: 101; /**Biblioteca Nazionale Centrale Vittorio Emanuele II, Rome**: 141; /**Derk Boddle Annual Customs & Festivals in Peking**, 1936: 87b; /**Ernst Boerschmann**: 58, 90, 224t, 224b, 273; /**The Bridgeman Art Library**: /Nottingham City Museums and Galleries: 232; /People's Republic of China, Lauros: 238; /Private Collection: 189; /**British Library**: 36–37, 125, 128–29; /**The British Museum**: /**© Copyright the Trustees of The British Museum**: 57, 102, 105, 112, 139, 235; /**Cambridge University Library**: 29, 109, /Permission of the Syndics: 261; /**China at Work by R.P. Hommel, 1937**: 116; /**Collection of Dr Lu Gwei-Djen**: 52; / **Corbis**: /Baldwin H. Ward & Kathyrn C. Ward: 200; / Bettmann: 157; /Stefano Bianchetti: 160; /Dean Conger: /Chris Daniels: 192–93; /Pat Jerrold/Papilio: 187; /Jose Fuste Raga/Zefa: 214, /Craig Lovell: 226–27; /Lawrence Manning: 22–23; /Reuters: 181; /Reuters/Bobby Yip: 253; /Tony Wharton/Frank Picture Agency: 174; /Nik Wheeler: 206; /Zefa: 24, 153; /**E.T.C. Werner Chinese Weapons, Royal Abiatic Society (North China Branch)**, 1933: 254; /**Eye Ubiquitous & Hutchinson**:18, 60–61, 63, 80–81; /**Frans Hals Museum**: 217; /**Getty Images**: /Henry Guttmann: 241; /Image Bank: 198; /**Ulrich Hausmann**: 48–49, 54–55, 184–85; /**Hupei Provincial Wuhan**: 222; /**Imperial War Museum**: /Cecil Beaton: 43, 59, 89, 91, 108 b, 135, 219; /**Joint Publishing Co, Hong Kong**: 65, 199; /**Kuangchow Historical Museum, Canton**: 204b; /**K.P. & S.B. Tritton**: 41; /**Lonely Planet Images**: 66–67, 202–03; /**MacQuitty International**:16, 21, 96–97; /**Multimedia Books**: 83, 168b; /John Webb: 45; /**Museum Rietberg, Zurich**: 165; /**Museum fur Volkerkunde, Berlin**: 130; /**National Galleries of Scotland**: 140; / **National Maritime Museum**: /Waters Collection: 204 tl, 205, 211; /**Dr Joseph Needham**: 51; /**Needham Science and Civilization in China**: 12, 13, 19, 20, 37, 47t; 121, 138b, 197, 204tr, 229; /**Joseph Needham 'Clerks and Craftsmen in China and West', Cambridge University Press, 1970**: 145; /**Needham 'The Development of Iron & Steel Technology in China, The Newcomen Society'**: 77; /**The Nelson-Atkins Museum of Art**: 50; /**Ontario Science Centre**: 94, 127, 132, 260, 263; /**Photolibrary.com**: 107; /**Panaroma Media (Beijing) Ltd.**: 78–79; /**Jerry Pavia**: 173; /**Private Collection**: 53, 64, 166–167,168t, 168 , 232; /**RMN**: 267; /**The Royal Artillery Museum**: 268, 272; /**Robert Temple**: 17, 27, 32–33, 34, 35, 38–39, 44, 72, 74, 75, 78, 82, 95, 98, 100, 103, 111, 126, 131, 137, 143, 151, 154t, 154bl, 154br, 158, 159, 161, 163t, 163b, 164, 171, 180, 183, 191bl, 191br, 196, 208–209, 210, 213, 218, 221, 225, 231, 233, 236, 243, 248, 252, 256, 259, 264, 265, 269, 270, 271, 274; /**Robin Yates**: 251; /**Science & Society**: /Science Museum Pictorial: 182; / Science Museum, London: 71, 117, 118, 166–167, 178, 179; /**Science Photo Library**: Mehau Kulyk: 176; /**Seattle Art Museum**: /Eugene Fuller Memorial Collection: 25; /**Simon Archery Foundation, Manchester Museum, University of Manchester**: 245, 246; /Smithsonian Institution: / Freer Gallery of Art: 195; /**Staatbibliothek Preussischer Kulturbesitz, Berlin**: 138t /**Topfoto.co.uk**: 86; /**Copyright © V&A Images/V&A Images. All rights reserved**: 8, 40, 47, 85, 88, 93, 99, 108t, 110, 114–115, 123, 223.

Every effort has been made to acknowledge correctly and contact the source and/or copyright holder of each picture and Carlton Books Limited apologizes for any unintentional errors or omissions, which will be corrected in future editions of this book.

鍼圖